PEARSON EDEXCEL INTERNATIONAL A LEVEL

CHEMISTRY

Student Book 2

Cliff Curtis
Jason Murgatroyd
with David Scott

Published by Pearson Education Limited, 80 Strand, London, WC2R 0RL.

www.pearsonglobalschools.com

Copies of official specifications for all Edexcel qualifications may be found on the website: https://qualifications.pearson.com

Designed by Tech-Set Ltd, Gateshead, UK
Typeset by Tech-Set Ltd, Gateshead, UK
Edited by Gerard Delaney, Sharon Jordan and Sarah Wright

Cover images: Front: **Getty Images:** David Malin/Science Faction
Inside front cover: **Shutterstock.com**/Dmitry Lobanov

First published 2019

25
11

British Library Cataloguing in Publication Data
A catalogue record for this book is available from the British Library
ISBN 978 1 2922 4472 3

Printed in Slovakia by Neografia

Acknowledgements

(Key: b-bottom; c-centre; l-left; r-right; t-top)

Images:
2-3 **Shutterstock:** Marilyn barbone/Shutterstock; 5 **Science Photo Library Ltd:** (tr) Andrew Lambert Photography/Science Photo Library Ltd; (cl) Martyn F.Chillmaid/Science Photo Library Ltd; (br) Trevor Clifford Photography/Science Photo Library Ltd; 30 **Getty Images:** Adri de Visser/Minden Pictures/Getty Images; 38 **Shutterstock:** Vlastas/Shutterstock; 50 **Pearson Education Ltd:** (tl1) Trevor Clifford/Pearson Education Ltd; **123RF:** (tl2) Anton Starikov/123RF; 54 **Toyota Motor Sales:** Courtesy of Toyota Motor Sales U.S.A., Inc.; 58-59 **Shutterstock:** Valentin Valkov/Shutterstock; 71 **Alamy Stock Photo:** PCN Photography/Alamy Stock Photo; 74 **Getty Images:** Reinhard Dirscherl/WaterFrame/Getty Images; 78-79 **Shutterstock:** Ikpro/Shutterstock; 80 **Alamy Stock Photo:** (cl1) Historic Images/Alamy Stock Photo; (cl2) **Thomas Martin Lowry:** T.M. Lowry; 91 **Science Photo Library Ltd:** Andrew Lambert Photography/Science Photo Library Ltd; 99 **Shutterstock:** Gennady/Shutterstock; 102 **123RF:** Federicofoto/123RF; 106-107 **Shutterstock:** WitR/Shutterstock; 116 **123RF:** Oksix/123RF; 118 **Science Photo Library Ltd:** (cl) Andrew Lambert Photography/Science Photo Library Ltd; (bl) Andrew Lambert Photography/Science Photo Library Ltd; 120 **Science Photo Library Ltd:** Andrew Lambert Photography/Science Photo Library Ltd; 123 **Shutterstock:** Gucio_55/Shutterstock; 125 **123RF:** Shane Maritch/123RF; 127 **Science Photo Library Ltd:** Andrew Lambert Photography/Science Photo Library Ltd; 129 **123RF:** Karandaev/123RF; 135 **Getty Images:** Ulrich Baumgarten/Getty Images;136 **Alamy Stock Photo:** CHASSENET/BSIP/Alamy Stock Photo; 138 **Getty Images:** BSIP/Universal Images Group/Getty Images; 140 **Alamy Stock Photo:** AC Images/Alamy Stock Photo; 141 **123RF:** Angellodeco/123RF; 142 **Shutterstock:** Image Source Trading Ltd/Shutterstock; 143 **Science Photo Library Ltd:** Mauro Fermariello/Science Photo Library Ltd; 158-159 **Shutterstock:** Assistant/Shutterstock; 173 **Getty Images:** Tramino/iStock/Getty Images; 174 **Science Photo Library Ltd:** Andrew Lambert Photography/Science Photo Library Ltd; 178 **123RF:** Sinisha Karich/123RF; 182-183 **Shutterstock:** Rost9/Shutterstock; 185 **Science Photo Library Ltd**; 190 **Science Photo Library Ltd:** Andrew Lambert Photography/Science Photo Library Ltd; 195 **Science Photo Library Ltd:** Ramon Andrade 3dciencia/Science Photo Library Ltd; 198 **Shutterstock:** Sashkin/Shutterstock; 199 **Science Photo Library Ltd:** Andrew Lambert Photography/Science Photo Library Ltd; 200 **Science Photo Library Ltd:** (tl) Martyn F.Chillmaid/Science Photo Library Ltd; (cl) Martyn F.Chillmaid/Science Photo Library Ltd; 201 **123RF:** 3Quarks/123RF;202 **Alamy Stock Photo:** Lehmann, Herbert/StockFood Ltd/Alamy Stock Photo; 206 **Science Photo Library Ltd:** Trevor Clifford Photography/Science Photo Library Ltd; 208 **Science Photo Library Ltd:** Andrew Lambert Photography/Science Photo Library Ltd; 212 **Science Photo Library Ltd:** Dick Luria/Science Photo Library Ltd; 213 **Science Photo Library Ltd:** Astrid & Hanns-Frieder Michler/Science Photo Library Ltd; 215 **Science Photo Library Ltd:** Martyn F.Chillmaid/Science Photo Library Ltd; 216 **Science Photo Library Ltd:** British Antarctic Survey/Science Photo Library Ltd; 220-221 **Shutterstock:** Diyana Dimitrova/Shutterstock; 222 **123RF:** Nila Newsom/123RF; 223 **Science Photo Library Ltd:** Trevor Clifford Photography/Science Photo Library Ltd; 227 **123RF:** Grzegorz Kula/123RF; 231 **Alamy Stock Photo:** IS981/Image Source Plus/Alamy Stock Photo; 232 **123RF:** Dmitriy Syechin/123RF; 236-237 **Shutterstock:** Estas/Shutterstock; 238 **123RF:** Dolgachov/123RF; 246 **Shutterstock:** (tl) Topseller/Shutterstock; **Science Photo Library Ltd:** (bl) Cordelia Molloy/Science Photo Library Ltd; 247 **Shutterstock:** Olga Miltsova/Shutterstock; 251 **Shutterstock:** Suriya yapin/Shutterstock; 258-259 **Getty Images:** Cultura/Phil Boorman/Getty Images; 265 **Alamy Stock Photo:** Chronicle/Alamy Stock Photo.

All other images © Pearson Education Limited 2019

We are grateful to the following for permission to reproduce copyright material:

Figures:
Figure on page 43 from 'Mass Number and Isotope', http://www.shimadzu.com, copyright © 2014 Shimadzu Corporation. All rights reserved; Figures on page 273 from Education Scotland © Crown copyright 2012; Figures on page 144 from 'Catalysts for a green industry' Education in Chemistry (Tony Hargreaves),July 2009, http://www.rsc.org/education/eic/issues/2009July /catalyst-green-chemistry-research-industry.asp, copyright © Royal Society of Chemistry; Figures on page 286 from 'Five rings good, four rings bad', Education in Chemistry (Dr Simon Cotton), March 2010, http://www.rsc.org/education/eic/issues/2010Mar/Five Rings Good Four Rings Bad.asp, copyright © Royal Society of Chemistry.

Text Credit(s):
26 **Royal Society of Chemistry:** Engineered Metalloenzyme Catalyses Friedal-Crafts Reaction by Debbie Houghton, 3 November, © 2014, Reproduced by permission of The Royal Society of Chemistry.; 54 **Toyota Motor Sales:** From Toyota The Official Blog of Toyota UK by Joe Clifford blog.toyota.co.uk/how-does-toyotas-fuel-cell-vehicle-work; 74 **Guardian News and Media Limited:** From an article in The Guardian, Scientists Study Ocean Absorption of Human Carbon Pollution by John Abraham, 16 February 2017, https://www.theguardian.com/environment/climate-consensus-97-per-cent/2017/feb/16/scientists-study-ocean-absorption-of-human-carbon-pollution; 102 **Miguel Llanos:** From The Daily Climate by Miguel Llanes https://www.dailyclimate.org/-2598789581.html; 154 **Royal Society of Chemistry:** Enriching the origin of life theory by Joanne Thomson, 26 November, © 2010, Reproduced by permission of The Royal Society of Chemistry.; 178 **Anthony Edward Hargreaves:** From After Battery Power by Tony Hargreaves. https://eic.rsc.org/section/feature/battery-power/2020096.article; 216 **National Geographic Society:** Blue Blood Helps Octopus Survive Brutally Cold Temperatures, 10 July © 2013, National Geographic Society. Used with permission; 232 **Royal Society of Chemistry:** From The Sweet Scent of Success by Emma Davies, February 2009. Reproduced by permission of The Royal Society of Chemistry; 278 **Royal Society of Chemistry:** 'Step change for organic synthesis' by Richard Van Noorden, 1 November 2007, © 2007, Reproduced by permission of The Royal Society of Chemistry.

UNIT 4: RATES, EQUILIBRIA AND FURTHER ORGANIC CHEMISTRY

UNIT 5: TRANSITION METALS AND ORGANIC NITROGEN CHEMISTRY

ABOUT THIS BOOK

This book is written for students following the Pearson Edexcel International Advanced Level (IAL) Chemistry specification. This book covers the full International Advanced Subsidiary (IAS) course and the first year of the International A Level (IAL) course.

The book contains full coverage of IAL units (or exam papers) 4 and 5. Each unit has five Topic areas that match the titles and order of those in the specification. You can refer to the Assessment Overview on pages xii–xiii for further information. Students can prepare for the written Practical Paper by using the IAL Chemistry Lab Book (see pages x–xi of this book).

Each Topic is divided into chapters and sections to break the content down into manageable chunks. Each section features a mix of learning and activities supported by the features explained below.

Learning objectives
Each chapter starts with a list of key learning objectives.

Specification reference
The exact specification points covered in the section are provided.

Learning tips
These help you focus your learning and avoid common errors. **Cross references** to previous or following Student Book content help you to navigate course content.

Worked examples
Worked examples show you how to work through questions, and set out calculations.

16A 2 ELECTROCHEMICAL CELLS

SPECIFICATION REFERENCE 16.7 16.8 CP12

LEARNING OBJECTIVES

- Be able to calculate a standard emf, E°_{cell}, by combining two standard electrode potentials.
- Be able to write cell diagrams using the conventional representation of half-cells.

ELECTROCHEMICAL CELLS

An electrochemical cell is a device for producing an electric current from chemical reactions. It is constructed from two half-cells. **Fig A** shows the apparatus used to construct an electrochemical cell from a Zn^{2+} | Zn half-cell and a Cu^{2+} | Cu half-cell under standard conditions.

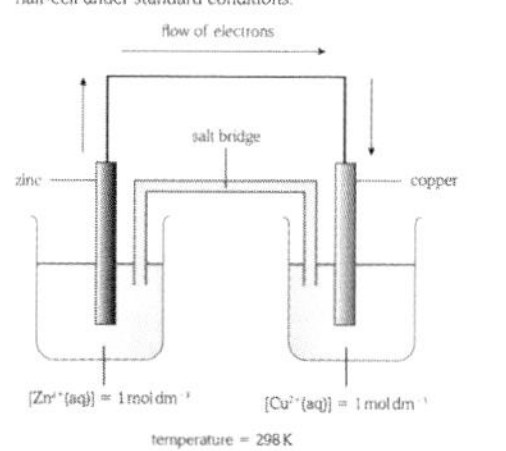

▲ **fig A** An electrochemical cell consisting of Zn^{2+}|Zn and Cu^{2+}|Cu half-cells

EXAM HINT

It is important that you can **describe** clearly how the charge flows in an electrochemical cell (such as the one shown in **fig A**).

Remember that it is the ions that flow in the salt bridge and not electrons. Make sure that you can explain why the K^+ ions move towards the copper half-cell and the nitrate ions move towards the zinc half-cell.

The relevant standard electrode potentials for the redox systems involved are:

$Zn^{2+}(aq) + 2e^- \rightleftharpoons Zn(s) \quad E^{\circ} = -0.76\,V$

$Cu^{2+}(aq) + 2e^- \rightleftharpoons Cu(s) \quad E^{\circ} = +0.34\,V$

The E° value for the Zn^{2+} | Zn half-cell is the more negative, so the zinc electrode will be the negative electrode of the cell. When the cell is in operation, i.e. it is generating an electric current, electrons will flow through the external circuit from the zinc electrode to the copper electrode.

The reactions taking place under these conditions are shown below.

At the negative electrode: $Zn(s) \rightarrow Zn^{2+}(aq) + 2e^-$

At the positive electrode: $Cu^{2+}(aq) + 2e^- \rightarrow Cu(s)$

The overall cell reaction is:

$Zn(s) + Cu^{2+}(aq) \rightarrow Zn^{2+}(aq) + Cu(s)$

CELL DIAGRAMS

It is not always convenient to draw a diagram of the full apparatus for a cell. For simplicity, chemists use a shorthand notation to represent half-cells. The half-cell made from zinc ions and zinc metal is written as:

$Zn^{2+}(aq) \mid Zn(s) \quad E^{\circ} = -0.76\,V$

The solid vertical line indicates a phase boundary, in this case between an aqueous phase, $Zn^{2+}(aq)$, and a solid phase, Zn(s).

Some other examples are:

$Mg^{2+}(aq) \mid Mg(s) \quad E^{\circ} = -2.37\,V$

$Cu^{2+}(aq) \mid Cu(s) \quad E^{\circ} = +0.34\,V$

$Al^{3+}(aq) \mid Al(s) \quad E^{\circ} = -1.66\,V$

$Ag^{+}(aq) \mid Ag(s) \quad E^{\circ} = +0.80\,V$

When the electrode, i.e. the electrical connection between the solution and the external circuit, is a piece of platinum foil, the following convention is used:

$Pt(s) \mid Fe^{3+}(aq), Fe^{2+}(aq) \quad E^{\circ} = +0.77\,V$

Notice that because there is no phase boundary between $Fe^{3+}(aq)$ and $Fe^{2+}(aq)$, a comma is used to separate them, not a solid vertical line.

The standard hydrogen electrode is represented as follows:

$H^{+}(aq) \mid \frac{1}{2}H_2(g) \mid Pt(s)$

These shorthand notations can now be used to represent a cell comprising two half-cells. Convention dictates two things:

1 The two reduced forms of the species are shown on the outside of the cell diagram.

2 The positive electrode is shown on the right-hand side of the cell diagram.

Applying these conventions produces the following cell diagram for a cell formed by combining the $Zn^{2+}(aq) \mid Zn(s)$ and the $Cu^{2+}(aq) \mid Cu(s)$ half-cells:

$Zn(s) \mid Zn^{2+}(aq) \vdots\vdots Cu^{2+}(aq) \mid Cu(s)$

TOPIC 16 16A.2 ELECTROCHEMICAL CELLS 167

The double vertical lines (⁞⁞) represent the salt bridge.

The emf, E°_{cell}, of this cell is simply the *difference* between the two standard electrode potentials of the two half-cells.

$Zn(s) \mid Zn^{2+}(aq) \vdots\vdots Cu^{2+}(aq) \mid Cu(s)$

$E^{\circ} = -0.76\,V \quad E^{\circ} = +0.34\,V$

The difference between these two numbers is 1.10, so the emf of the cell is 1.10 V. To indicate that the right-hand electrode, i.e. the copper, is the positive electrode of the cell, the emf is given a positive (+) sign.

So, the complete cell diagram is:

$Zn(s) \mid Zn^{2+}(aq) \vdots\vdots Cu^{2+}(aq) \mid Cu(s) \quad E^{\circ}_{cell} = +1.10\,V$

LEARNING TIP

Note that the sign for the E° of the Zn^{2+}|Zn half-cell is still given as negative, despite the half-cell being written in reverse. As mentioned in **Section 16A.1**, the sign of a standard electrode potential remains the same no matter how the half-cell is represented.

DID YOU KNOW?

It is also possible to calculate the emf of the cell from a cell diagram by subtracting the E° value of the left-hand half-cell from that of the right-hand half-cell:

$$E^{\circ}_{cell} = E^{\circ}_{right} - E^{\circ}_{left}$$

It is important to remember, however, that you must not change the sign of the E° value of the left-hand half-cell, even though the reaction is written as an oxidation and not a reduction.

WHAT IS MEANT BY THE 'DIFFERENCE' BETWEEN TWO STANDARD ELECTRODE POTENTIALS?

The easiest way to explain this is to represent the two values on a number scale, such as the one shown in **fig B**.

+ 1.00

+ 0.34

0.00 difference = 1.10

− 0.76

− 1.00

▲ **fig B** Example of a number scale to calculate an emf

To move from one number to the other on the number scale involves a change of 1.10 units.

BREAKING THE CELL CONVENTION

Chemists allow themselves to break the rules if it suits their purpose. This is exactly what we do when we write cell diagrams to represent the measurement of a standard electrode potential. In this case, the standard hydrogen electrode is always written on the left-hand side.

The cell diagrams for the set-up used when measuring the standard electrode potentials of the $Zn^{2+}(aq) \mid Zn(s)$ and the $Cu^{2+}(aq) \mid Cu(s)$ half-cells are therefore:

$Pt(s) \mid \frac{1}{2}H_2(g) \mid H^{+}(aq) \vdots\vdots Zn^{2+}(aq) \mid Zn(s) \quad E^{\circ}_{cell} = -0.76\,V$

$Pt(s) \mid \frac{1}{2}H_2(g) \mid H^{+}(aq) \vdots\vdots Cu^{2+}(aq) \mid Cu(s) \quad E^{\circ}_{cell} = +0.34\,V$

As before, the sign of E°_{cell} indicates the polarity of the right-hand electrode. Zinc is the negative electrode of the cell formed in combination with the standard hydrogen electrode. Copper is the positive electrode of the cell formed in combination with the standard hydrogen electrode.

CHECKPOINT

SKILLS CREATIVITY, ADAPTIVE LEARNING

1. (a) Draw a labelled diagram of the apparatus that can be used to construct a cell, under standard conditions, from a Zn^{2+}|Zn half-cell and a Fe^{3+}, Fe^{2+}|Pt half-cell.
   (b) The standard electrode potentials for the half-cells are:
   $Zn^{2+}|Zn \quad E^{\circ} = -0.76\,V$
   $Fe^{3+}, Fe^{2+}|Pt \quad E^{\circ} = +0.77\,V$
   Explain the direction of electron flow in the external circuit when this cell is in use.
   (c) Write the cell diagram for this cell and calculate the emf (E°_{cell}) of the cell.
2. Chlorine may be prepared in the laboratory by reacting dilute hydrochloric acid with potassium manganate(VII). The standard electrode potentials that relate to this reaction are:
   $\frac{1}{2}Cl_2(g) + e^- \rightleftharpoons Cl^-(aq) \quad E^{\circ} = +1.36\,V$
   $MnO_4^-(aq) + 8H^+(aq) + 5e^- \rightleftharpoons Mn^{2+}(aq) + 4H_2O(l) \quad E^{\circ} = +1.51\,V$
   (a) Calculate the emf, E°_{cell}, of a cell constructed from these two redox systems.
   (b) Explain the direction of electron flow that would take place in the external circuit of this cell when in use.
   (c) Write the cell diagram for this cell.

SUBJECT VOCABULARY

thermodynamically feasible reaction a reaction that should take place without any intervention by us, if we consider the enthalpy and entropy changes involved
kinetically stable the reaction does not take place, or is very slow, because the activation energy for the reaction is very high
disproportionation a reaction in which an element is both oxidised and reduced at the same time

Exam hints
Tips on how to answer exam-style questions and guidance for exam preparation, including requirements indicated by particular **command words**.

Did you know?
Interesting facts help you remember the key concepts.

Checkpoint
Questions at the end of each section check understanding of the key learning points in each chapter. Certain questions allow you to develop **skills** that will be valuable for further study and in the workplace.

Subject Vocabulary
Key terms are highlighted in blue in the text. Clear definitions are provided at the end of each section for easy reference, and are also collated in a **glossary** at the back of the book.

You should be able to put every stage of your learning in context, chapter by chapter.

- Links to other areas of Chemistry include previous knowledge that is built on in the topic, and future areas of knowledge and application that you will cover later in your course.
- Maths knowledge required is detailed in a handy checklist. If you need to practise the maths you need, you can use the **Maths Skills** reference at the back of the book as a starting point.

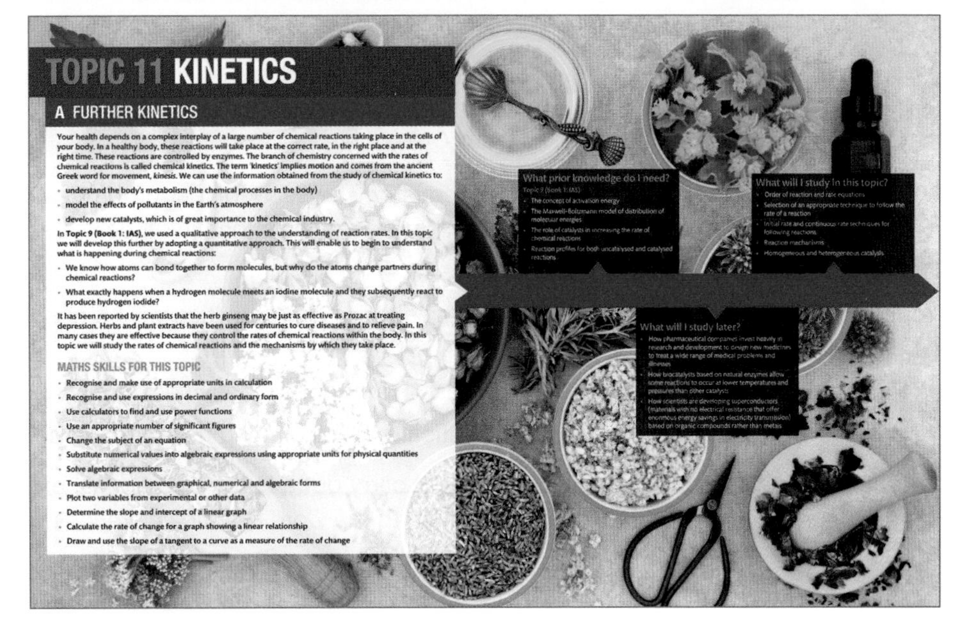

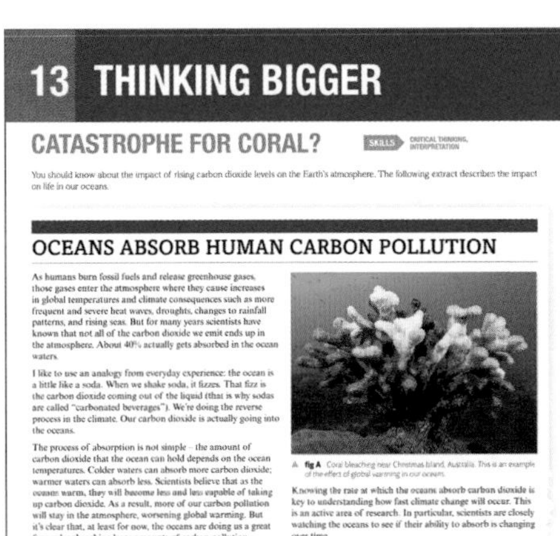

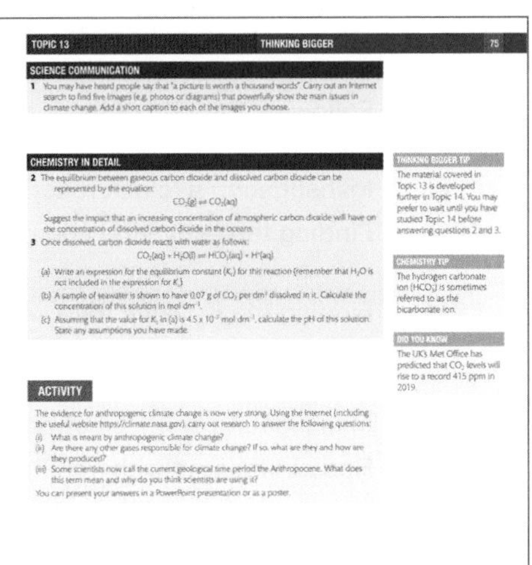

Thinking Bigger
At the end of each topic there is an opportunity to read and work with real-life research and writing about science.

The activities help you to read real-life material that's relevant to your course, analyse how scientists write, think critically and consider how different aspects of your learning piece together.

These Thinking Bigger activities focus on key transferable skills, which are an important basis for key academic qualities.

Exam Practice
Exam-style questions at the end of each chapter are tailored to the Pearson Edexcel specification to allow for practice and development of exam writing technique. They also allow for practice responding to the command words used in the exams (see the **command words glossary** at the back of this book).
You can also refer to the **Preparing for Your Exams** section at the back of the book, for sample exam answers with commentary.

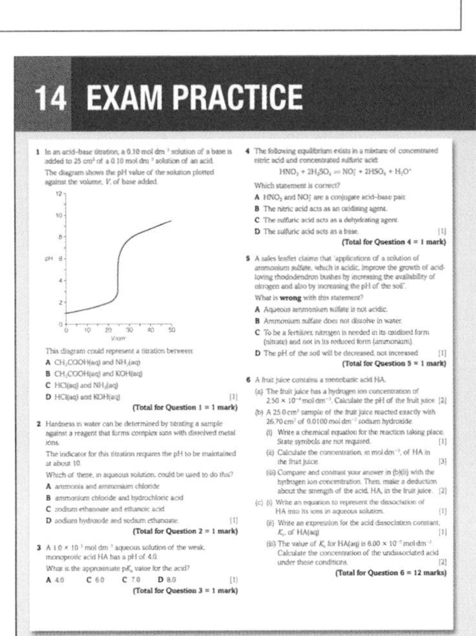

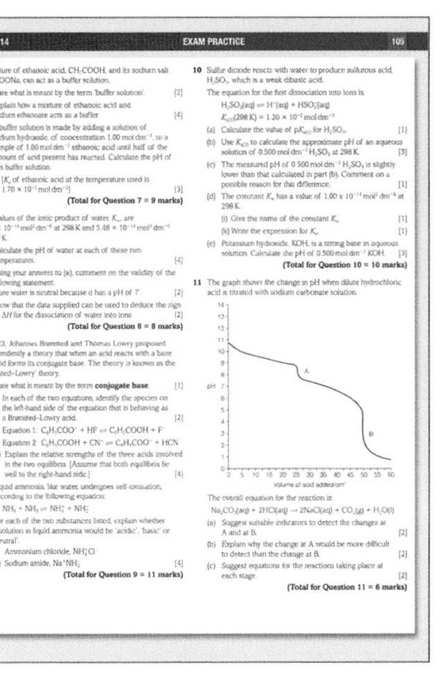

PRACTICAL SKILLS

Practical work is central to the study of chemistry. The second year of the Pearson Edexcel International Advanced Level (IAL) Chemistry course includes eight Core Practicals that link theoretical knowledge and understanding to practical scenarios.

Your knowledge and understanding of practical skills and activities will be assessed in all examination papers for the IAL Chemistry qualification.

- Papers 4 and 5 will include questions based on practical activities, including novel scenarios.
- Paper 6 will test your ability to plan practical work, including risk management and selection of apparatus.

In order to develop practical skills, you should carry out a range of practical experiments related to the topics covered in your course. Further suggestions in addition to the Core Practicals are included in the specification, which is available online.

STUDENT BOOK TOPIC	IAL CORE PRACTICALS	
TOPIC 11 **KINETICS**	**CP9a**	Following the rate of the iodine–propanone reaction by a titrimetric method
	CP9b	Investigating a 'clock reaction' (Harcourt–Esson, iodine clock)
	CP10	Finding the activation energy of a reaction
TOPIC 14 **ACID–BASE EQUILIBRIA**	**CP11**	Finding the K_a value for a weak acid
TOPIC 16 **REDOX EQUILIBRIA**	**CP12**	Investigating some electrochemical cells
	CP13a	Redox titrations with iron(II) ions and potassium manganate(VII)
	CP13b	Redox titrations with sodium thiosulfate and iodine
TOPIC 17 **TRANSITION METALS AND THEIR CHEMISTRY**	**CP14**	Preparation of a transition metal complex
TOPIC 19 **ORGANIC NITROGEN COMPOUNDS: AMINES, AMIDES, AMINO ACIDS AND PROTEINS**	**CP15**	Analysis of some inorganic and organic unknowns
TOPIC 20 **ORGANIC SYNTHESIS**	**CP16**	Preparation of aspirin

17B 1 DIFFERENT TYPES OF REACTIONS

SPECIFICATION REFERENCE 17.11 17.24(i) 17.24(ii) CP14

In the **Student Book**, the Core Practical specifications are supplied in the relevant sections.

LEARNING OBJECTIVES

- Understand that colour changes in transition metal ions may arise as a result of changes in:
  - (i) oxidation number of the ion
  - (ii) ligand
  - (iii) coordination number of the complex.
- Understand that ligand exchange, and an accompanying colour change, occurs in the formation of:
  - (i) $[Cu(NH_3)_4(H_2O)_2]^{2+}$ from $[Cu(H_2O)_6]^{2+}$ via $Cu(OH)_2(H_2O)_4$
  - (ii) $[CuCl_4]^{2-}$ from $[Cu(H_2O)_6]^{2+}$.
- Be able to write ionic equations to show the meaning of amphoteric behaviour, deprotonation and ligand exchange reactions.

TYPES OF REACTIONS

So far, we have considered the origin of colour in transition metal ions. We can now consider why there are often colour changes when transition metal ions take part in reactions. Four main types of reactions can occur:

- redox – the oxidation number of the transition metal ion changes
- **deprotonation** – one or more of the ligands gains or loses a hydrogen ion (proton)
- **ligand exchange** – one or more of the ligands around the transition metal ion is replaced by a different ligand
- coordination number change – the number of ligands changes.

Any one of these types of reactions can cause a change in the colour of the complex. Some reactions involve more than one of these types of reactions.

CHANGE IN OXIDATION NUMBER

An aqueous solution containing $Fe^{2+}(aq)$ ions is pale green, but when it is exposed to air it gradually turns yellow or brown, as the oxidation number of iron increases from +2 to +3. The type and number of ligands remain unchanged in this oxidation reaction, so the formulae of the two complexes are $[Fe(H_2O)_6]^{2+}$ and $[Fe(H_2O)_6]^{3+}$. Colour changes such as the one in this reaction are best illustrated using solid samples containing the ions (see **Fig A**).

Equations are not usually written for oxidation reactions in which the only change is the oxidation number of the transition metal ion.

Iron(II) sulphate (FeSO^4)

Iron(III) sulphate (Fe2(SO^4)^3)

fig A Solid samples clearly show colour differences between ions.

FORMATION OF $[Cu(NH_3)_4(H_2O)_2]^{2+}$: DEPROTONATION AND LIGAND EXCHANGE REACTIONS

Consider the reaction that occurs when aqueous sodium hydroxide is added to copper(II) sulfate solution. The observation is that a pale blue solution forms a blue precipitate. The equation for this reaction is:

$$[Cu(H_2O)_6]^{2+} + 2OH^- \rightarrow [Cu(H_2O)_4(OH)_2] + 2H_2O$$

You might think that this is a ligand substitution reaction – that two hydroxide ions have replaced two water molecules. In fact, it is a deprotonation reaction – the two hydroxide ions have removed hydrogen ions from two of the water ligands and converted them into water molecules. The two water ligands that have lost hydrogen ions are now hydroxide ligands.

PRACTICAL SKILLS CP14

Tetramminecopper(II) sulfate-1-water, $[Cu(NH_3)_4SO_4].H_2O$, can be prepared by adding aqueous ammonia to an aqueous solution of copper(II) sulfate.

The overall equation for the reaction is:

$$[Cu(H_2O)_6]^{2+} + 4NH_3 + SO_4^{2-} \rightarrow [Cu(NH_3)_4]SO_4.H_2O + 5H_2O$$

This is an example of a ligand exchange reaction. This is the reaction you might investigate in **CP14: The preparation of a transition metal complex**.

Practical Skills
Practical skills boxes explain techniques used in the Core Practicals, and also detail useful skills and knowledge gained in other related investigations.

Exactly the same observations can be made during the careful addition of aqueous ammonia instead of aqueous sodium hydroxide. The equation for this reaction is:

$$[Cu(H_2O)_6]^{2+} + 2NH_3 \rightarrow [Cu(H_2O)_4(OH)_2] + 2NH_4^+$$

DID YOU KNOW?

The formula for the copper(II)-ammine complex in aqueous solution is sometimes given as $[Cu(NH_3)_4]^{2+}$. This is *not* correct. The correct formula is shown in the text. The confusion arose because the bonds from the Cu^{2+} ion to the water ligands are longer than the bonds from the Cu^{2+} ion to the ammonia ligands. This is the result of something called the Jahn–Teller effect. The explanation for this effect is beyond the aims of this book.

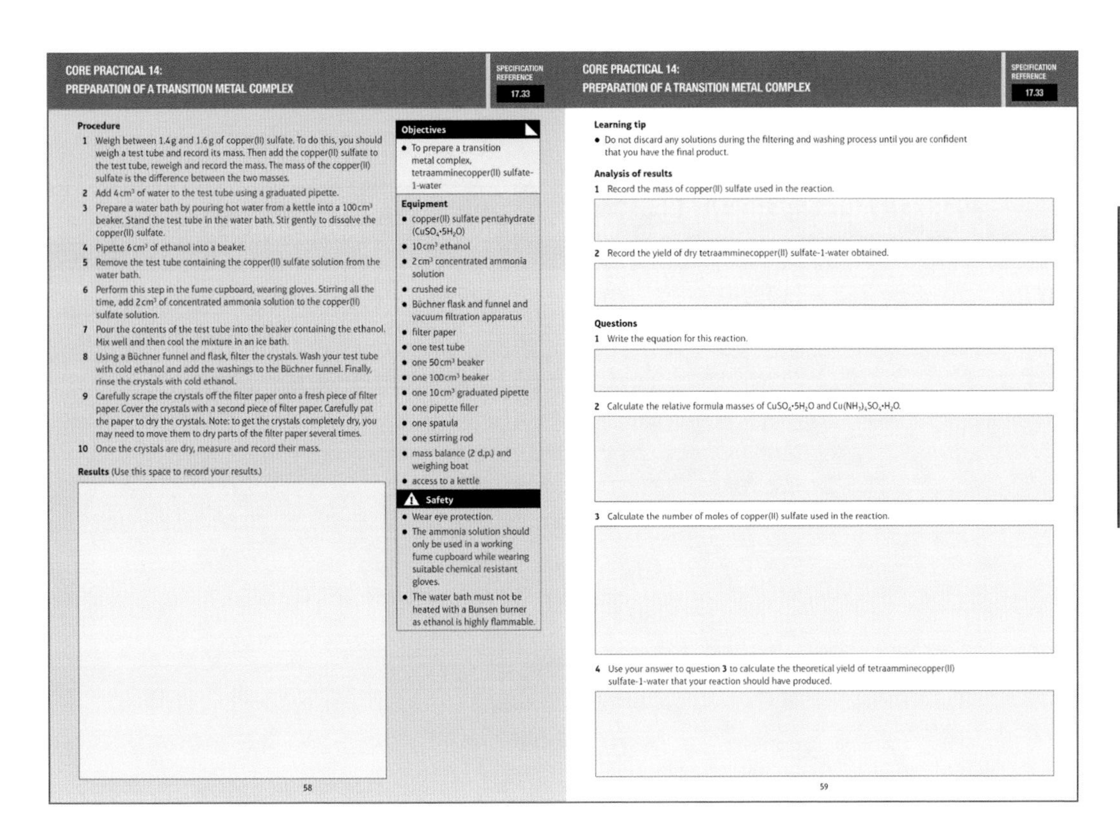
CORE PRACTICAL 14:
PREPARATION OF A TRANSITION METAL COMPLEX

SPECIFICATION REFERENCE 17.33

Procedure

1. Weigh between 1.4 g and 1.6 g of copper(II) sulfate. To do this, you should weigh a test tube and record its mass. Then add the copper(II) sulfate to the test tube, reweigh and record the mass. The mass of the copper(II) sulfate is the difference between the two masses.
2. Add 4 cm^3 of water to the test tube using a graduated pipette.
3. Prepare a water bath by pouring hot water from a kettle into a 100 cm^3 beaker. Stand the test tube in the water bath. Stir gently to dissolve the copper(II) sulfate.
4. Pipette 6 cm^3 of ethanol into a beaker.
5. Remove the test tube containing the copper(II) sulfate solution from the water bath.
6. Perform this step in the fume cupboard, wearing gloves. Stirring all the time, add 2 cm^3 of concentrated ammonia solution to the copper(II) sulfate solution.
7. Pour the contents of the test tube into the beaker containing the ethanol. Mix well and then cool the mixture in an ice bath.
8. Using a Büchner funnel and flask, filter the crystals. Wash your test tube with cold ethanol and add the washings to the Büchner funnel. Finally, rinse the crystals with cold ethanol.
9. Carefully scrape the crystals off the filter paper onto a fresh piece of filter paper. Cover the crystals with a second piece of filter paper. Carefully pat the paper to dry the crystals. Note: to get the crystals completely dry, you may need to move them to dry parts of the filter paper several times.
10. Once the crystals are dry, measure and record their mass.

Results (Use this space to record your results.)

Objectives

- To prepare a transition metal complex, tetraamminecopper(II) sulfate-1-water

Equipment

- copper(II) sulfate pentahydrate ($CuSO_4 \cdot 5H_2O$)
- 10 cm^3 ethanol
- 2 cm^3 concentrated ammonia solution
- crushed ice
- Büchner flask and funnel and vacuum filtration apparatus
- filter paper
- one test tube
- one 50 cm^3 beaker
- one 100 cm^3 beaker
- one 10 cm^3 graduated pipette
- one pipette filler
- one spatula
- one stirring rod
- mass balance (2 d.p.) and weighing boat
- access to a kettle

Safety

- Wear eye protection.
- The ammonia solution should only be used in a working fume cupboard while wearing suitable chemical resistant gloves.
- The water bath must not be heated with a Bunsen burner as ethanol is highly flammable.

58

CORE PRACTICAL 14:
PREPARATION OF A TRANSITION METAL COMPLEX

SPECIFICATION REFERENCE 17.33

Learning tip

- Do not discard any solutions during the filtering and washing process until you are confident that you have the final product.

Analysis of results

1. Record the mass of copper(II) sulfate used in the reaction.
2. Record the yield of dry tetraamminecopper(II) sulfate-1-water obtained.

Questions

1. Write the equation for this reaction.
2. Calculate the relative formula masses of $CuSO_4 \cdot 5H_2O$ and $Cu(NH_3)_4SO_4 \cdot H_2O$.
3. Calculate the number of moles of copper(II) sulfate used in the reaction.
4. Use your answer to question **3** to calculate the theoretical yield of tetraamminecopper(II) sulfate-1-water that your reaction should have produced.

59

This Student Book is accompanied by a **Lab Book**, which includes instructions and writing frames for the Core Practicals for you to record your results and reflect on your work. Practical skills practice questions and answers are also provided. The Lab Book records can be used as preparation for the Practical Skills Paper.

ASSESSMENT OVERVIEW

The following tables give an overview of the assessment for Pearson Edexcel International Advanced Level course in Chemistry. You should study this information closely to help ensure that you are fully prepared for this course and know exactly what to expect in each part of the examinations. More information about this qualification, and about the question types in the different papers, can be found in *Preparing for your exams* on page 286 of this book.

PAPER / UNIT 4	PERCENTAGE OF IA2	PERCENTAGE OF IAL	MARK	TIME	AVAILABILITY
RATES, EQUILIBRIA AND FURTHER ORGANIC CHEMISTRY Written exam paper Paper code WCH14/01 Externally set and marked by Pearson Edexcel Single tier of entry	40%	20%	90	1 hour 45 minutes	January, June and October First assessment: January 2020

PAPER / UNIT 5	PERCENTAGE OF IA2	PERCENTAGE OF IAL	MARK	TIME	AVAILABILITY
TRANSITION METALS AND ORGANIC NITROGEN CHEMISTRY Written exam paper Paper code WCH15/01 Externally set and marked by Pearson Edexcel Single tier of entry	40%	20%	90	1 hour 45 minutes	January, June and October First assessment: June 2020

PAPER / UNIT 6	PERCENTAGE OF IA2	PERCENTAGE OF IAL	MARK	TIME	AVAILABILITY
PRACTICAL SKILLS IN CHEMISTRY II Written exam paper Paper / Unit code WCH16/01 Externally set and marked by Pearson Edexcel Single tier of entry	20%	10%	50	1 hour 20 minutes	January, June and October First assessment: June 2020

ASSESSMENT OBJECTIVES AND WEIGHTINGS

ASSESSMENT OBJECTIVE	DESCRIPTION	% IN IAS	% IN IA2	% IN IAL
A01	Demonstrate knowledge and understanding of science.	34–36	29–31	32–34
A02	(a) Application of knowledge and understanding of science in familiar and unfamiliar contexts.	34–36	33–36	33–36
	(b) Analysis and evaluation of scientific information to make judgements and reach conclusions.	9–11	14–16	11–14
A03	Experimental skills in science, including analysis and evaluation of data and methods.	20	20	20

RELATIONSHIP OF ASSESSMENT OBJECTIVES TO UNITS

UNIT NUMBER	ASSESSMENT OBJECTIVE (%)			
	A01	A02 (A)	A02 (B)	A03
UNIT 1	17–18	17–18	4.5–5.5	0.0
UNIT 2	17–18	17–18	4.5–5.5	0.0
UNIT 3	0.0	0.0	0.0	20
TOTAL FOR INTERNATIONAL ADVANCED SUBSIDIARY	**33–36**	**34–36**	**9–11**	**20**

UNIT NUMBER	ASSESSMENT OBJECTIVE (%)			
	A01	A02 (A)	A02 (B)	A03
UNIT 1	8.5–9.0	8.5–9.0	2.2–2.8	0.0
UNIT 2	8.5–9.0	8.5–9.0	2.2–2.8	0.0
UNIT 3	0.0	0.0	0.0	10
UNIT 4	7.3–7.8	8.4–8.9	3.6–4.0	0.0
UNIT 5	7.3–7.8	8.4–8.9	3.6–4.0	0.0
UNIT 6	0.0	0.0	0.0	10
TOTAL FOR INTERNATIONAL ADVANCED LEVEL	**32–34**	**33–36**	**11–14**	**20**

TOPIC 11 KINETICS

A FURTHER KINETICS

Your health depends on a complex interplay of a large number of chemical reactions taking place in the cells of your body. In a healthy body, these reactions will take place at the correct rate, in the right place and at the right time. These reactions are controlled by enzymes. The branch of chemistry concerned with the rates of chemical reactions is called chemical kinetics. The term 'kinetics' implies motion and comes from the ancient Greek word for movement, *kinesis*. We can use the information obtained from the study of chemical kinetics to:

- understand the body's metabolism (the chemical processes in the body)
- model the effects of pollutants in the Earth's atmosphere
- develop new catalysts, which is of great importance to the chemical industry.

In **Topic 9 (Book 1: IAS)**, we used a qualitative approach to the understanding of reaction rates. In this topic we will develop this further by adopting a quantitative approach. This will enable us to begin to understand what is happening during chemical reactions:

- We know how atoms can bond together to form molecules, but why do the atoms change partners during chemical reactions?
- What exactly happens when a hydrogen molecule meets an iodine molecule and they subsequently react to produce hydrogen iodide?

It has been reported by scientists that the herb ginseng may be just as effective as Prozac at treating depression. Herbs and plant extracts have been used for centuries to cure diseases and to relieve pain. In many cases they are effective because they control the rates of chemical reactions within the body. In this topic we will study the rates of chemical reactions and the mechanisms by which they take place.

MATHS SKILLS FOR THIS TOPIC

- Recognise and make use of appropriate units in calculation
- Recognise and use expressions in decimal and ordinary form
- Use calculators to find and use power functions
- Use an appropriate number of significant figures
- Change the subject of an equation
- Substitute numerical values into algebraic expressions using appropriate units for physical quantities
- Solve algebraic expressions
- Translate information between graphical, numerical and algebraic forms
- Plot two variables from experimental or other data
- Determine the slope and intercept of a linear graph
- Calculate the rate of change for a graph showing a linear relationship
- Draw and use the slope of a tangent to a curve as a measure of the rate of change

What prior knowledge do I need?

Topic 9 (Book 1: IAS)

- The concept of activation energy
- The Maxwell-Boltzmann model of distribution of molecular energies
- The role of catalysts in increasing the rate of chemical reactions
- Reaction profiles for both uncatalysed and catalysed reactions

What will I study in this topic?

- Order of reaction and rate equations
- Selection of an appropriate tecÚique to follow the rate of a reaction
- Initial rate and continuous rate tecÚiques for following reactions
- Reaction mechanisms
- Homogeneous and heterogeneous catalysis

What will I study later?

- How pharmaceutical companies invest heavily in research and development to design new medicines to treat a wide range of medical problems and illnesses
- How biocatalysts based on natural enzymes allow some reactions to occur at lower temperatures and pressures than other catalysts
- How scientists are developing superconductors (materials with no electrical resistance that offer enormous energy savings in electricity transmission) based on organic compounds rather than metals

11A 1 TECHNIQUES FOR MEASURING THE RATE OF REACTION

SPECIFICATION REFERENCE 11.1(i) 11.3(i) 11.3(ii) 11.3(iii) 11.3(iv) 11.3(v)

LEARNING OBJECTIVES

- Understand the term 'rate of reaction'.
- Select and justify a suitable experimental tecÜique to obtain rate data for a given reaction, including:
  (i) titration
  (ii) colorimetry
  (iii) mass change
  (iv) volume of gas evolved
  (v) other suitable tecÜique(s) for a given reaction.

RATE OF REACTION

The **rate of a reaction** can be expressed in two ways:

(1) How the concentration of a product *increases* with time.

$$\text{rate} = \frac{\text{change in concentration of product}}{\text{time}}$$

(2) How the concentration of a reactant *decreases* with time.

$$\text{rate} = -\frac{\text{change in concentration of reactant}}{\text{time}}$$

The negative sign in the second expression shows that the concentration of the reactant is decreasing and therefore gives a positive value for the rate.

Rate is measured in units of concentration per unit time, and the most common units are mol dm^{-3} s^{-1}.

The expressions in calculus notation are:

$$\text{rate} = \frac{d[\text{product}]}{dt}$$

$$\text{rate} = -\frac{d[\text{reactant}]}{dt}$$

This **rate of reaction** is sometimes called the 'overall rate of reaction'.

TECHNIQUES FOR MEASURING THE RATE OF REACTION

Before investigating the rate of a particular reaction, it is necessary to know the overall equation, including state symbols, for the reaction so that we can decide what technique to use to follow the reaction.

There are various techniques available to use, such as:

1 measuring the volume of a gas evolved
2 measuring the change in mass of a reaction mixture
3 monitoring the change in intensity of colour of a reaction mixture (colorimetry)
4 measuring the change in concentration of a reactant or product using titration
5 measuring the change in pH of a solution
6 measuring the change in electrical conductivity of a reaction mixture.

The technique chosen to follow the reaction will depend on the nature of the reactants and products, as well as the conditions under which the reaction is carried out.

For example, the reaction between calcium carbonate and dilute hydrochloric acid,

$$CaCO_3(s) + 2HCl(aq) \rightarrow CaCl_2(aq) + H_2O(l) + CO_2(g)$$

could conveniently be followed by measuring the volume of gas (technique 1) given off at regular time intervals, or by measuring the change in mass of the reaction mixture with time (technique 2).

However, the reaction between propanone and iodine in aqueous solution,

$$CH_3COCH_3(aq) + I_2(aq) \rightarrow CH_3COCH_2I(aq) + H^+(aq) + I^-(aq)$$

could not be followed by measuring the change in mass because all products of the reaction remain in solution. It would be possible, however, to follow the reaction by monitoring the decrease in intensity of colour of the reaction mixture, since $I_2(aq)$ is the only coloured **species** present (technique 3).

TECHNIQUE 1: MEASURING THE VOLUME OF A GAS EVOLVED

The two most common techniques for collecting and measuring the volume of a gas evolved during a reaction are:

1 collection over water into a measuring cylinder (**fig A**), and
2 collection using a gas syringe (**fig B**).

The technique chosen will depend partly on the level of precision required. The gas syringe has a greater degree of precision, but if a large volume of gas is being measured, the difference in the degree of measurement uncertainty becomes so small that either instrument is sufficiently precise.

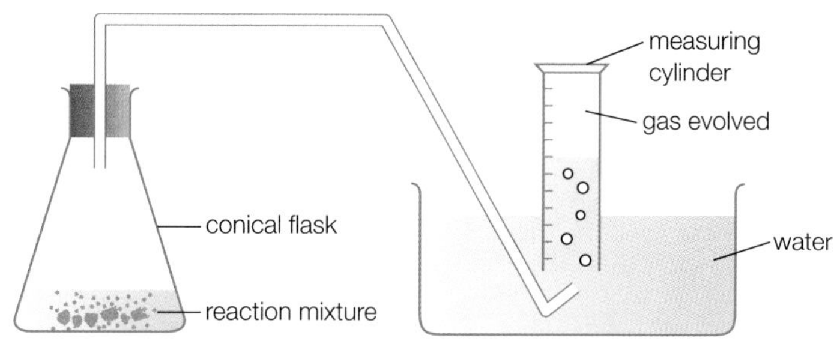

▲ **fig A** Collecting a gas over water.

EXAM HINT

Reactions producing gases that are very soluble in water, such as sulfur dioxide, cannot use the gas collection over water tecÚique.

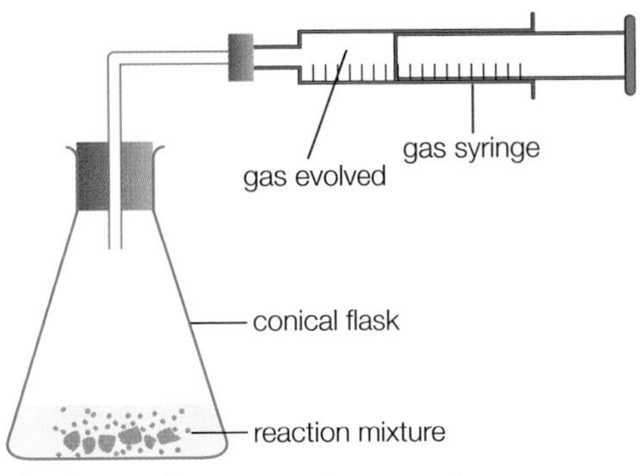

▲ **fig B** Collecting a gas in a gas syringe.

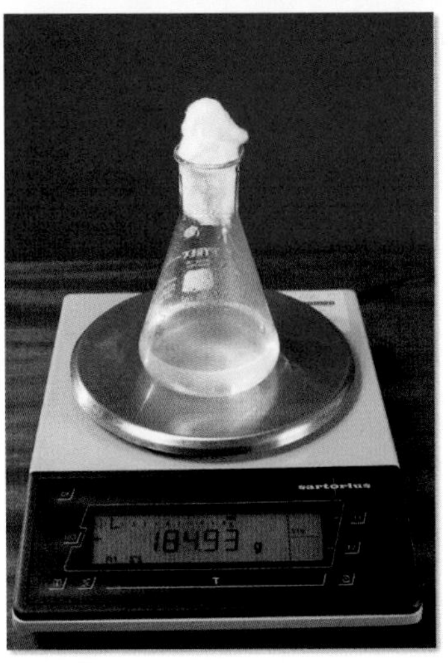

▲ **fig C** Cotton wool is placed in the neck of the flask to prevent the loss of liquid spray.

TECHNIQUE 2: MEASURING THE CHANGE IN MASS OF A REACTION MIXTURE

This is another technique applicable to reactions in which a gas is evolved.

The reaction flask and contents are placed on a digital balance and the decrease in mass is measured as the reaction proceeds (**fig C**).

This technique is most precise when the gas given off has a relatively high density, such as with carbon dioxide. With a low-density (i.e. low relative molecular mass) gas such as hydrogen, the mass changes are so small that the measurement uncertainties become significant.

TECHNIQUE 3: MONITORING A COLOUR CHANGE (COLORIMETRY)

Colour change can sometimes be monitored using observation only. However, using a colorimeter gives more precise results (**fig D**). A colorimeter can detect far more subtle changes than the human eye can observe, and provides a quantitative (rather than a subjective) measurement.

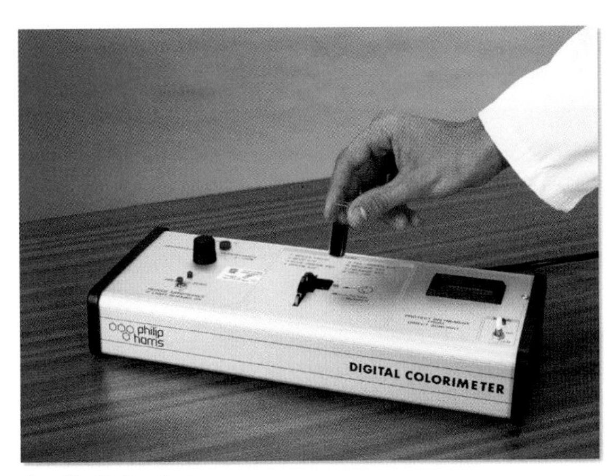

▲ **fig D** A colorimeter.

TECHNIQUE 4: ANALYSIS BY TITRATION

This technique involves using a pipette to remove small samples (aliquots) from a reaction mixture at regular intervals. The reaction in the aliquot can either be stopped by adding another substance to it or slowed down to almost zero by immersing it in an ice bath. The aliquot is then titrated to determine the concentration of a reactant or product species.

The process of stopping or slowing down the reaction in an aliquot is known as 'quenching'.

For example, if the reaction involves an acid, the aliquot, after quenching, could be titrated against a standard solution of sodium hydroxide to determine the concentration of the acid. This technique is used to investigate the reaction between iodine and propanone, which is catalysed by acid. Sodium hydrogen carbonate is added to the aliquot to remove the acid catalyst and, as a result, effectively stops the reaction. The remaining iodine is then titrated against a standard solution of sodium thiosulfate (**fig E**).

$$CH_3COCH_3(aq) + I_2(aq) \rightarrow CH_3COCH_2I(aq) + H^+(aq) + I^-(aq)$$

$$I_2(aq) + 2S_2O_3^{2-}(aq) \rightarrow 2I^-(aq) + S_4O_6^{2-}(aq)$$

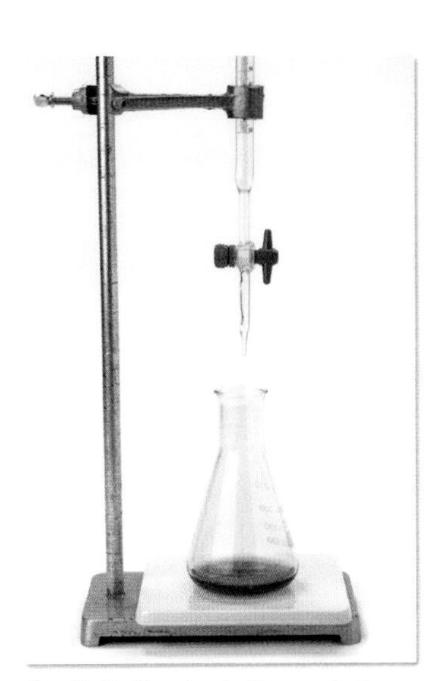

▲ **fig E** Titrating iodine against sodium thiosulfate.

TECHNIQUE 5: MEASURING THE ELECTRICAL CONDUCTIVITY

If the total number, or type, of ions in solution changes during a reaction, it might be possible to follow the reaction by measuring changes in the electrical conductivity of the solution using a conductivity meter. For example, it could be used to follow this reaction:

$$5Br^-(aq) + BrO_3^-(aq) + 6H^+(aq) \rightarrow 3Br_2(aq) + 3H_2O(l)$$

TECHNIQUE 6: MEASURING ANY OTHER PHYSICAL PROPERTY THAT SHOWS A SIGNIFICANT CHANGE

Possible physical properties that have not already been mentioned include changes in the volume of liquid ('dilatometry'), chirality and refractive index.

CHECKPOINT

1. State suitable tecÚiques to collect rate data for each of the following reactions.

(a) Magnesium with dilute sulfuric acid:

$$Mg(s) + 2H^+(aq) \rightarrow Mg^{2+}(aq) + H_2(g)$$

(b) Ethyl ethanoate with sodium hydroxide:

$$CH_3COOCH_2CH_3(l) + OH^-(aq) \rightarrow CH_3COO^-(aq) + CH_3CH_2OH(aq)$$

(c) Hydrogen gas with iodine gas:

$$H_2(g) + I_2(g) \rightarrow 2HI(g)$$

SKILLS CREATIVITY

2. Why would the tecÚique of measuring the change in mass of a reaction vessel and contents not work well in the reaction between magnesium and dilute sulfuric acid?

3. The reaction between calcium carbonate and dilute hydrochloric acid can be followed by collecting and measuring the volume of gas produced. The gas could be collected over water in a measuring cylinder or in a gas syringe. Which tecÚique would be the more suitable for this reaction? Explain your answer.

SUBJECT VOCABULARY

(overall) **rate of reaction** the change in concentration of a species divided by the time it takes for the change to occur. All reaction rates are positive

(chemical) **species** an atom, a molecule or an ion that is taking part in a chemical reaction

11A 2 RATE EQUATIONS, RATE CONSTANTS AND ORDERS OF REACTION

SPECIFICATION REFERENCE 11.1(ii) 11.1(iii) 11.1(iv) 11.1(v) 11.1(vii)

LEARNING OBJECTIVES

- Understand the terms:
  (i) rate equation, rate = $k[A]^m[B]^n$, where m and n are 0, 1 or 2
  (ii) order with respect to a substance in a rate equation
  (iii) overall order of a reaction
  (iv) rate constant
  (v) rate-determining step.

RATE EQUATION

WHAT IS A RATE EQUATION?

The usual relationship between the rate of reaction and the concentration of a reactant is that the rate of reaction is directly proportional to the concentration. In other words, as the concentration is doubled, the rate of reaction doubles.

Unfortunately, this is not always the case. Sometimes the rate will double, but sometimes it will increase by a factor of four. In some cases, the rate of reaction does not increase at all, or it increases in an unexpected way.

Let us consider the simple relationship where the rate is directly proportional to the concentration of a reactant, say A. We can represent this by the expression:

$$\text{rate} \propto [A]$$

or:

$$\text{rate} = k[A]$$

where k is the proportionality constant.

This is called the *first order* **rate equation**. The constant k is called the *rate constant.*

- Every reaction has its own particular rate equation and its own rate constant.
- Rate constants will only change their value with a change in temperature.

Other common rate equations with respect to an individual reactant are:

second order rate equation: rate = $k[A]^2$

zero order rate equation: rate = $k[A]^0$ or rate = k

Zero order reactions do not occur very often, and it might be difficult at this stage for you to imagine why they should occur at all. However, you will find out shortly why they can occur.

If two or more reactants are involved, then it is possible to have a third order rate equation:

third order rate equation: rate = $k[A]^2[B]$

THE UNITS OF RATE CONSTANTS

Table A shows the units for rate constants, using mol dm^{-3} as the unit of concentration and seconds as the unit of time. You find the units by rearranging the rate equation.

ORDER	UNIT
Zero	mol dm^{-3} s^{-1}
First	s^{-1}
Second	dm^3 mol^{-1} s^{-1}
Third	dm^6 mol^{-2} s^{-1}

table A

EXAM HINT

The values for orders of reaction in rate equations will only ever be 0, 1 or 2 at International A Level.

The units are obtained by rearranging the rate equation. For example, for a second order reaction:

$$k = \frac{\text{rate}}{[A]^2}$$

Inserting the units we obtain:

$$\frac{\text{mol dm}^{-3}\,\text{s}^{-1}}{\text{mol dm}^{-3} \times \text{mol dm}^{-3}}$$

This cancels down to:

$$\frac{\text{s}^{-1}}{\text{mol dm}^{-3}}$$

which equates to $dm^3\ mol^{-1}\ s^{-1}$.

The majority of reactions involve two or more reactants. If we call the reactants A, B and C, then the reaction may be first **order** with respect to A, first order with respect to B and second order with respect to C. The *overall* rate equation will be:

$$\text{rate} = k[A][B][C]^2$$

and the **overall order** of the reaction is four (1 + 1 + 2). Note that you are adding the powers.

For a general reaction in which the orders are m, n and p, we have:

$$\text{rate} = k[A]^m[B]^n[C]^p$$

The overall order of the reaction is $m + n + p$.

REACTION MECHANISMS

Many reactions can be represented by a stoichiometric equation containing many reactant particles. For example, the reaction between manganate(VII) ions and iron(II) ions in acidic solution can be represented by:

$$MnO_4^-(aq) + 8H^+(aq) + 5Fe^{2+}(aq) \rightarrow Mn^{2+}(aq) + 4H_2O(l) + 5Fe^{3+}(aq)$$

If this reaction actually proceeded in a single step, then the reaction would be very slow indeed. The probability of 14 particles simultaneously colliding, all with the correct orientation and energy, is so small that we can say it is effectively zero. The reaction is, however, very fast indeed even at room temperature. It must, therefore, proceed via a series of steps, all of which follow on quickly from one another. It is important to recognise that a reaction involving simultaneous collision of more than two particles is very rare.

The orders of reaction of the individual reactants can help us to suggest a possible mechanism for a reaction. The mechanism cannot be inferred from the stoichiometric equation, because the mathematical relationship between the rate of reaction and the concentration of reactants (i.e. the orders of reaction) can only be determined *through experiments.*

Consider the reaction:

$$A + B + C \rightarrow D + E$$

for which the experimentally determined rate equation is:

$$\text{rate} = k[A][B]$$

This could mean that C was present in such a large excess that changes in its concentration were negligible and therefore had no measurable effect on the rate of reaction.

If, however, this is not the case, and changes in [C] really do not have any effect on the overall rate of reaction, then a different explanation must be sought for why [C] does not appear in the rate equation.

In this case, there must be a step involving a reaction between A and B that has an effect on the rate of reaction. There must also be another step in which C reacts, but this reaction has *no effect* on the overall rate of reaction. This could be explained by assuming that the reaction between A and B takes place *before* C has a chance to react, and that the reaction between A and B is significantly *slower* than the reaction involving C. If this were the case, then the mechanism for the reaction could be:

Step 1: $A + B \rightarrow Z$ SLOW

Step 2: $Z + C \rightarrow D + E$ FAST

Since only Step 1 is **rate-determining**, then only changes in [A] and [B] will affect the overall rate of reaction. Changes in the rate at which Step 2 occurs, owing to changes in [C], will be negligible.

Important points to remember are that:

- *The slowest step in a reaction determines the overall rate of the reaction.*
- The slowest step is known as the *rate-determining step* of the reaction.

We will return to the concept of reaction mechanisms, and consider them in much more detail, in **Section 11A.4**.

WORKED EXAMPLE

A useful way of visualising the idea of a rate-determining step is to imagine that three students are arranging some sheets of notes into sets (**fig A**).

Step 1: the notes are arranged into ten piles and the first student collects a sheet from each of the piles.

Step 2: the second student takes the set of ten papers and shuffles them so that they are tidy and ready for stapling.

Step 3: the third student staples the set of notes together.

The overall rate of this process (i.e. the rate at which the final sets of notes are prepared) depends on the rate of Step 1, the collecting of the sheets of notes, since this is by far the slowest step.

It does not matter, within reason, how quickly the tidying up for stapling is done. For most of the time the second and third students will be doing nothing while they wait for the first student to collect the sheets. The mechanism of the process is therefore:

Step 1: Student 1 collects sheets SLOW

Step 2: Student 2 tidies set of sheets FAST

Step 3: Student 3 staples set of sheets FAST

fig A Analogy for rate-determining step.

LEARNING TIP

All reactant species involved either in, or before, the rate-determining step have an effect on the rate of the reaction and will appear in the rate equation.

CHECKPOINT

1. The rate equation for the reaction between peroxydisulfate ions and iodide ions is:

rate = $k[S_2O_8^{2-}][I^-]$

(a) What is the order of reaction with respect to (i) peroxydisulfate ions and (ii) iodide ions?

(b) What is the overall order of reaction?

2. The rate equation for the reaction between P and Q is:

rate = $k[P][Q]^2$

What will be the increase in rate if:

(a) [P] is doubled, while [Q] is kept constant?

(b) [Q] is doubled, while [P] is kept constant?

(c) [P] and [Q] are both doubled?

3. In the reaction between R, S and T:

- when the concentration of R is doubled, the rate increases by four times
- when the concentration of S is doubled, the rate does not change
- when the concentration of T is doubled, the rate doubles.

(a) Deduce the orders of reaction with respect to R, S and T.

(b) Write the rate equation for the reaction.

(c) What is the overall order of reaction?

SKILLS PROBLEM-SOLVING

SUBJECT VOCABULARY

rate equation an equation expressing the mathematical relationship between the rate of reaction and the concentrations of the reactants

order (of a reactant species) the power to which the concentration of the species is raised in the rate equation

overall order (of a reaction) the sum of all the individual orders

rate-determining step (of a reaction) the slowest step in the mechanism for the reaction

11A 3 DETERMINING ORDERS OF REACTION

SPECIFICATION REFERENCE

11.1(ii) 11.1(vi) 11.2 11.4(i) 11.4(ii) 11.5(i) 11.5(ii) 11.5(iii)

LEARNING OBJECTIVES

- Understand the terms:
  - (i) rate equation, rate = $k[A]^m[B]^n$, where m and n are 0, 1 or 2
  - (ii) half-life.
- Be able to calculate the half-life of a reaction, using data from a suitable graph, and identify a reaction with a constant half-life as being first order.
- Understand experiments that can be used to investigate reaction rates by:
  - (i) an initial-rate method
  - (ii) a continuous monitoring method.
- Be able to deduce the order (0, 1 or 2) with respect to a substance in a rate equation, using data from:
  - (i) a concentration–time graph
  - (ii) a rate–concentration graph
  - (iii) an initial-rate method.

HOW CAN WE DETERMINE THE RATE EQUATION?

This question can equally be phrased 'How can we determine the order of reaction with respect to each reactant?'

There are two methods for determining orders of reaction. They are both experimental. Indeed, orders of reaction can only be determined by experiment.

The first is sometimes called the 'continuous method'.

The second is sometimes called the 'initial-rate method'.

THE CONTINUOUS METHOD

In this method, *one* reaction mixture is made up and samples of the reaction mixture are withdrawn at regular time intervals. The reaction in the sample is stopped, if necessary, by quenching. The concentration of the reactant is then determined by an appropriate experimental technique, such as titration.

- The first step is to draw a 'concentration–time' graph.
- The second step is to find out the **half-life** for the reaction at different concentrations.

If the half-life has a constant value, then the reaction is first order with respect to the reactant.

The 1st half-life, for the change in concentration from 120 to 60 units, is 100 s (**fig A**).

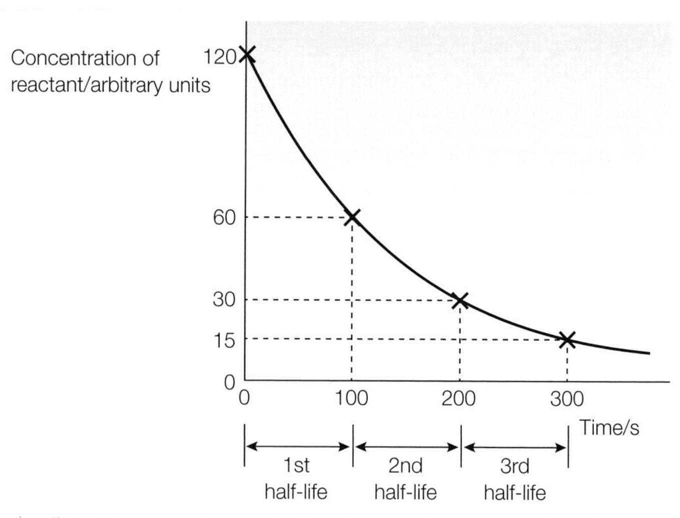

fig A A graph of concentration of reactant against time for a first order reaction.

The 2nd half-life, for the change in concentration from 60 to 30 units, is 100 s.

The 3rd half-life, for the change in concentration from 30 to 15 units, is 100 s.

Since all three half-lives are the same, the reaction is first order with respect to the concentration of the reactant plotted.

If the half-life doubles as the reaction proceeds, then the reaction is second order.

If the graph is a straight line with a negative gradient, then the rate of reaction is constant no matter what the concentration of the reactant (**fig B**). In other words, the reaction is zero order with respect to the reactant.

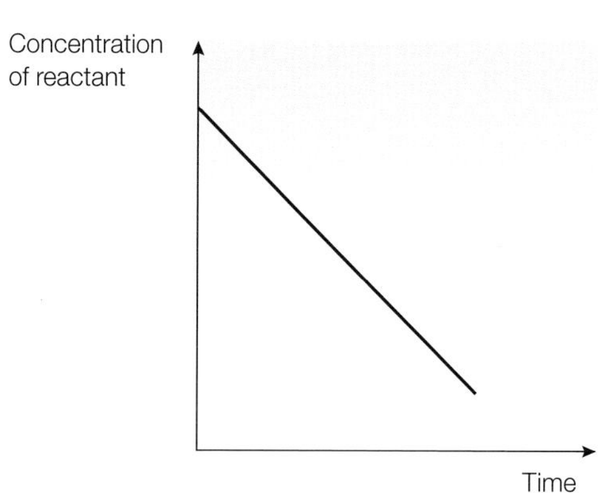

fig B A graph of concentration against time for a zero order reaction.

A typical set of results is shown in **table A** for the reaction:

$$2N_2O_5(g) \rightarrow 4NO_2(g) + O_2(g)$$

TIME/min	0	10	20	30	40	50	60	70	80	90
$[N_2O_5]/10^{-3}$ mol dm^{-3}	22.90	16.27	12.29	9.35	6.89	4.88	3.68	2.74	2.16	1.85

table A

If you plot a graph of $[N_2O_5]$ against time using the above data, you will find that the line is curved and the half-life is constant; as a result, the reaction is first order with respect to N_2O_5.

CALCULATING RATE FROM A CONCENTRATION–TIME GRAPH

The rate of reaction at any given time can be determined from a concentration–time or volume–time graph by drawing a tangent to the curve at the given time and calculating the gradient of the tangent.

Fig C shows the change in concentration of a reactant with time.

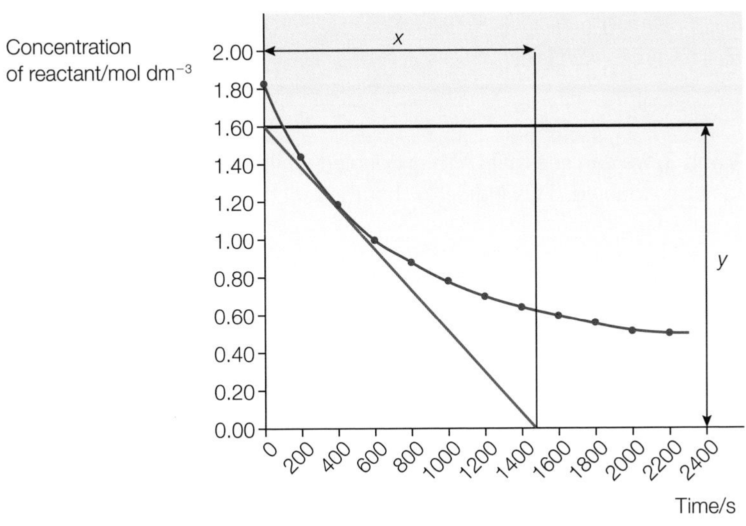

fig C Determining an instantaneous rate of reaction from a graph of concentration of reactant against time.

A tangent to the curve has been drawn at time = 400 s.

To find the rate at this time point, draw as large a triangle as possible and then measure x and y.

$x = 1470$ s and $y = -1.60$ mol dm^{-3}.

$$\text{rate} = -\frac{\text{change in concentration of reactant}}{\text{time}}$$

$$= -\frac{-1.60\ \text{mol dm}^{-3}}{1470\ \text{s}} = 1.09 \times 10^{-3}\ \text{mol dm}^{-3}\ \text{s}^{-1}$$

CALCULATING RATE FROM A VOLUME–TIME GRAPH

The procedure is exactly the same for a volume–time graph for a gas evolved (**Section 11A.1**). This time the curve will slope upwards not downwards (**fig D**).

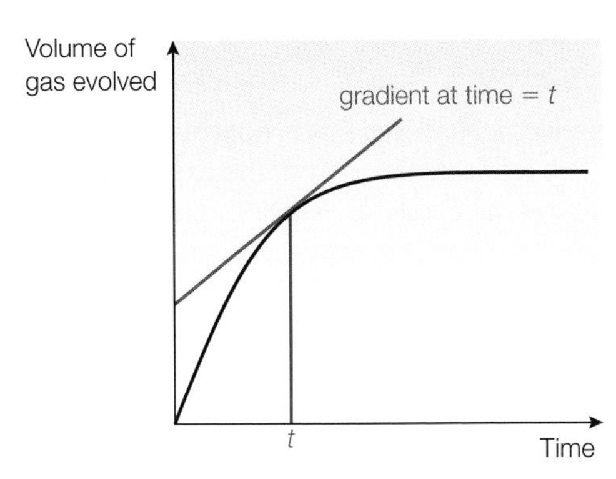

fig D Determining an instantaneous rate of reaction from a graph of volume of gas evolved against time.

The rate of reaction obtained in this way is sometimes called the **instantaneous reaction rate**.

THE INITIAL-RATE METHOD FOR DETERMINING THE RATE EQUATION

In this method, *several* reaction mixtures are made up and the initial rate (i.e. the time taken for a fixed amount of reactant to be used up or for a fixed amount of product to be formed) is measured. From these times, it is then possible to calculate the mathematical relationship between the rate of the reaction and the concentration of the reactant.

A typical set of results is shown in **table B** for the reaction:

$$A + B \rightarrow C$$

EXPERIMENT	[A]/mol dm^{-3}	[B]/mol dm^{-3}	INITIAL RATE OF FORMATION OF C/mol dm^{-3} s^{-1}
1	0.1	0.1	0.02
2	0.1	0.2	0.04
3	0.2	0.1	0.04
4	0.2	0.2	0.08

table B

If we now look at experiments 1 and 3, we can see that [A] has doubled while [B] has remained constant. The rate of reaction has also doubled. This indicates that the reaction is first order with respect to A and, as a result, we can write:

$$\text{rate} \propto [A]$$

If we now look at experiments 1 and 2, or experiments 3 and 4, we can see that [B] has doubled while [A] has remained constant. The initial rate of reaction has also doubled. This indicates that the rate of reaction is directly proportional to [B]. That is, the reaction is first order with respect to B. As a result, we can write:

$$\text{rate} \propto [B]$$

The overall order of the reaction is two, and the complete rate law (rate equation) is:

$$\text{rate} = k[A][B]$$

Note that the numbers in **table B** were deliberately kept simple to illustrate a point. Real experimental results rarely fit such an exact pattern, and you may have to look for the nearest whole numbers to obtain orders.

DETERMINING ORDER FROM A RATE–CONCENTRATION GRAPH

The method described above to determine initial rates, with changing concentrations, results in only approximate values for the initial rates. However, since most orders of reaction with respect to individual reactants have integer values, this approximation is usually acceptable.

In an initial rates experiment, the time measured is that for a fixed amount of product to be formed, or that for a fixed amount of reactant to be used up. Since the amount of product formed, or reactant used up, is kept constant, the initial rate of reaction is proportional to the reciprocal of the time, t, measured. That is:

$$\text{rate} \propto \frac{1}{t}$$

This assumes that the rate is constant for the whole time period, t. However, this is not true because as soon as the reaction starts the rate begins to decrease. The rate calculated from the expression is, therefore, the *mean (average)* rate over time, t, and not the *true* initial rate of reaction. The approximation becomes poorer with larger values of t. However, the approximation is good enough to determine integer order.

It might be possible to determine orders of reaction from just a few measurements. However, it is usual to record a range of results and plot a graph of $\frac{1}{t}$ against concentration of reactant.

The shape of the graph in **table C** indicates the order of reaction.

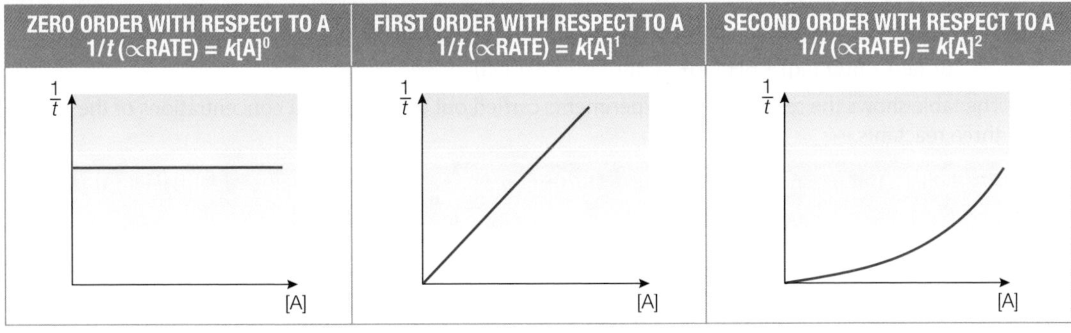

table C

It is impossible to determine directly by sight from a rate–concentration graph that the reaction is second order. If the graph is a curve as shown, then it is necessary to then plot $\frac{1}{t}$ against $[A]^2$. If this produces a straight line passing through the origin, then the reaction is second order with respect to A.

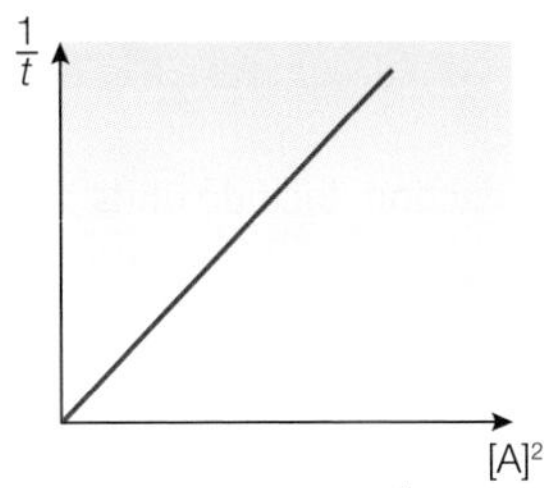

▲ **fig E** A graph of $\frac{1}{t}$ against the square of the concentration of a reactant for a second order reaction.

EXAM HINT

Be careful to check the units on the axes of graphs. Graphs with the same shape will have different meanings if there are different units on the axes!

CHECKPOINT

SKILLS INTERPRETATION

1. A compound P decomposes when heated. The graph shows the change in concentration when a sample of P is heated.

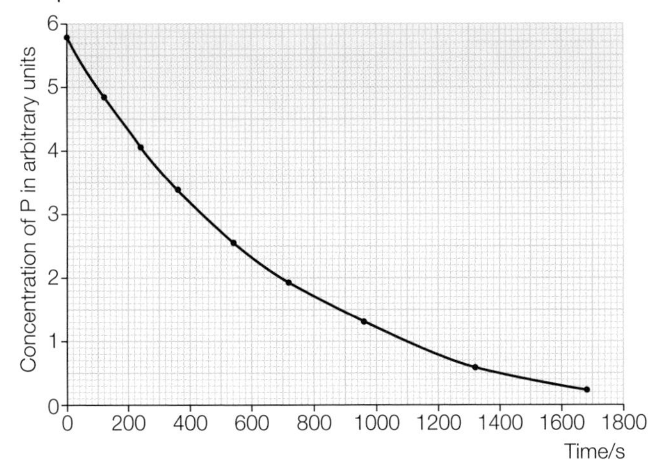

(a) State what is meant by the term *half-life of reaction*.

(b) Use the graph to show that the decomposition of P is a first order reaction.

(c) Explain the effect on the half-life of doubling the initial concentration of P.

(d) Calculate the rate constant, k, for this reaction using the following expression:

$$k = \frac{0.693}{\text{half-life}}$$

Include units in your answer.

(e) Write the rate equation for this reaction.

(f) (i) Use the graph to determine the concentration of P at 800 s.

(ii) Use the rate equation from (e) and your answers to (d) and (f)(i) to calculate the rate of reaction at 800 s. Include units in your answer.

(g) Describe how could you determine the reaction rate at 800 s directly from the graph.

SKILLS INTERPRETATION

2. The equation for the reaction between bromide ions and bromate(V) ions in acidified aqueous solution is:

$$5Br^-(aq) + BrO_3^-(aq) + 6H^+(aq) \rightarrow 3Br_2(aq) + 3H_2O(l)$$

The table shows the results of four experiments carried out using different concentrations of the three reactants.

EXPERIMENT	$[Br^-(aq)]$/ mol dm^{-3}	$[BrO_3^-(aq)]$/ mol dm^{-3}	$[H^+(aq)]$/ mol dm^{-3}	INITIAL RATE/ mol dm^{-3} s^{-1}
1	0.10	0.10	0.10	1.2×10^{-3}
2	0.10	0.20	0.10	2.4×10^{-3}
3	0.30	0.10	0.10	3.6×10^{-3}
4	0.10	0.20	0.20	9.6×10^{-3}

(a) Deduce the order of reaction with respect to:

(i) Br^-

(ii) BrO_3^-

(iii) H^+

(b) Write the rate equation for the reaction.

(c) Using the results from experiment 1, calculate the rate constant, *k*, for the reaction. Include units in your answer.

SUBJECT VOCABULARY

half-life (of a reaction) the time taken for the concentration of the reactant to fall to one-half of its initial value

instantaneous reaction rate the gradient of a tangent drawn to the line of the graph of concentration against time. The instantaneous rate varies as the reaction proceeds (except for a zero order reaction)

11A 4 RATE EQUATIONS AND MECHANISMS

SPECIFICATION REFERENCE

11.6	11.7	11.8
11.9	CP9A	CP9B

LEARNING OBJECTIVES

- Deduce the rate-determining step from a rate equation and vice versa.
- Deduce a reaction mechanism, using knowledge of the rate equation and the stoichiometric equation for a reaction.
- Understand that knowledge of the rate equations for the hydrolysis of halogenoalkanes can be used to provide evidence for S_N1 or S_N2 mechanisms for tertiary and primary halogenoalkane hydrolysis.
- Understand how to:
  (i) obtain data to calculate the order with respect to the reactants (and the hydrogen ion) in the acid-catalysed iodination of propanone
  (ii) use these data to make predictions about species involved in the rate-determining step
  (iii) deduce a possible mechanism for the reaction.

REACTION MECHANISMS

You will remember from **Book 1** that the basic view as to how a reaction takes place at a particulate level is that particles (atoms, molecules, ions or radicals) first have to collide in the correct orientation and with sufficient energy for products to be formed.

For the following reaction:

$$P + Q \rightarrow \text{products}$$

we expect the rate law to be:

$$\text{rate} = k[P][Q]$$

In other words, we might expect the reaction to be first order with respect to each reactant and second order overall.

ELEMENTARY REACTIONS

A reaction taking place in this manner (a single collision between the two reactant particles) is described as being *elementary*.

If we know that a reaction is elementary, then we can deduce the rate law directly from the stoichiometric equation. For example, the following reaction is known to be elementary:

$$NO(g) + O_3(g) \rightarrow NO_2(g) + O_2(g)$$

so the rate equation is:

$$\text{rate} = k[NO][O_3]$$

The reaction takes place when a molecule of NO collides with a molecule of O_3.

If the reaction is not elementary, it is *not* possible to deduce the rate equation by simply looking at the stoichiometric equation for the reaction.

For example, the decomposition of dinitrogen pentoxide into nitrogen dioxide and oxygen is first order with respect to dinitrogen pentoxide, not second.

$$2N_2O_5(g) \rightarrow 4NO_2(g) + O_2(g)$$

The experimentally determined rate equation is:

$$\text{rate} = k[N_2O_5]$$

A reaction that is not elementary takes place via a series of interconnected elementary reactions that are collectively called the *mechanism* for the reaction. You will have already come across a number of such mechanisms in your study of organic chemistry. For example, the radical substitution reaction between methane and chlorine to form chloromethane (CH_3Cl) is thought to have the following mechanism:

Step 1: $Cl_2 \rightarrow 2Cl\bullet$

Step 2: $Cl\bullet + CH_4 \rightarrow HCl + \bullet CH_3$

Step 3: $Cl_2 + \bullet CH_3 \rightarrow Cl\bullet + CH_3Cl$

The overall stoichiometric equation for the reaction is:

$$CH_4 + Cl_2 \rightarrow CH_3Cl + HCl$$

The species $Cl\bullet$ and $\bullet CH_3$ are called *intermediates*. They do not appear in the overall equation for the reaction, but are involved in reactions that ultimately result in the reactants being converted into the products.

If the experimentally determined rate equation does not match the overall stoichiometry, then it is almost certain that the reaction is not elementary. For example, the rate equation for the following reaction:

$$NO_2(g) + CO(g) \rightarrow NO(g) + CO_2(g)$$

is:

$$\text{rate} = k[NO_2]^2$$

This suggests that only molecules of NO_2 are involved in the rate-determining step, and that *two* molecules of NO_2 are involved in this step.

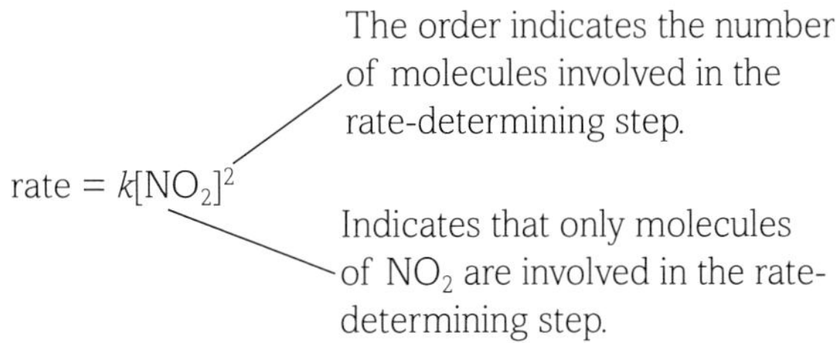

fig A Molecules involved in the rate-determining step.

Using our knowledge of molecules that do exist, two possible rate-determining steps are:

1 $NO_2 + NO_2 \rightarrow N_2O_4$, or

2 $NO_2 + NO_2 \rightarrow 2NO + O_2$

Both of these reactions are equally valid, and we have no way of knowing, without carrying out further investigations, which is the more likely to be taking place.

We also have no way of knowing what is involved in the remaining steps, but we do know that the sum of all the steps must add up to the overall stoichiometric equation.

Using the second of the two possible rate-determining steps, the following mechanism is consistent with the data:

$NO_2 + NO_2 \rightarrow 2NO + O_2$ SLOW

$CO + O_2 \rightarrow CO_2 + O$ FAST

$NO + O \rightarrow NO_2$ FAST

This shows how the particles in the proposed mechanism cancel to produce the overall equation:

$$NO_2 + \cancel{NO_2} \rightarrow \cancel{NO} + NO + \cancel{O_2}$$
$$CO + \cancel{O_2} \rightarrow CO_2 + \cancel{O}$$
$$\cancel{NO} + \cancel{O} \rightarrow \cancel{NO_2}$$
$$NO_2 + CO \rightarrow NO + CO_2$$

As it happens, further investigations into this reaction have identified the mechanism as:

$NO_2 + NO_2 \rightarrow NO_3 + NO$ SLOW

$NO_3 + CO \rightarrow NO_2 + CO_2$ FAST

This is not what we would immediately suspect as the mechanism because the existence of NO_3 is not something with which we would be familiar.

Even if the experimentally determined rate equation is simple, it does not necessarily follow that the reaction proceeds in a single elementary step. For example, the rate expression for the following gas-phase reaction:

$$2N_2O_5(g) \rightarrow 4NO_2(g) + O_2(g)$$

is:

$$\text{rate} = k[N_2O_5]$$

Yet, the reaction is thought to involve several steps and a number of intermediates.

ALKALINE HYDROLYSIS OF HALOGENOALKANES

The hydrolysis of halogenoalkanes by hydroxide ions is a reaction we introduced in **Topic 10 (Book 1: IAS)**. The hydroxide ion acts as a nucleophile and replaces the halogen in the halogenoalkanes. The reaction, therefore, can also be described as *nucleophilic substitution*.

HYDROLYSIS OF A TERTIARY HALOGENOALKANE

The equation for the alkaline hydrolysis of 2-chloromethylpropane is:

$$(CH_3)_3CCl + OH^- \rightarrow (CH_3)_3COH + Cl^-$$

The experimentally determined rate equation for this reaction is:

$$\text{rate} = k[(CH_3)_3CCl]$$

The reaction is first order with respect to 2-chloromethylpropane, but zero order with respect to the hydroxide ion.

The sensible conclusion to reach is that the 2-chloromethylpropane undergoes slow ionisation as the rate-determining step. This is then followed by a very fast step involving attack by the hydroxide ion on the carbocation formed in Step 1.

Step 1: $(CH_3)_3CCl \rightarrow (CH_3)_3C^+ + Cl^-$ SLOW

Step 2: $(CH_3)_3C^+ + OH^- \rightarrow (CH_3)_3COH$ FAST

This type of mechanism is named S_N1, i.e. *S*ubstitution *N*ucleophilic *uni*molecular.

The rate-determining step is said to be *uni*molecular because there is only *one* reactant particle present, $(CH_3)_3CCl$.

The carbocation, $(CH_3)_3C^+$, formed in Step 1 is an *intermediate* (see **Section 11A.6**).

HYDROLYSIS OF A PRIMARY HALOGENOALKANE

The equation for the alkaline hydrolysis of 1-chlorobutane is:

$$CH_3CH_2CH_2CH_2Cl + OH^- \rightarrow CH_3CH_2CH_2CH_2OH + Cl^-$$

The experimentally determined rate equation for this reaction is:

$$\text{rate} = k[CH_3CH_2CH_2CH_2Cl][OH^-]$$

This time the reaction is first order with respect to each reactant, so it is reasonable to suggest that one particle of each reactant is present in the rate-determining step of the mechanism.

The accepted mechanism for the reaction is:

$CH_3CH_2CH_2$–C(H)(H)–Cl + OH^- ⟶ $[HO \cdots C(CH_2CH_2CH_3)(H)(H) \cdots Cl]^-$ (transition state) ⟶ HO–C($CH_2CH_2CH_3$)(H)(H) + Cl^-

▲ **fig B** S_N2 mechanism of the alkaline hydrolysis of 1-chlorobutane.

This type of mechanism is named S_N2, i.e. *S*ubstitution *N*ucleophilic *bi*molecular.

The rate-determining step is *bi*molecular because there are *two* reactant particles present.

It is a continuous single one-step reaction. The complex shown in square brackets is not an intermediate (like the carbocation formed in the S_N1 mechanism), but is a *transition state* (see **Section 11A.6**).

PRACTICAL SKILLS CP9A

A study of the acid-catalysed iodination of propanone

The reaction between propanone and iodine in aqueous solution may be acid catalysed:

$$I_2(aq) + CH_3COCH_3(aq) + H^+(aq) \rightarrow CH_3COCH_2I(aq) + 2H^+(aq) + I^-(aq)$$

The influence of the iodine on the reaction rate may be studied if the concentrations of propanone and hydrogen ions effectively remain constant during the reaction. This is achieved by using a *large excess* of both propanone and acid in the original reaction mixture. You will investigate this in **CP9A Following the rate of iodine-propanone reaction by a titrimetric method**.

Procedure

1 Mix 25 cm^3 of 1 mol dm^{-3} aqueous propanone with 25 cm^3 of 1 mol dm^{-3} sulfuric acid.

2 Start the stop clock *the moment you add* 50 cm^3 of 0.02 mol dm^{-3} iodine solution. Shake well.

3 Using a pipette, withdraw a 10 cm^3 sample and place it in a conical flask. Stop the reaction by adding a 'spatula-measure' of sodium hydrogen carbonate. Note the exact time at which the sodium hydrogen carbonate is added.

4 Titrate the remaining iodine present in the sample with 0.01 mol dm^{-3} sodium thiosulfate(VI) solution, using starch indicator.

5 Withdraw further 10 cm^3 samples at suitable time intervals (approx. 5 to 7 minutes) and treat them similarly, always noting the exact time at which the sodium hydrogen carbonate is added. Waste should be contained in a fume cupboard.

Treatment of results

The Lab Book can give additional guidance in the treatment and analysis of results.

- Plot a graph of titre against time. (The titre is proportional to the concentration of iodine.)
- Deduce from the graph the order of reaction with respect to iodine.

Analysis of results

The graph produced shows that the reaction is zero order with respect to iodine.

Similar experiments show that the reaction is first order with respect to both propanone and hydrogen ions.

This gives us the following rate equation:

$$\text{rate} = k[CH_3COCH_3][H^+]$$

This would suggest the following reaction as the first step of the reaction:

$$CH_3-C(=O)-CH_3 + H^+ \rightarrow CH_3-C(=O^+-H)-CH_3$$

Similar reactions in other mechanisms are very fast, so this reaction is unlikely to be the rate-determining step of this reaction.

EXAM HINT

Examiners may ask you to **evaluate** different methods for investigating rates of reaction. One limitation of the method described here is the time delay between withdrawing a reaction sample by pipette and quenching it with sodium hydrogen carbonate.

The second step probably controls the rate of the reaction and produces H^+ that, being a catalyst, is not used up in the reaction:

$$CH_3{-}C({=}{}^+O{-}H){-}CH_3 \rightarrow CH_3{-}C(O{-}H){=}CH_2 + H^+ \quad \text{SLOW}$$

This rearrangement is likely to be very slow and hence is probably the rate-determining step of the reaction.

Iodine can now react in a fast step as follows:

$$CH_3{-}C(O{-}H){=}CH_2 + I_2 \rightarrow CH_3{-}C({=}O){-}CH_2I + I^- + H^+ \quad \text{FAST}$$

Testing the mechanism

We have a proposed mechanism for the reaction that is consistent with the kinetic data obtained from experiment. This is not the same as saying that the mechanism is the correct one, or indeed the most likely one.

We now need to carry out further experiments to confirm, or deny, our proposed mechanism. Here are three tecÚiques that can be employed in this case.

- *Use a wider range of concentrations*

  The experimentally determined rate equation may hold over only a limited range of concentrations. For example, the mechanism we have proposed for the iodination of propanone predicts that, at very low concentrations of iodine, the order of reaction with respect to iodine will change from zero to first. This is because the rate of the final step in the mechanism is given by the expression:

  Rate = $k[I_2][CH_3C(OH){=}CH_2]$

  and so the rate of the step will decrease as $[I_2]$ decreases. This has been shown to be the case, so supports the proposed mechanism.

- *Use instrumental analysis*

  This may be able to detect the presence of intermediates that have been proposed. Nuclear magnetic resonance can be used to show that acidified propanone contains about one molecule in a million in the form $CH_3C(OH){=}CH_2$. This is an intermediate stated in the proposed mechanism.

- *Carry out the reaction with deuterated propanone, CD_3COCD_3*

  Deuterium ${}^2_1H$ (or simply D) behaves slightly differently to hydrogen ${}^1_1H$ (or simply H).

  A C–D bond is slightly stronger, and therefore harder to break, than a C–H bond. If this is the bond that breaks in the rate-determining step, as suggested by our proposed mechanism, then the iodination of CD_3COCD_3 will be slower than that of CH_3COCH_3. This is found to be the case.

CHECKPOINT

1. The equation for the reaction between ethanal, CH_3CHO, and hydrogen cyanide, HCN, is:

   $CH_3CHO + HCN \rightarrow CH_3CH(OH)CN$

   Two mechanisms that have been proposed for this reaction are:

   Mechanism 1

   Step 1: $CH_3CHO + H^+ \rightarrow [CH_3CHOH]^+$

   Step 2: $[CH_3CHOH]^+ + CN^- \rightarrow CH_3CH(OH)CN$

   Mechanism 2

   Step 1: $CH_3CHO + CN^- \rightarrow [CH_3CHOCN]^-$

   Step 2: $[CH_3CHOCN]^- + H^+ \rightarrow CH_3CH(OH)CN$

   The rate equation for the reaction is:

   rate = $k[CH_3CHO][CN^-][H^+]^0$

   (a) Explain which of the two mechanisms is consistent with the rate equation.

   (b) Which step in this mechanism is the rate-determining step?

2. The equation for the reaction between hydrogen and nitrogen monoxide is:

   $2H_2(g) + 2NO(g) \rightarrow 2H_2O(g) + N_2(g)$

   The rate equation for the reaction is:

   rate = $k[H_2][NO]^2$

   A proposed mechanism for this reaction is:

   Step 1: $2NO(g) \rightarrow N_2O_2(g)$ FAST

   Step 2: $N_2O_2(g) + H_2(g) \rightarrow H_2O(g) + N_2O(g)$ SLOW

   Step 3: $N_2O(g) + H_2(g) \rightarrow N_2(g) + H_2O(g)$ FAST

   Is this mechanism consistent with the rate equation? Explain your answer.

3. Assume the following proposed reaction mechanism is correct:

   $Cl_2 \rightarrow 2Cl\bullet$ SLOW

   $H_2 + Cl\bullet \rightarrow HCl + H\bullet$ FAST

   $H\bullet + Cl\bullet \rightarrow HCl$ FAST

   (a) Write the overall equation for the reaction.

   (b) Write a rate equation for the reaction that is consistent with the mechanism.

   (c) What would be the effect on the rate of reaction of doubling the concentration of Cl_2?

   (d) What would be the effect on the rate of reaction of doubling the concentration of H_2?

4. The nucleophilic substitution reaction between equimolar quantities of CH_3Cl and OH^- is second order overall. However, if the reaction is carried out using a large excess of OH^-, the reaction becomes first order overall. Suggest an explanation for these observations.

5. Bromine can be formed by the oxidation of hydrogen bromide with oxygen.

   A proposed mechanism for this reaction is:

   Step 1: $HBr + O_2 \rightarrow HBrO_2$

   Step 2: $HBrO_2 + HBr \rightarrow 2HBrO$

   Step 3: $HBrO + HBr \rightarrow Br_2 + H_2O$

   Step 4: $HBrO + HBr \rightarrow Br_2 + H_2O$ (a repeat of Step 3)

   The rate equation for this reaction is:

   rate = $k[HBr][O_2]$

   (a) Explain which of the above four steps is the rate-determing step for this reaction.

   (b) Write the overall equation for the reaction.

11A 5 ACTIVATION ENERGY AND CATALYSIS

SPECIFICATION REFERENCE

11.1(viii) 11.1(ix) 11.11 CP10

LEARNING OBJECTIVES

- Understand the terms:
  (i) activation energy
  (ii) heterogeneous and homogeneous catalyst
  (iii) autocatalysis.
- Understand the use of a solid (heterogeneous) catalyst for industrial reactions, in the gas phase, in terms of providing a surface for the reaction.

ACTIVATION ENERGY, E_a

In **Topic 9** (**Book 1: IAS**), we defined activation energy, E_a, as the minimum energy that colliding particles must possess for a reaction to occur.

The activation energy represents the energy that the colliding particles must obtain in order to reach the energy level of the *transition state* (see **Section 11A.6** for more details). Once the energy level of the transition state has been reached, the particles can react to form the products and release energy as they do so. The energy profile diagram for an exothermic reaction is shown in **fig A**.

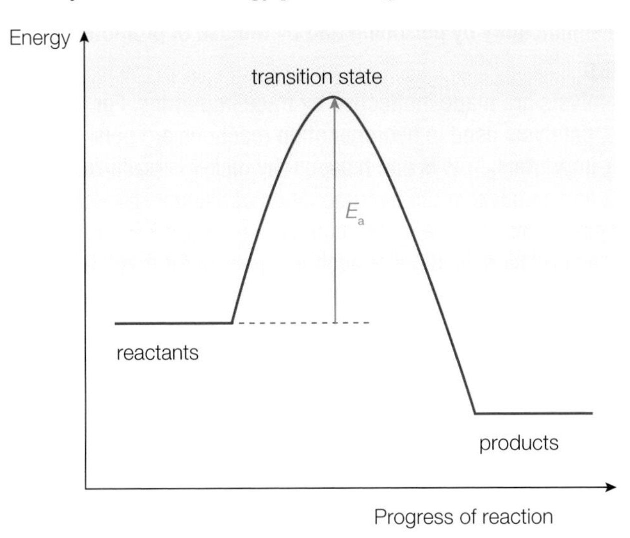

▲ **fig A** A reaction profile for an elementary exothermic reaction.

CATALYSTS

In **Topic 9** (**Book 1: IAS**), we defined a catalyst as a substance that increases the rate of a chemical reaction but is chemically unchanged at the end of the reaction.

We then went on to explain that a catalyst works by providing an alternative route for the reaction, and that this alternative route has a lower activation energy than the original route. Although the original route is still available for the reactants, most collisions resulting in reaction will occur by the alternative route, since the fraction of particles possessing this lower activation energy will be greater.

Catalysts can be divided into two groups:

- homogeneous catalysts
- heterogeneous catalysts.

HOMOGENEOUS CATALYSTS

A *homogeneous catalyst* is in the same phase (solid, liquid, solution or gas) as the reactants.

Many reactions in aqueous solution are catalysed by the hydrogen ion, $H^+(aq)$. An example is the iodination of propanone, a reaction we discussed in detail in **Section 11A.4**:

$$CH_3COCH_3(aq) + I_2(aq) \rightarrow CH_3COCH_2I(aq) + H^+(aq) + I^-(aq)$$

The production of chlorine radicals from chlorofluorocarbons (CFCs) is responsible for the destruction of ozone in the upper atmosphere. Ultraviolet radiation from the Sun produces chlorine radicals, Cl•, from CFCs such as dichlorodifluoromethane, CCl_2F_2:

$$CCl_2F_2(g) \xrightarrow{\text{ultraviolet radiation}} \bullet CClF_2(g) + Cl\bullet(g)$$

The chlorine radicals then take part in a chain reaction with ozone:

$$Cl\bullet(g) + O_3(g) \rightarrow ClO\bullet(g) + O_2(g)$$
$$ClO\bullet(g) + O_3(g) \rightarrow 2O_2(g) + Cl\bullet(g)$$

The chlorine radical is regenerated and so is acting as a catalyst. Since the catalyst is in the same phase, i.e. the gas phase, as the reacting species, it is classified as a homogeneous catalyst.

The overall reaction is:

$$2O_3(g) \rightarrow 3O_2(g)$$

Another reaction involving a homogeneous catalyst is the one between peroxydisulfate ions and iodide ions in aqueous solution:

$$S_2O_8^{2-}(aq) + 2I^-(aq) \rightarrow 2SO_4^{2-}(aq) + I_2(aq)$$

This reaction is catalysed by either $Fe^{2+}(aq)$ or $Fe^{3+}(aq)$.

With $Fe^{2+}(aq)$, the reaction mechanism is:

Step 1: $S_2O_8^{2-}(aq) + 2Fe^{2+}(aq) \rightarrow 2SO_4^{2-}(aq) + 2Fe^{3+}(aq)$

Step 2: $2Fe^{3+}(aq) + 2I^-(aq) \rightarrow 2Fe^{2+}(aq) + I_2(aq)$

With $Fe^{3+}(aq)$ as the catalyst, Steps 1 and 2 occur in the reverse order

LEARNING TIP

The action of $Fe^{2+}(aq)$ or $Fe^{3+}(aq)$ as a catalyst in the reaction between $S_2O_8^{2-}(aq)$ and $I^-(aq)$ is sometimes explained in terms of standard electrode potentials ($E^\ominus$ values).

However, it is important to note that standard electrode potentials can predict only the thermodynamic feasibility of a reaction, and not the kinetics. The reason why the reactions are faster in the presence of either Fe^{2+} or Fe^{3+} is that the activation energies for both Steps 1 and 2 are lower than the activation energy for the overall reaction.

HETEROGENEOUS CATALYSTS

A *heterogeneous* catalyst is in a different phase to that of the reactants.

Two important uses of heterogeneous catalysts in industry are in the Haber process and the Contact process. The use of solid vanadium(V) oxide (V_2O_5), in which the vanadium changes its oxidation number, in the Contact process is described in the section on transition elements (**Topic 17**).

We will now describe the action of solid iron as a catalyst in the reaction between nitrogen gas and hydrogen gas to form ammonia gas in the Haber process.

The equation for the formation of ammonia in the Haber process is:

$$N_2(g) + 3H_2(g) \rightleftharpoons 2NH_3(g)$$

Iron is able to act as a catalyst because it can form an *interstitial hydride* with hydrogen molecules. In this hydride, hydrogen atoms are held in spaces between the metal ions in the lattice. The atoms are then able to react with nitrogen molecules that are adsorbed onto the metal surface nearby. There are three stages in catalysis involving surface **adsorption**. These are:

1. *Adsorption* – the reactants are first adsorbed onto the surface of the catalyst.
2. *Reaction* – the reactant molecules are held in positions that enable them to react together.
3. *Desorption* – the product molecules leave the surface.

The rate of reaction is controlled by how fast the reactants are adsorbed and how fast the products are desorbed. As mentioned in **Topic 9** (**Book 1: IAS**), once the surface of the iron is covered with molecules, there is no further increase in reaction rate even if the pressure of the reactants is increased.

DID YOU KNOW?

The efficiency of a heterogeneous catalyst depends on the surface of the catalyst. In particular, the efficiency of the catalyst can be affected significantly by poisoning and by the use of promoters.

POISONING

Many catalysts are made ineffective by trace impurities. For example, catalysts used in hydrogenation reactions are poisoned by sulfur impurities. This is one reason why nickel is preferred to platinum as a catalyst in the hydrogenation of alkenes. Nickel is relatively inexpensive, so a large quantity can be used. If some of it is deactivated by poisoning, enough will remain for it still to be effective. On the other hand, platinum is expensive, and so small quantities would have to be used. In this case, it is possible for all of the catalyst to become poisoned.

PROMOTERS

The spacing on the surface of the catalyst is important. For example, only some surfaces of the iron crystals act as effective catalysts in the Haber process. The addition of traces of potassium oxide and aluminium oxide act as promoters by producing *active sites* where the reaction takes place most readily.

DID YOU KNOW?

An interstitial hydride, sometimes called a metallic hydride, is not strictly a compound. It is more like an alloy than a compound. The hydrogen absorbs into the metal and can exist in the form of either atoms or diatomic molecules. For example, palladium absorbs up to 900 times its own volume of hydrogen at room temperature, and forms palladium hydride. This material has been considered as a method to carry hydrogen in fuel cells for use in vehicles (see **Section 14.2.1**).

A heterogeneous catalyst is used in a three-way catalytic converter, which is employed in the exhaust of cars. This converts unburned hydrocarbons into water and carbon dioxide, and carbon and carbon monoxide into carbon dioxide. It also converts oxides of nitrogen into oxygen and nitrogen. The reactions involved are discussed in more detail in **Topic 17**.

DID YOU KNOW?

The electrophilic addition reaction between ethene and bromine vapour (**Topic 5 (Book 1: IAS)**) was once thought to be entirely a homogeneous process in which the molecules collided in the gas phase and reacted.

It is now known that the reaction takes place almost entirely on the surface of the walls of the reaction vessel. The reaction takes place very quickly on a glass surface and even more quickly when the surface of the vessel is covered with a polar substance such as cetyl alcohol (hexadecan-1-ol, $CH_3(CH_2)_{15}OH$). It takes place even more quickly still on a surface of stearic acid (octadecanoic acid, $CH_3(CH_2)_{16}COOH$), presumably because stearic acid molecules are more polar than molecules of cetyl alcohol.

However, if the walls of the reaction vessel are coated with a non-polar substance such as paraffin wax, very little reaction, if any, occurs. So, a process once thought to be a homogeneous gas phase reaction is now known to be heterogeneously catalysed.

It is sometimes stated that a catalyst does not change the products of a reaction, but only increases the rate at which the reaction takes place. This is not entirely true, as demonstrated by the different products obtained when hot ethanol vapour is passed over different solid catalysts.

With copper or nickel as the catalyst, at 400 °C, a dehydrogenation reaction takes place in which the products are ethanal and hydrogen:

$$CH_3CH_2OH(g) \longrightarrow CH_3CHO(g) + H_2(g)$$

If, at the same temperature, aluminium oxide is used as the catalyst, the more familiar dehydration reaction into ethene and water takes place:

$$CH_3CH_2OH(g) \longrightarrow CH_2{=}CH_2(g) + H_2O(g)$$

It is interesting to note that the dehydrogenation catalysts, copper and nickel, adsorb hydrogen very strongly, whereas the dehydration catalyst, aluminium oxide, adsorbs water in preference.

These two examples illustrate that there is a lot more to heterogeneous catalysis than is mentioned at International A Level. This might encourage you to do your own research into the topic and learn even more about it.

OXIDATION OF ETHANEDIOIC ACID BY MANGANATE(VII) IONS

In **autocatalysis**, the reaction is catalysed by one of its products. One of the simplest examples of this is the oxidation of ethanedioic acid by acidified potassium manganate(VII). The equation for the reaction is:

$$5(COOH)_2(aq) + 2MnO_4^-(aq) + 6H^+(aq) \longrightarrow 10CO_2(g) + 2Mn^{2+}(aq) + 8H_2O(l)$$

The reaction is very slow at room temperature, but is catalysed by manganese(II) ions, Mn^{2+}. The Mn^{2+} ions are not present initially, so the reaction starts off extremely slowly at room temperature. However, Mn^{2+} is a product of the reaction. As soon as it is produced in a catalytic amount, the reaction rate increases.

You can show this effect by plotting the concentration of one of the reactants against time. The graph obtained is unlike the normal rate curve for a reaction.

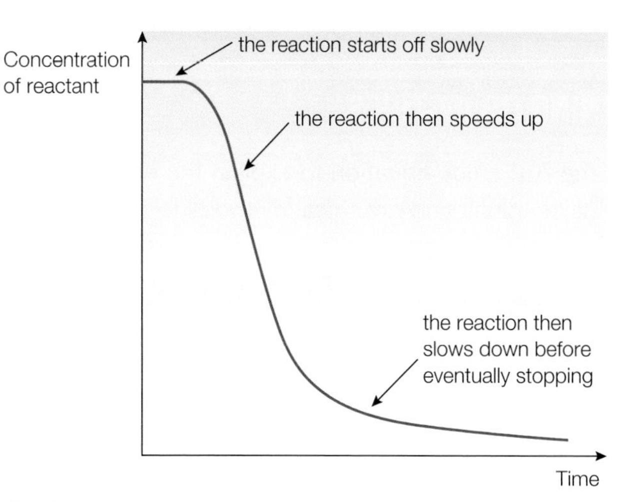

fig B A graph of concentration against time for an autocatalysed reaction.

In **fig B** you can see the slow (uncatalysed) reaction at the beginning. As catalyst begins to be formed in the mixture, the reaction speeds up – it gets faster and faster as more and more catalyst is formed. Eventually, the rate decreases in the normal way as the concentrations of the reactants decrease.

LEARNING TIP

Do not assume that a rate curve similar to the curve in **fig B** always shows an example of autocatalysis. There are other effects that might produce a similar graph.

For example, if the reaction involves a solid reacting with a liquid, liquid might have to penetrate a substance on the surface of the solid before the expected reaction can happen.

Another possibility is that the reaction is strongly exothermic and the temperature is not being controlled. The heat evolved during the reaction speeds up the reaction.

CHECKPOINT

1. Explain why only a small quantity of a homogeneous catalyst is required in order for it to be effective.
2. Cars in some parts of the world still run on leaded petrol. Leaded petrol contains a compound called tetraethyl lead, $(CH_3CH_2)_4Pb$. Tetraethyl lead reacts in the engine with oxygen and forms lead and lead(II) oxide. These substances remove radical intermediates in the combustion reactions, thus increasing the octane rating of the petrol. Excess lead and lead(II) oxide is removed by adding 1,2-dibromoethane. This forms volatile lead(II) bromide that escapes through the exhaust.

   Suggest why a catalytic converter could not be used in a car running on leaded petrol.

SUBJECT VOCABULARY

adsorption the adhesion of atoms, molecules or ions to the surface of a solid
autocatalysis when a reaction product acts as a catalyst for the reaction

11A 6 EFFECT OF TEMPERATURE ON THE RATE CONSTANT

SPECIFICATION REFERENCE 11.10

LEARNING OBJECTIVES

- Use the Arrhenius equation to explain the effect of temperature on the rate constant of a reaction.
- Use calculations and graphical methods to find the activation energy for a reaction from experimental data.

RELATIONSHIP BETWEEN TEMPERATURE AND RATE OF REACTION

In **Book 1** we discussed qualitatively why an increase in temperature increased the rate of a reaction. There are two reasons for this:

- an increase in the fraction of molecules with energy equal to or greater than the activation energy for the reaction
- an overall increase in the frequency of collisions between the reacting molecules.

The second effect is considerably less significant than the first and means that we can effectively ignore the overall increase in the frequency of collisions.

THE ARRHENIUS EQUATION

EXAM HINT

You will always be given the Arrhenius equation if it is needed in an exam question.

In 1889, Svante Arrhenius, a Swedish chemist, proposed a quantitative relationship between temperature and the rate constant, k, for a reaction. This is described by the *Arrhenius equation* and is usually expressed in the form:

$$k = Ae^{\left(-\frac{E_a}{RT}\right)}$$

where:

- A is a constant known as the pre-exponential factor, which is a measure of the rate at which collisions occur irrespective of their energy. It also includes other factors, the most important of which is that reactions can only occur when the molecules are correctly orientated at the time of collision.
- E_a is the activation energy of the reaction.
- R is the gas constant.
- T is the absolute temperature (i.e. the temperature in kelvin).

If we take natural logarithms (i.e. logarithms to the base 'e') of the Arrhenius equation. we obtain:

$$\ln k = -\frac{E_a}{R}\frac{1}{T} + \ln A$$

If a graph of $\ln k$ is plotted against $\frac{1}{T}$, a straight line is obtained with a gradient of $-\frac{E_a}{R}$

This provides an experimental method for determining the activation energy of a reaction.

The intercept with the vertical axis gives $\ln A$.

WORKED EXAMPLE

Table A shows some data related to the reaction between peroxydisulfate and iodide ions in aqueous solution:

$$S_2O_8^{2-}(aq) + 2I^-(aq) \rightarrow 2SO_4^{2-}(aq) + I_2(aq)$$

TEMPERATURE/K	MAGNITUDE OF RATE CONSTANT, k	$\ln k$	$1/T$/K^{-1}
300	0.00513	−5.27	0.00333
310	0.00833	−4.79	0.00323
320	0.0128	−4.36	0.00313
330	0.0201	−3.91	0.00303
340	0.0301	−3.50	0.00294

table A

Fig A shows ln$k$ plotted against $\frac{1}{T}$.

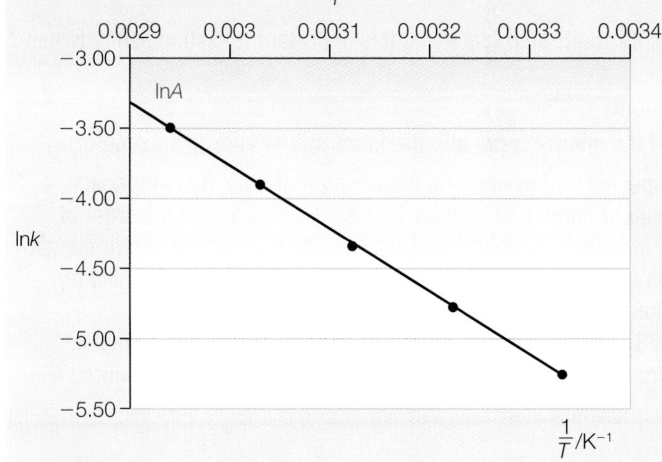

fig A A graph of ln$k$ against $1/T$.

The gradient of the line, $-\frac{E_a}{R}$

$$= -\frac{(-3.50 - -5.27)}{(0.00333 - 0.00294)}$$

$= -4538$

[$R = 8.31\,\text{J mol}^{-1}\,\text{K}^{-1}$]

$E_a = -(-4538 \times 8.31) = 37\,700\,\text{J mol}^{-1} = +37.7\,\text{kJ mol}^{-1}$ (to 3 s.f.)

DID YOU KNOW?

When we state that the relationship between ln$k$ and $-\frac{1}{T}$ is linear, we have assumed that both E_a and A remain constant over a range of temperatures. This is not strictly true. However, the change in the values of E_a and A with temperature are insignificant compared with the effect of temperature on the rate constant, and so can be ignored.

DID YOU KNOW?

The significance of the factor $e^{-\frac{E_a}{RT}}$

The factor $e^{-\frac{E_a}{RT}}$ represents the fraction of collisions that have energy equal to or greater than the activation energy, E_a. For example, a reaction with an E_a of 60 kJ mol^{-1} gives a fraction, at 298 K, of:

$$e^{-(60\,000/8.31 \times 298)} = 3 \times 10^{-11}$$

So, only one collision in 3×10^{11} has sufficient energy to react.

It is often assumed that the fraction of collisions with energy equal to or greater than E_a is the same as the fraction of molecules with this energy. This is not strictly true, but the difference is insignificant at high energies, where the fraction is very small.

It is, therefore, reasonable to draw the molecular energy distribution curve, instead of the collision frequency curve, when demonstrating the effect of temperature on collision frequency. (See **Topic 9** **(Book 1: IAS)**).

LEARNING TIP

The effect of E_a on reaction rate:

- reactions with a large E_a are slow, but the rate increases rapidly with an increase in temperature
- reactions with a small E_a are fast, but the rate does not increase as rapidly with an increase in temperature
- catalysed reactions have small values of E_a.

DID YOU KNOW?

Reaction profiles

When drawing reaction profile diagrams for multi-step reactions, it is important to distinguish between an *intermediate* and a *transition state*:

- an intermediate has an energy minimum
- a transition state occurs at the top of the energy curve and therefore has an energy maximum.

An intermediate is a definite chemical species that exists for a finite length of time. By comparison, a transition state has no significant permanent lifetime of its own: it exists for a period of the order of 10^{-15} seconds, when the molecules are in contact with one another. Even a very reactive intermediate, with a lifetime of only 10^{-6} seconds, has a long lifetime in comparison with the period that colliding molecules are in contact with one another.

A simple one-step reaction, such as the S_N2 hydrolysis of a primary halogenoalkane, has a single maximum energy. The reaction profile for the hydrolysis of a primary halogenoalkane, represented as RHal, is therefore:

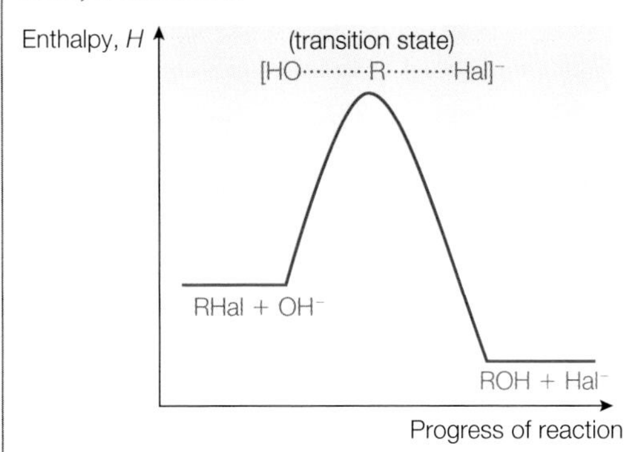

▲ **fig B** The reaction profile diagram for the hydrolysis of a primary halogenoalkane.

The reaction profile for the S_N1 hydrolysis of a tertiary halogenoalkane, R_3CHal, which is a two-step reaction involving the intermediate R_3C^+, is:

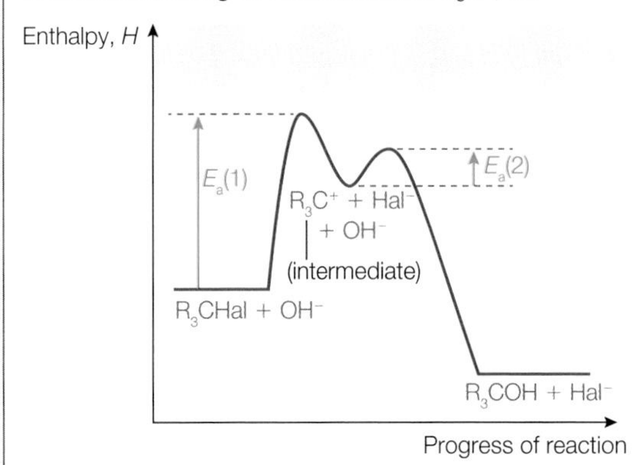

▲ **fig C** The reaction profile diagram for the hydrolysis of a tertiary halogenoalkane.

LEARNING TIP

In the S_N1 hydrolysis of a tertiary halogenoalkane, the first step of the mechanism is the rate-determining step of the reaction, and it therefore has the higher activation energy. That is, $E_a(1) > E_a(2)$.

If the second step in a reaction mechanism is rate-determining, $E_a(1) < E_a(2)$.

DID YOU KNOW?

The relative rates of S_N1 and S_N2 reactions depend on structure

You learned in **Book 1** that alkyl groups donate electrons by the inductive effect. This means that the stability of carbocations increases in the order 1° < 2° < 3°, as the number of alkyl groups donating electrons towards the positive carbon atom increases.

1° 2° 3°

As the stability of the carbocation increases, the activation energy for the reaction leading to its formation also decreases.

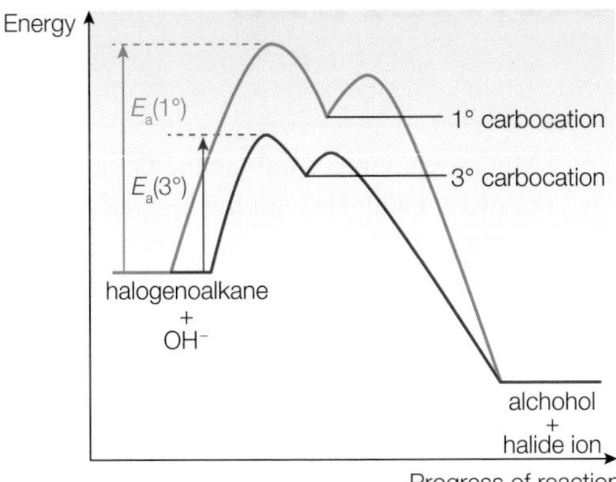

▲ **fig D** Reaction profiles for the S_N1 hydrolysis of primary and tertiary halogenoalkanes.

We therefore expect the rate of the S_N1 reaction to increase in the order 1° < 2° < 3°.

No carbocations are formed during the S_N2 reaction. In an S_N2 reaction, the transition state has five groups arranged around the central carbon atom. It is therefore more crowded than either the starting halogenoalkane or the alcohol product, each of which have only four groups around the central carbon atom.

Alkyl groups are much larger than hydrogen atoms. Therefore, the more alkyl groups around the central carbon atom, the more crowded will be the transition state, and the higher the activation energy for its formation.

We therefore expect the rate of the S_N2 reaction to increase in the order 3° < 2° < 1°.

The two effects reinforce one another. The S_N1 reaction is fastest with tertiary halides and slowest with primary halides, while the S_N2 reaction is fastest with primary halides and slowest with tertiary halides. Overall, primary halides react predominantly via the S_N2 mechanism and tertiary halides react predominantly via the S_N1 mechanism. Secondary halides tend to react via a mixture of the two mechanisms.

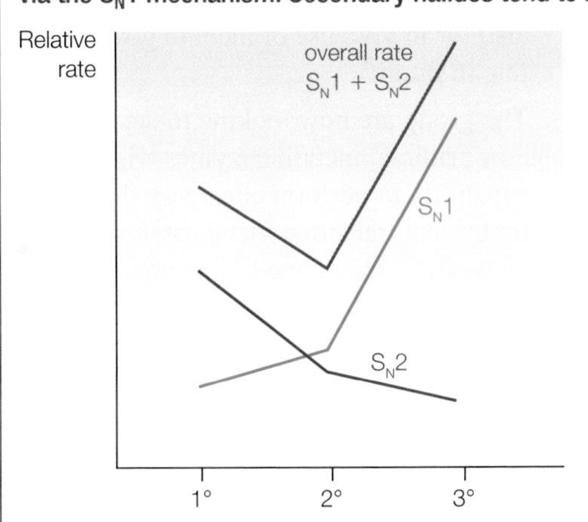

▲ **fig E** S_N1 and S_N2 hydrolysis reactions for primary, secondary and tertiary halogenoalkanes.

CHECKPOINT

SKILLS CREATIVITY

1. Use the Arrhenius equation to explain why an increase in temperature results in an increase in the rate of reaction.
2. The rate constant, k, for a reaction increases from $10.0\ s^{-1}$ to $100.0\ s^{-1}$ when the temperature is increased from 300 K to 400 K.

   Calculate the activation energy, E_a, for this reaction. [$R = 8.31\ J\ mol^{-1}\ K^{-1}$].

CATALYST CRAFT

SKILLS CREATIVITY

In this article you will look at an example of how an engineered metalloenzyme can improve both the kinetics and specificity of a reaction. The extract is from *Chemistry World*, the print and online magazine of the Royal Society of Chemistry.

ENGINEERED METALLOENZYME CATALYSES FRIEDEL–CRAFTS REACTION

Reprogramming the genetic code of bacteria to include an unnatural amino acid has allowed scientists in the Netherlands to create a new metalloenzyme capable of catalysing an enantioselective reaction.

'Nature is extremely good at catalysing reactions with very high rate accelerations and very high selectivity. But it does so, from our perspective, with a relatively limited set of reactions,' explains Gerard Roelfes from the University of Groningen in the Netherlands, who led the study. His group is looking at existing reactions that use traditional catalysts, but fail to achieve the same rate acceleration and selectivity as enzyme catalysed reactions.

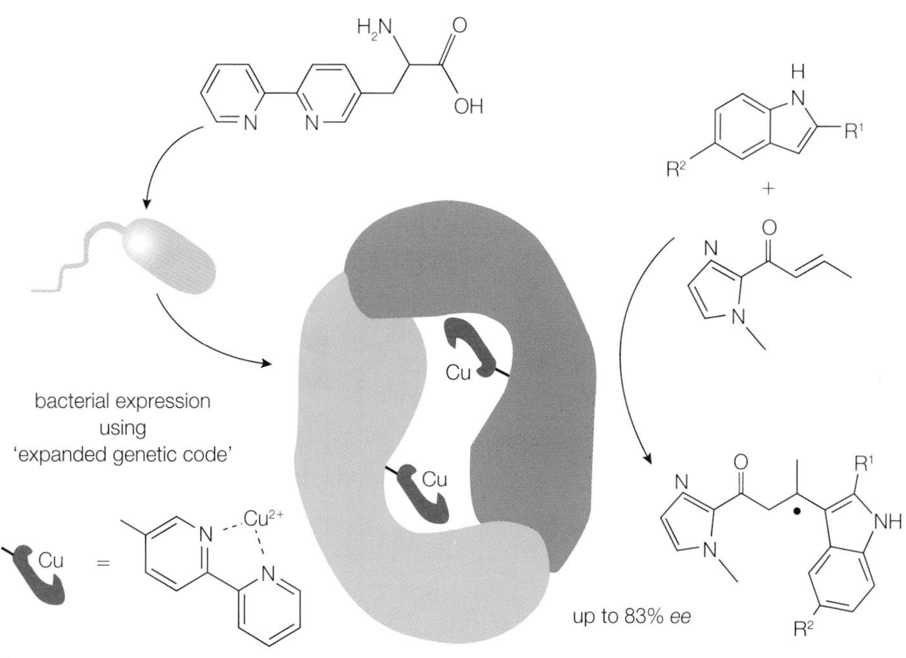

▲ **fig A** The artificial metalloenzymes were applied in a catalytic asymmetric Friedel–Crafts alkylation reaction.

Metalloenzymes combine the flexibility of metal catalysts with the high activity and selectivity of enzymes. Artificial metalloenzymes are produced by inserting a catalytically active transition metal complex into a biomolecular scaffold, like a protein.

Roelfes' team engineered *Escherichia coli* cells to include a copper-binding amino acid into one of its proteins. This method requires no further chemical modification or purification steps, giving it an advantage over existing methods. The resulting metalloenzyme was tested on a catalytic asymmetric Friedel–Crafts alkylation reaction, achieving an enantiomeric excess of up to 83%.

Takafumi Ueno, who researches the mechanisms of chemical reactions in living cells at the Tokyo Institute of Technology in Japan, is impressed by the work. 'It could be applied not only for rapid screening of artificial metalloenzymes but also for in vivo use of them to govern cell fate in future'.

The group are now looking to develop new artificial metalloenzymes with the capability to perform chemistry that traditional transition metal catalysts cannot. Ultimately, they hope to integrate these enzymes into biosynthetic pathways.

Adapted (with permission) from 'Engineered Metalloenzyme Catalyses Friedal–Crafts Reaction' by Debbie Houghton, *Chemistry World*, 3 November 2014

SCIENCE COMMUNICATION

1 (a) After reading the article for the first time, write a one-paragraph summary of what you think the article is about.

(b) Now write down a list of any words from the article of which you do not know the meaning. Do some research into what they mean. Is there anything about your summary from (a) that you would change?

INTERPRETATION NOTE

You may find that it is easier than you think to get an overall understanding of what a detailed scientific article is about. You do not need to understand every single word to learn something new and interesting.

CHEMISTRY IN DETAIL

2 Assuming the reaction is exothermic, sketch an energy profile for the catalysed reaction showing:

(a) the overall reaction enthalpy change

(b) the activation energy for the formation of the enzyme–substrate complex.

3 Under certain conditions the rate equation for an enzyme-catalysed reaction takes the form: Rate = $k[E]^n$, where [E] is the enzyme concentration and k and n are experimentally determined constants. What does this rate equation suggest about the reaction mechanism under these conditions?

4 Name the type of bond that is formed between the copper ion and the unusual amino acid in the metalloenzyme shown in **fig A**.

5 Suggest why the use of a metalloenzyme of this type might be a better alternative than the traditional Friedel-Crafts methods for alkylation.

6 Suggest why this method might be expected to increase the enantioselectivity of this reaction.

7 Suggest why the Zn^{2+} ion may not be a suitable replacement for the Cu^{2+} ion in this enzyme.

THINKING BIGGER TIP

When you are asked to draw something, make sure you use a sharp HB pencil so that your drawing is clear, and use a ruler if you are drawing straight lines.

ACTIVITY

Although the metalloenzyme detailed above is genetically engineered, there are many naturally occurring metalloenzymes that carry out a range of reactions in organisms. Most of these can be found in a freely accessible online database called the PDB (alternatively, the PDBe).

Choose one of the following metals: Fe, Zn, Cu, Mo, Co.

- Find a protein that contains the metal.
- Find out what the protein does.
- Find out how much of the metal is present in a typical healthy adult.

Give a five-minute presentation on your chosen metal, featuring no more than five slides. You will have the opportunity to display your chosen molecules using some of the freely available software packages such as Jmol.

DID YOU KNOW?

Although a typical adult has only about 1.5 mg of copper per kg of body mass, an inability to regulate copper has disastrous effects. For example, the inability to excrete excess copper from the body (Wilson's disease) can be fatal if left untreated. Menkes' syndrome manifests itself if the body is unable to retain (hold on to) copper. Although not fatal, it can lead to developmental delay and neurological problems.

11 EXAM PRACTICE

1 An experiment is set up to measure the rate of hydrolysis of methyl ethanoate, CH_3COOCH_3.

$$CH_3COOCH_3 + H_2O \rightleftharpoons CH_3COOH + CH_3OH$$

The hydrolysis is very slow in neutral aqueous solution. When dilute hydrochloric acid is added, the reaction is faster.

What is the function of the hydrochloric acid?

A to increase the reaction rate by acting as a catalyst
B to make sure that the reaction reaches equilibrium
C to maintain a constant pH during the reaction
D to dissolve the methyl ethanoate [1]

(Total for Question 1 = 1 mark)

2 For the gaseous reaction $2X(g) + Y(g) \rightarrow Z(g)$, the rate equation is

rate = $k[X]^2[Y]^0$

If the pressure in the reaction vessel is doubled but the temperature remains constant, by what factor does the rate of reaction increase?

A 2 **B** 3 **C** 4 **D** 8 [1]

(Total for Question 2 = 1 mark)

3 The alkaline hydrolysis of RBr, where RBr = $(CH_3)_3Br$, takes place in two steps:

$RBr \rightarrow R^+ + Br^-$ SLOW

$R^+ + OH^- \rightarrow ROH$ FAST

Which of the following rate equations is consistent with this scheme?

A rate = $k[OH^-]$
B rate = $k[RBr]$
C rate = $k[RBr][OH^-]$
D rate = $k[R^+][OH^-]$ [1]

(Total for Question 3 = 1 mark)

4 The table gives data for the reaction between X and Y at constant temperature.

Experiment	[X] / mol dm^{-3}	[Y] / mol dm^{-3}	Initial rate / mol dm^{-3} s^{-1}
1	0.3	0.2	4.0×10^{-4}
2	0.6	0.4	1.6×10^{-3}
3	0.6	0.8	6.4×10^{-3}

What is the rate equation for the reaction?

A rate = $k[X][Y]^2$
B rate = $k[X]^2[Y]$
C rate = $k[X]^2$
D rate = $k[Y]^2$ [1]

(Total for Question 4 = 1 mark)

5 The reaction of acidified aqueous potassium iodide with aqueous hydrogen peroxide

$$2I^-(aq) + H_2O_2(aq) + 2H^+(aq) \rightarrow I_2(aq) + 2H_2O(l)$$

is thought to involve three steps:

$H_2O_2 + I^- \rightarrow H_2O + OI^-$ SLOW

$OI^- + H^+ \rightarrow HOI$ FAST

$HOI + H^+ + I^- \rightarrow I_2 + H_2O$ FAST

Which of the following conclusions **cannot** be deduced from this information?

A The acid is a catalyst.
B The reaction is first order with respect to the iodide ion.
C The rate determining step is $H_2O_2 + I^- \rightarrow H_2O + OI^-$
D The rate equation for the reaction is: rate = $k[H_2O_2][I^-]$ [1]

(Total for Question 5 = 1 mark)

6 Propanone and iodine react in aqueous acidic solution according to the following overall equation:

$$CH_3COCH_3 + I_2 + H^+ \rightarrow CH_3COCH_2I + 2H^+ + I^-$$

The experimentally determined rate equation for this reaction is:

rate = $k[CH_3COCH_3][H^+]$

(a) With initial concentrations as shown, the initial rate of reaction was 1.43×10^{-6} mol dm^{-3} s^{-1}.

	Initial concentration/mol dm^{-3}
CH_3COCH_3	0.400
H^+	0.200
I_2	4.00×10^{-4}

Calculate a value for the rate constant, k, for the reaction. [3]

(b) Explain the effect on the rate of reaction of doubling the concentration of iodine, but keeping the concentrations of propanone and hydrogen ions constant. [2]

(c) The proposed mechanism for the overall reaction is:

Step 1 $CH_3COCH_3 + H^+ \rightarrow CH_3–C^+(OH)–CH_2–H$

Step 2 $CH_3–C^+(OH)–CH_2–H \rightarrow CH_3–C(OH){=}CH_2 + H^+$

Step 3 $CH_3–C(OH){=}CH_2 + I_2 \rightarrow CH_3–C({=}O^+–H)–CH_2I + I^-$

Step 4 $CH_3–C({=}O^+–H)–CH_2I \rightarrow CH_3–C({=}O)–CH_2I + H^+$

Explain which of the four steps could be the rate-determing step. [3]

(Total for Question 6 = 8 marks)

7 Most chemical reactions involve two or more steps. The experimentally determined rate equation indicates which species are involved either before or in the rate-determining step.

(a) State what is meant by the term **rate-determining step**. [1]

(b) Hydrogen reacts with iodine monochloride in a two-step mechanism according to the following overall equation:

$H_2(g) + 2ICl(g) \rightarrow 2HCl(g) + I_2(g)$

The experimentally determined rate equation for this reaction is:

rate = $k[H_2(g)][ICl(g)]$

The rate-determining step is the first step in the mechanism for the reaction.

(i) Write an equation for the rate-determining step. [1]

(ii) Write an equation for the second step. [1]

(c) A series of experiments were carried out on the reaction A + B + C → products. The results are shown in the table.

Experiment	Initial concentrations/ mol dm^{-3}			Initial rate/ mol dm^{-3} s^{-1}
	[A]	[B]	[C]	
1	0.100	0.100	0.100	6.20×10^{-4}
2	0.100	0.200	0.100	6.20×10^{-4}
3	0.100	0.100	0.200	2.48×10^{-3}
4	0.200	0.100	0.100	1.24×10^{-3}

(i) Determine the order of reaction with respect to A, B and C. Show how you obtain your answers. [3]

(ii) Write the rate equation for the reaction. [1]

(iii) Which reactant is unlikely to be in the rate determining step? [1]

(Total for Question 7 = 8 marks)

8 Nitrogen pentoxide decomposes on heating to form nitrogen tetroxide and oxygen. The equation for the reaction is:

$N_2O_5(g) \rightarrow N_2O_4(g) + \frac{1}{2}O_2(g)$

The progress of the reaction can be followed by measuring the concentration of N_2O_5 present. The graph shows the results obtained in an experiment conducted at constant temperature.

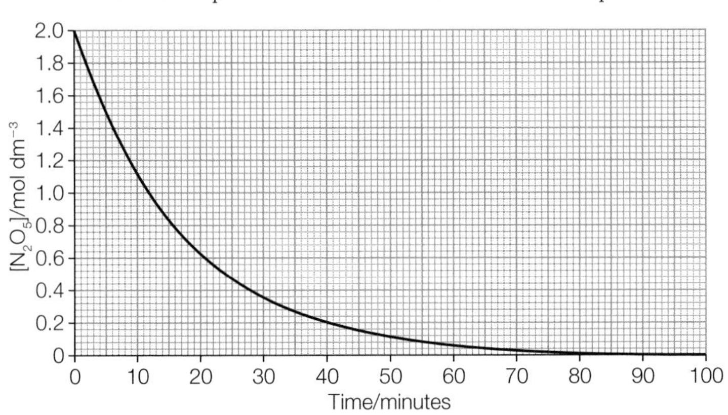

(a) (i) Use the graph to predict the rate of reaction after 20 minutes and after 90 minutes. [3]

(ii) Deduce the rate of production of oxygen after 20 minutes? [1]

(b) (i) Plot on the graph two successive half-lives for this reaction. [2]

(ii) Deduce the order of reaction with respect to N_2O_5. Justify your answer. [2]

(iii) Write a rate equation for the reaction. [1]

(c) (i) Calculate a value for the rate constant, k, for this reaction. [3]

(ii) Deduce the initial rate of reaction when the initial concentration of N_2O_5 is 1.50 mol dm^{-3}. Assume the reaction is carried out under the same conditions of temperature and pressure. [2]

(Total for Question 8 = 14 marks)

9 Nitrogen dioxide, NO_2, is decomposed on heating into nitrogen monoxide, NO, and oxygen. The proposed mechanism for this reaction is:

$NO_2 + NO_2 \rightarrow NO_3 + NO$ SLOW

$NO_3 \rightarrow NO + O_2$ FAST

(a) (i) Write an overall equation for the decomposition of nitrogen dioxide. [1]

(ii) Write a rate equation for the reaction. [1]

(b) The rate constant, k, for the decomposition was determined at several different temperatures. The results obtained were used to plot the graph shown below.

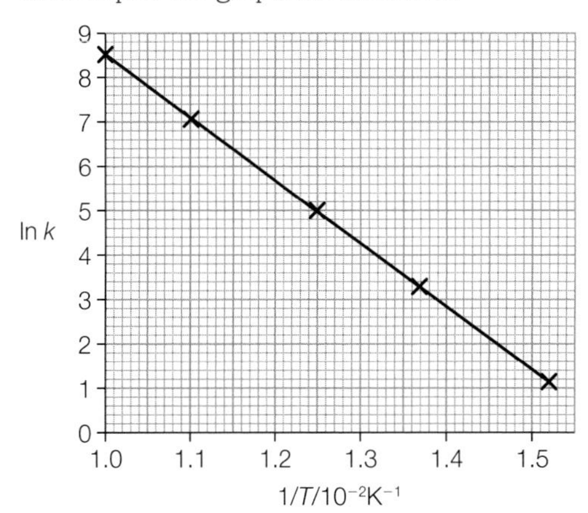

(i) Use the graph and the expression

$$\ln k = -\frac{E_a}{R}\frac{1}{T} + \ln A$$

to calculate the activation energy, E_a, for the thermal decomposition of nitrogen dioxide. Give your answer to an appropriate number of significant figures. [4]

(ii) A vessel of volume 2.00 dm^3 is filled with 4.00 mol of ntitrogen dioxide at a temperature of 650 K. The initial rate of reaction was found to be 12.64 mol dm^{-3} s^{-1}.

Calculate a value for the rate constant, k, at this temperature. [3]

(Total for Question 9 = 9 marks)

TOPIC 12 ENTROPY AND ENERGETICS

A ENTROPY | B LATTICE ENERGY

People lived very simply in the Stone Age. Their energy needs were supplied by the Sun, and by burning plants and trees. Because the plants and trees were continuously renewed, and populations were small, energy supplies during the Stone Age period were plentiful. However, this simple lifestyle has almost completely vanished, and people today cannot survive without abundant fuel.

Today, most of the world's energy is supplied from fossil fuels, which consist of natural gas, coal and products from crude oil, such as gasoline, diesel and kerosene. Fossil fuels are not renewable and will one day run out. To manage these energy resources and develop new fuels we need to understand how energy is released or used in chemical reactions, including the reactions in plants, animals and our own bodies.

Energy is an essential part of chemistry as well as of civilisation as we know it. In chemistry, energy determines which reactions can occur and which compounds can exist. You have already studied a large number of chemical reactions. All of these reactions either absorb or release energy. In this topic we will develop an understanding of how to measure and report these energy changes.

Have you ever wondered why water evaporates? Why hot objects cool? Why hydrogen combines with oxygen? Why green leaves turn red in the autumn? Why anything happens? Part of the answer is related to energy. We need energy to think, to move and to live. Every chemical reaction makes use of energy to rearrange the bonding between elements in compounds. Thermodynamics deals with questions like these.

In the first half of this topic we will deal with the second law of thermodynamics, which governs the direction of natural change. The second law enables us to predict whether or not a reaction has a tendency to occur, and to what extent it will occur. The second law is of fundamental importance in chemistry because it provides a basis for discussing, explaining and predicting equilibria, the subject of **Topics 13** and **14** in this book. The second law is also the foundation of the whole field of electrochemistry, the subject of **Topic 16**.

MATHS SKILLS FOR THIS TOPIC

- Recognise and make use of appropriate units in calculations
- Recognise and use expressions in decimal and ordinary form
- Use calculators to find and use power, exponential and logarithmic functions
- Use an appropriate number of significant figures
- Change the subject of an equation
- Substitute numerical values into algebraic equations using appropriate units for physical quantities
- Solve algebraic equations
- Use logarithms in relation to quantities that range over several orders of magnitude

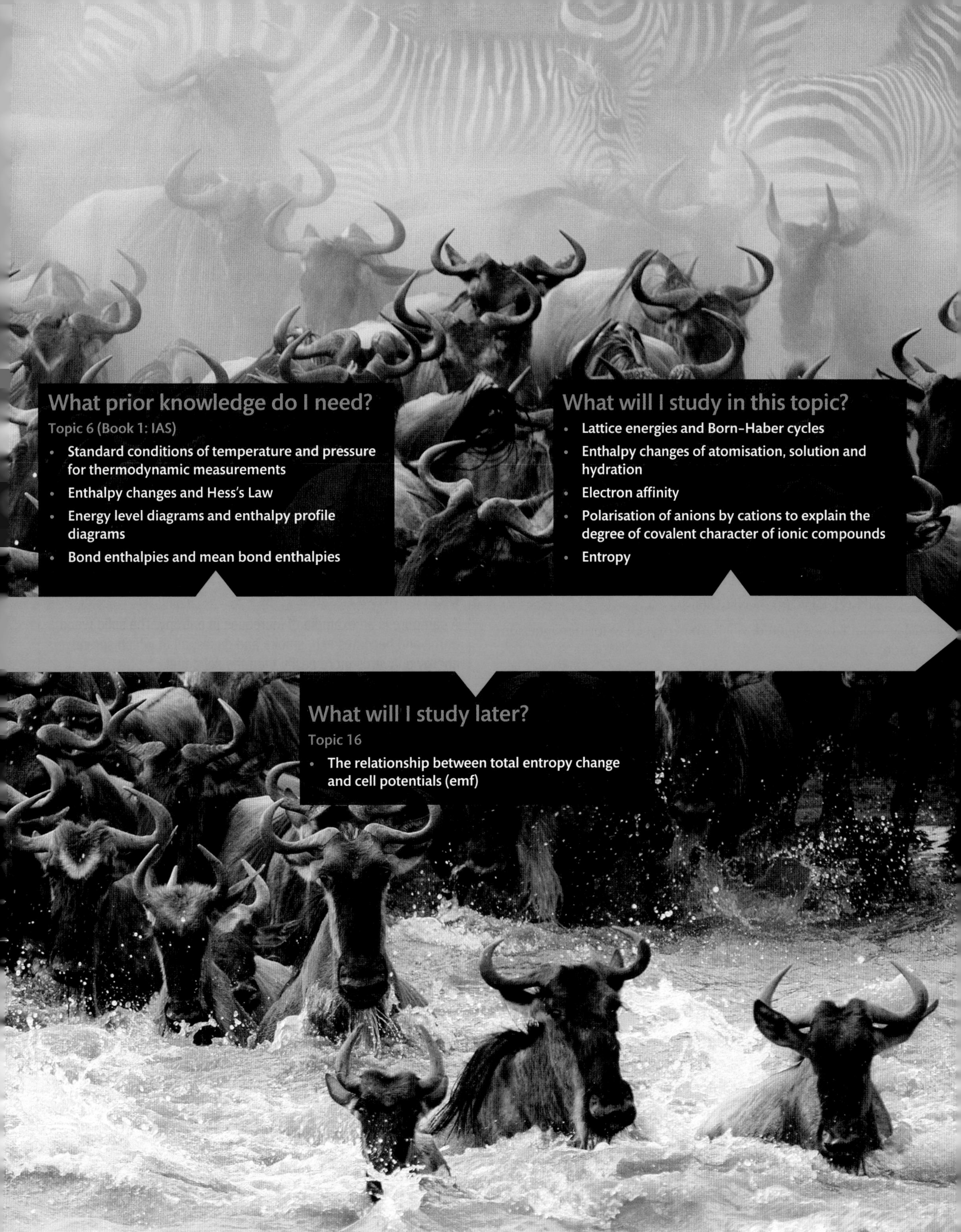

What prior knowledge do I need?

Topic 6 (Book 1: IAS)

- Standard conditions of temperature and pressure for thermodynamic measurements
- Enthalpy changes and Hess's Law
- Energy level diagrams and enthalpy profile diagrams
- Bond enthalpies and mean bond enthalpies

What will I study in this topic?

- Lattice energies and Born–Haber cycles
- Enthalpy changes of atomisation, solution and hydration
- Electron affinity
- Polarisation of anions by cations to explain the degree of covalent character of ionic compounds
- Entropy

What will I study later?

Topic 16

- The relationship between total entropy change and cell potentials (emf)

12A 1 INTRODUCTION TO ENTROPY

SPECIFICATION REFERENCE 12.1 12.2 12.3 12.4

LEARNING OBJECTIVES

- Understand that, since endothermic reactions can occur spontaneously at room temperature, enthalpy changes alone do not control whether reactions occur.
- Understand entropy as a measure of disorder of a system in terms of the random dispersal of molecules and of energy quanta between molecules.
- Understand that the entropy of a substance increases with temperature, that entropy increases as solid → liquid → gas and that perfect crystals at zero kelvin have zero entropy.
- Be able to interpret the natural direction of change as being in the direction of increasing total entropy (positive entropy change), including gases spread spontaneously through a room.

WHAT MAKES A REACTION OCCUR?

Perhaps one of the most important questions to ask in chemistry is 'Will a reaction occur?'

We know that, once started, some chemical reactions simply 'go' with no further, continuous help from us.

For example, ammonia gas and hydrogen chloride gas react together at room temperature to form the white solid, ammonium chloride:

$$NH_3(g) + HCl(g) \rightarrow NH_4Cl(s)$$

Magnesium, once ignited, will burn in oxygen to form magnesium oxide:

$$Mg(s) + \frac{1}{2}O_2(g) \rightarrow MgO(s)$$

In other reactions, rather than the reactants changing completely into the products, a position of equilibrium is reached, with the final mixture containing a measurable amount of both reactants and products.

For example, ethanoic acid dissociates in water. In a 0.1 mol dm^{-3} solution of ethanoic acid, only about 1% of the ethanoic acid molecules are present as ions:

$$CH_3COOH(aq) \rightleftharpoons CH_3COO^-(aq) + H^+(aq)$$

Another example is the dimerisation of nitrogen dioxide in the gas phase:

$$2NO_2(g) \rightleftharpoons N_2O_4(g)$$

At a temperature of 298 K and a pressure of 100 kPa, the equilibrium mixture contains about 70% N_2O_4.

There are other types of reactions that simply do not occur at all, at least not without some help. For example:

- ammonium chloride does not spontaneously decompose into ammonia and hydrogen chloride
- magnesium oxide does not break apart to form magnesium and oxygen without some continuous intervention from us in the form of heating.

If we consider all of these types of reaction together, we see that the real difference between them is the *position* of equilibrium that is established. For some reactions, the position of equilibrium is so far over to the products side that, to all intents and purposes, the reaction has gone to completion. For some other reactions, significant amounts of both reactants and products are present at equilibrium. For other reactions, the equilibrium lies so far to the left that they appear not to take place at all.

So, perhaps the better question to ask is *not* 'Will a chemical reaction occur?' but 'What will be the position of equilibrium?' This is the question we hope to answer in this topic.

DID YOU KNOW?

A campfire is an example of increases in entropy. The solid wood burns and becomes ash, smoke and gases, all of which spread energy outwards more easily than the solid fuel.

EXOTHERMIC AND ENDOTHERMIC REACTIONS

The reaction between magnesium and oxygen is exothermic and can be represented by the enthalpy level diagram shown in **fig A**.

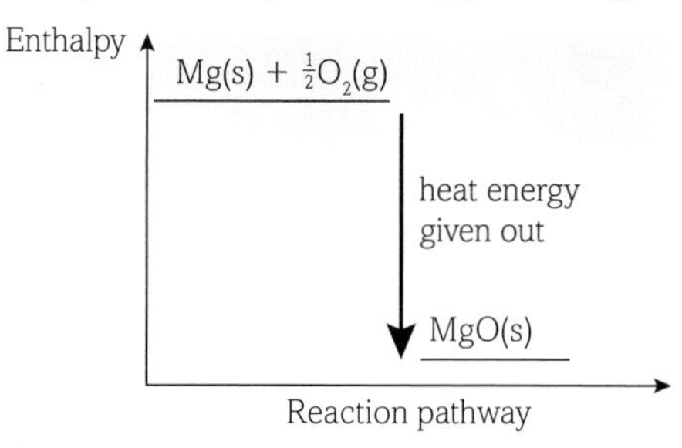

▲ **fig A** Enthalpy level diagram for the reaction between magnesium and oxygen.

As the products have less energy than the reactants, we often say that the products are more *energetically stable* than the reactants.

It is very tempting to conclude that this reaction occurs *because* the magnesium oxide is energetically more stable than its elements, magnesium and oxygen. However tempting this argument may be, we must discard it, because experience tells us that many endothermic reactions occur at room temperature.

Let us consider the dimerisation of $NO_2(g)$ at 298 K:

$$2NO_2(g) \rightleftharpoons N_2O_4(g)$$

The enthalpy level diagram for this reaction is shown in **fig B**.

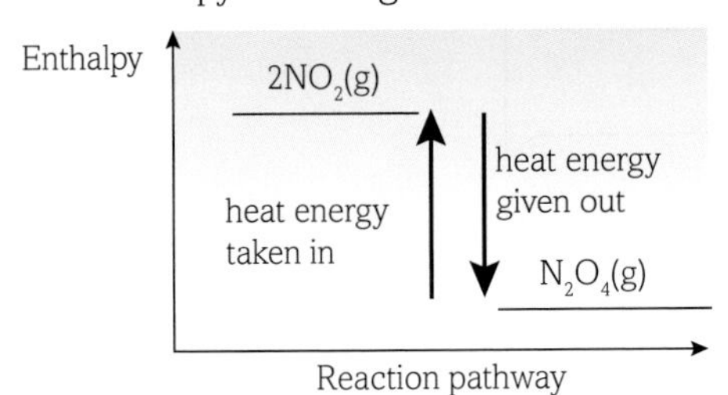

▲ **fig B** Enthalpy level diagram for the dimerisation of NO_2.

This reaction is readily reversible and reaches a position of equilibrium at 298 K. The forward reaction leading to the formation of $N_2O_4(g)$ is exothermic. However, the backward reaction is endothermic.

As we discovered in **Topic 9** (**Book 1: IAS**), the position of equilibrium can be reached from either direction. This means that if we place some $N_2O_4(g)$ in a sealed container at room temperature, some of it will decompose to form $NO_2(g)$. So, an endothermic reaction is taking place at room temperature without us doing anything to it. We say that the reaction is *spontaneous* (see later in this section).

Clearly, it is not only exothermic reactions that can take place spontaneously. The driving force for spontaneous endothermic reactions cannot be the formation of more energetically favourable, i.e. lower-energy, products but must involve another factor we have yet to consider.

That factor is a quantity known as **entropy**, which is governed by the *Second Law of Thermodynamics*.

But first, we need to appreciate what is meant by the term **spontaneous process**.

SPONTANEOUS PROCESSES

A spontaneous process is one that takes place without continuous intervention by us.

A good example of a spontaneous process is the freezing of water to form ice. If water is placed in an environment at −20 °C it will turn into ice. However, the reverse will never happen: at −20 °C ice will never melt to form water.

Another example of a spontaneous process is the mixing of gases. If two unreactive gases are present in a container, they will mix completely in a process known as diffusion.

We can bring about the reverse of spontaneous processes by intervening. For example, we can melt ice by heating it, and we can separate a mixture of gases by liquefying followed by distillation.

The key point, however, is that the reverse of a spontaneous process never happens on its own. A mixture of gases will *never* separate on their own.

At the beginning of this section we stated that, in chemistry, we are concerned with whether or not chemical reactions will occur, and we suggested that perhaps the most important question to ask in chemistry is 'Why does a reaction occur?'

From the point of view of thermodynamics, all reactions are reversible, so we suggested that an even better question to ask is:

> When a reaction reaches a point of equilibrium, what determines whether the equilibrium will favour the reactants or the products, and to what extent?

This question is answered using the Second Law of Thermodynamics, in which the concept of entropy is introduced.

ENTROPY

The Second Law of Thermodynamics allows us to predict whether a process is likely to occur. It is, therefore, the key to understanding what drives chemical reactions and what determines the position of equilibrium.

The Second Law of Thermodynamics introduces a term called entropy, which is a property of matter, just like density or energy. The simplest way to describe entropy is to say that it is a quantity associated with randomness or disorder.

This definition is rather basic and has to be used with care. Entropy refers not only to the distribution of molecules but also to the ways of distributing the energy of the system in all of the energy levels available.

To help you to understand entropy, it helps to think about what happens during the diffusion of gases.

DISTRIBUTION OF MOLECULES

Fig C shows two gas jars. The bottom one contains bromine gas and the top one contains air.

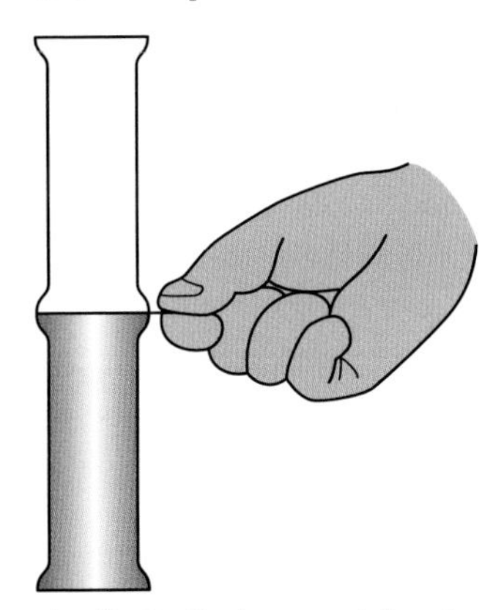

▲ **fig C** Gas jars containing air and bromine

When the cover slip between the two gas jars is removed, the bromine and air diffuse so that the molecules of each gas are spread out evenly throughout both gas jars.

To understand why this happens we can look at a much simpler example where there are only five molecules. **Fig D** shows two gas jars. The left-hand jar contains five molecules of bromine and the right-hand jar contains no molecules.

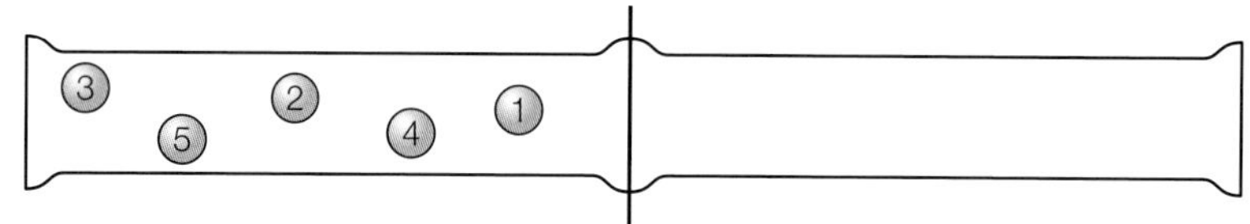

Five molecules of bromine start off in the left-hand jar

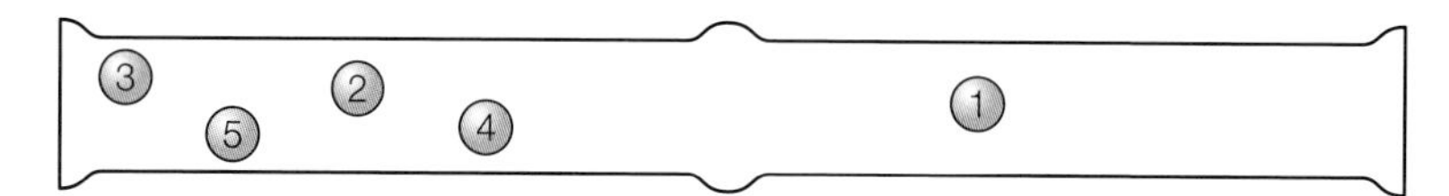

As soon as the cover slip is removed, the molecules are free to move between the jars

▲ **fig D** Diffusion in a gas: one of the ways the molecules might rearrange. Remember that air molecules are also present but have not been shown.

Each molecule has two possible arrangements. It can be in either the left-hand jar or the right-hand jar. We can write the number of possible arrangements for molecule 1 as W_1, and the number of possible arrangements for molecule 2 as W_2, and so on.

As there are two possible arrangements for molecule 1, $W_1 = 2$. For molecule 2, $W_2 = 2$, and so on.

The equation for the *total* number of ways, W, that the five molecules can be arranged is given by:

$$\begin{aligned} W &= W_1 \times W_2 \times W_3 \times W_4 \times W_5 \\ &= 2 \times 2 \times 2 \times 2 \times 2 \\ &= 2^5 \\ &= 32 \end{aligned}$$

Most of the 32 possible arrangements have molecules in both jars. Only one has all five molecules in the left-hand jar. It would be surprising if this arrangement happened very often.

If we increase the number of molecules to 100, W becomes 2^{100}. The left-hand gas jar would contain all 100 molecules only once in every 3.2×10^{22} years! We would have to wait for a very long time indeed to see all 100 molecules in the left-hand jar!

You can see that as the number of molecules increases, the number of possible arrangements also increases very rapidly. To make the numbers easier to work with we need to use an equation to help us. You do not need to remember the equation, but it helps us to see how the spreading out of the molecules is related to an increase in entropy.

Entropy, S, is linked to W by the equation

$$S = k \ln W$$

where W is the number of ways of arranging the molecules, k is a constant called the Boltzmann constant, with a value $1.38 \times 10^{-23}\,\mathrm{J\,K^{-1}}$ and ln is the natural logarithm (i.e. the logarithm to the base e).

Entropy is measured in units of $\mathrm{J\,K^{-1}\,mol^{-1}}$.

EXAM HINT

It is satisfying to make links between topics in chemistry. If you multiply the Boltzmann constant k (1.38 × 10–23) by the Avogadro constant L (6.02 × 1023),
you will get 8.31. This is also known as the gas constant, R, which you should remember from PV = nRT in **Topic 1 (Book 1: IAS)**.

Entropy measures the degree of randomness, so you would expect a gas to have a larger entropy than its liquid form, and a liquid to have a larger entropy than its solid form. The figures in **table A** show that this is the case for water vapour, water and ice.

SUBSTANCE	ENTROPY, $S/\mathrm{J\,K^{-1}\,mol^{-1}}$
water vapour	189
water	70
ice	48

table A Entropies of water vapour, water and ice.

DISTRIBUTION OF ENERGY

The spreading out of molecules in diffusion is an increase in entropy, because the molecules become more randomly dispersed. In the same way, the spreading out of heat energy also represents an increase in entropy.

Energy exists in 'packets' called quanta. You can have a whole number of quanta, but not halves or fractions of quanta.

The distribution of the quanta between molecules is random. The more quanta there are, the more ways there are of arranging them between the molecules. Also, the more molecules there are, the more ways there are of sharing the quanta.

As an example, **table B** shows that there are five possible arrangements of four quanta between two molecules. **Table C** shows that there are six possible arrangements of five quanta between two molecules. **Table D** shows that there are 10 possible arrangements of three quanta between three molecules.

table B

MOLECULE 1	MOLECULE 2
4	0
3	1
2	2
1	3
0	4

table C

MOLECULE 1	MOLECULE 2
5	0
4	1
3	2
2	3
1	4
0	5

table D

MOLECULE 1	MOLECULE 2	MOLECULE 3
3	0	0
0	3	0
0	0	3
2	1	0
2	0	1
1	1	1
1	2	0
1	0	2
0	1	2
0	2	1

table B 2 molecules, 4 quanta, 5 arrangements
table C 2 molecules, 5 quanta, 6 arrangements
table D 3 molecules, 3 quanta, 10 arrangements

The higher the temperature, the more energy a substance has. The more energy a substance has, the more ways there are to distribute the energy. The more ways there are to distribute the energy, the higher the entropy. So, increasing the temperature will increase the entropy.

PERFECT CRYSTALS

A 'perfect' crystal is one in which the internal lattice structure is the same at all times. In other words, the particles (atoms, molecules or ions) are fixed and are not moving in any way, rotating or vibrating.

The Third Law of Thermodynamics states that the entropy of a perfect crystal at the temperature of absolute zero (i.e. 0 K) is zero.

THE SECOND LAW OF THERMODYNAMICS

The Second Law of Thermodynamics is sometimes misquoted. You may have come across the statement that in a spontaneous change, entropy always increases. This cannot be true: if it were, water could never freeze because this change involves a decrease in entropy.

Another example is the reaction between ammonia gas and hydrogen chloride gas:

$$NH_3(g) + HCl(g) \rightarrow NH_4Cl(s)$$

Clearly, there is a reduction in the entropy of the system, since two gases are forming a solid, yet this reaction occurs spontaneously at room temperature.

The statement *entropy always increases* is almost correct. It needs to be expressed more carefully to give the Second Law of Thermodynamics:

In a spontaneous process, the *total* entropy increases.

In **Section 12A.2**, we explain what is meant by the term 'total entropy'.

CHECKPOINT

1. For each of the following, predict whether the change is accompanied by an increase or a decrease in total entropy.
   (a) $H_2O(g) \rightarrow H_2O(l)$ (b) $I_2(s) \rightarrow I_2(g)$ (c) $Na(l) \rightarrow Na(s)$
2. Arrange the following in order of entropy, with the lowest first:
   steam at 110 °C; ice at −10 °C; water at 30 °C

SUBJECT VOCABULARY

entropy a property of matter that is associated with the degree of disorder, or degree of randomness, of particles (i.e. atoms, molecules or ions), and also with the distribution of the quanta of energy between the particles

spontaneous process a process that takes place without continuous intervention by us

12A 2 TOTAL ENTROPY

SPECIFICATION REFERENCE

12.6 12.7 12.8 12.9(i) 12.9(ii) 12.10 12.11

LEARNING OBJECTIVES

- Understand that the total entropy change of any reaction is the sum of the entropy change of the system and the entropy change of the surroundings, summarised by the expression:
  $\Delta S_{total} = \Delta S_{system} + \Delta S_{surroundings}$
- Be able to calculate the entropy change of the system for a reaction, ΔS_{system}, given the entropies of the reactants and products.
- Be able to calculate the entropy change of the surroundings, and hence ΔS_{total}, using the expression
  $\Delta S_{surroundings} = -\frac{\Delta H}{T}$
- Understand that the feasibility of a reaction depends on:
  (i) the balance between ΔS_{system} and $\Delta S_{surroundings}$, so that even endothermic reactions can occur spontaneously at room temperature
  (ii) temperature, as higher temperatures decrease the magnitude of $\Delta S_{surroundings}$, so its contribution to ΔS_{total} is less.
- Understand that reactions can occur as long as ΔS_{total} is positive even if one of the other entropy changes is negative.
- Understand and distinguish between the concepts of thermodynamic stability and kinetic stability.

TOTAL ENTROPY CHANGE, ΔS_{total}

The total entropy change of a process comprises two components:

- the entropy change of the system, ΔS_{system}, and
- the entropy change of the surroundings, $\Delta S_{surroundings}$.

In a chemical reaction, the system is the species that are taking part in the reaction.

The surroundings is everything else. In practice, this usually means the reaction vessel, e.g. test tube or beaker, and the air in the laboratory.

The total entropy change is defined as the sum of the entropy change of the system and the entropy change of the surroundings. That is:

$$\Delta S_{total} = \Delta S_{system} + \Delta S_{surroundings}$$

For a reaction to be spontaneous, ΔS_{total} must be positive. This is another way of expressing the Second Law of Thermodynamics.

DID YOU KNOW?

The official definition of the Second Law of Thermodynamics does not use the term 'total entropy change'. Instead, it refers to the 'entropy change of the universe'. These two terms have the same meaning. We will use the term 'total entropy change' throughout this book.

ENTROPY CHANGE OF THE SYSTEM

ΔS_{system} is calculated using the expression:

$$\Delta S_{system} = \Sigma S\,(\text{products}) - \Sigma S\,(\text{reactants})$$

where S represents entropy and Σ represents 'the sum of'.

The standard entropy values of some substances are shown in **table A**.

Standard refers to conditions of 100 kPa and 298 K.

GAS	ENTROPY $S^\ominus$ /J K^{-1} mol^{-1}
$H_2O(g)$	188.7
H_2	130.6
O_2	205.0
N_2	191.6
Cl_2	165.0
CO_2	213.6
NH_3	192.3

LIQUID	ENTROPY $S^\ominus$ /J K^{-1} mol^{-1}
$H_2O(l)$	69.9
CH_3OH	239.7
CH_3CH_2OH	160.7
C_6H_6	172.8

SOLID	ENTROPY $S^\ominus$ /J K^{-1} mol^{-1}
$H_2O(s)$	47.9
C(diamond)	2.4
C(graphite)	5.7
CaO	39.7
$CaCO_3$	92.9

table A

EXAM HINT

An examiner may ask you to **suggest** why the standard entropy value of diamond is lower than that of graphite. Diamond is ordered in three dimensions (a 3D lattice), whereas graphite is ordered in two dimensions and has greater degrees of freedom between the layers.

WORKED EXAMPLE 1

Use the values in **table A** to calculate $\Delta S^\ominus_{system}$ for the following reaction:

$$CaCO_3(s) \rightarrow CaO(s) + CO_2(g)$$

$$\Delta S^\ominus_{system} = S^\ominus[CaO(s)] + S^\ominus[CO_2(g)] - S^\ominus[CaCO_3(s)]$$

$$= 39.7 + 213.6 - 92.9$$

$$= +160.4\text{ J K}^{-1}\text{ mol}^{-1}$$

WORKED EXAMPLE 2

Use the values in **table A** to calculate $\Delta S^\ominus_{system}$ for the following reaction:

$$N_2(g) + 3H_2(g) \rightarrow 2NH_3(g)$$

$$\Delta S^\ominus_{system} = 2 \times S^\ominus[NH_3(g)] - S^\ominus[N_2(g)] - 3 \times S^\ominus[H_2(s)]$$

$$= (2 \times 192.3) - 191.6 - (3 \times 130.6)$$

$$= -198.8\text{ J K}^{-1}\text{ mol}^{-1}$$

ENTROPY CHANGE OF THE SURROUNDINGS

The entropy change of the surroundings, $\Delta S_{surroundings}$, is related to the enthalpy change of the reaction, ΔH, by the expression:

$$\Delta S_{surroundings} = -\frac{\Delta H}{T}$$

where T is the temperature in kelvin.

For an *exothermic* reaction, where ΔH is negative, $\Delta S_{surroundings}$ will always be positive, so the entropy of the surroundings *increases*.

Conversely, for an *endothermic* reaction, $\Delta S_{surroundings}$ will always be negative, so the entropy of the surroundings *decreases*.

WORKED EXAMPLE 3

Calculate $\Delta S_{surroundings}$ at 298 K when one mole of hydrogen gas is burned in oxygen.

$$H_2(g) + \tfrac{1}{2}O_2(g) \rightarrow H_2O(l) \qquad \Delta H = -286\text{ kJ mol}^{-1}$$

$$\Delta S_{surroundings} = -\frac{-286}{298}$$

$$= +0.960\text{ kJ K}^{-1}\text{ mol}^{-1} \text{ or } +960\text{ J K}^{-1}\text{ mol}^{-1}$$

LEARNING TIP

Note that entropy values are usually quoted in J K^{-1} mol^{-1}, whereas enthalpy changes are usually quoted in kJ mol^{-1}. So, in any equation linking ΔS and ΔH, one of the energy terms will need a conversion of units.

CALCULATING THE TOTAL ENTROPY CHANGE, ΔS_{total}

Now that we know how to calculate both ΔS_{system} and $\Delta S_{surroundings}$, we are in a position to calculate the total entropy change, ΔS_{total}, for a reaction.

WORKED EXAMPLE 4

Using the information in **table A**, calculate the total entropy change at 298 K for the following reaction:

$$H_2(g) + \frac{1}{2}O_2(g) \rightarrow H_2O(l) \quad \Delta H = -286\ \text{kJ mol}^{-1}$$

$$\Delta S_{system} = S[H_2O(l)] - S[H_2(g)] - \frac{1}{2} \times S[O_2(g)]$$
$$= 69.9 - 130.6 - (\frac{1}{2} \times 205)$$
$$= -163.2\ \text{J K}^{-1}\ \text{mol}^{-1}$$
$$\Delta S_{total} = -163.2\ \text{J K}^{-1}\ \text{mol}^{-1} + 960\ \text{J K}^{-1}\ \text{mol}^{-1}$$
$$= +796.8\ \text{J K}^{-1}\ \text{mol}^{-1}$$

SUMMARY

ΔS_{total} will be positive if:

- both $\Delta S_{surroundings}$ and ΔS_{system} are positive
- $\Delta S_{surroundings}$ is positive and ΔS_{system} is negative, but the magnitude of $\Delta S_{surroundings}$ > the magnitude of ΔS_{system}
- $\Delta S_{surroundings}$ is negative and ΔS_{system} is positive, but the magnitude of $\Delta S_{surroundings}$ < the magnitude of ΔS_{system}.

THERMODYNAMIC AND KINETIC STABILITY

The reaction between methane and oxygen is exothermic, and the total entropy change has a positive value:

$$CH_4(g) + 2O_2(g) \rightarrow CO_2(g) + 2H_2O(l) \quad \Delta H_r^{\ominus} = -890.3\ \text{kJ mol}^{-1}$$
$$\Delta S_{total} = +2.95\ \text{kJ mol}^{-1}$$

Since ΔS_{total} is positive, the reaction is thermodynamically feasible at 298 K. However, the reaction does not take place until the reaction mixture is ignited, usually with a flame.

The added heat energy is required to overcome the high activation energy of the reaction.

A reaction mixture with a high activation energy, where a reaction will not take place spontaneously at room temperature, is said to be kinetically stable.

You can see here that a reaction mixture (such as methane and oxygen) can be thermodynamically unstable but kinetically stable, so a reaction does not take place under standard conditions.

THE ROLE OF TEMPERATURE

The increase in entropy obtained by supplying a certain amount of heat energy to an object depends on the temperature of the system.

If an object is very cold, the molecules are not moving around very much. Supplying some heat energy to the object will make the molecules move around more, so the entropy increases.

If we supply the same amount of heat energy to a much hotter object, the entropy will still increase, but not by as much as with the cold object. This is because in the hot object the molecules are already moving around vigorously and the increased degree of movement is less for the hot object.

▲ **fig A** Why does water freeze?

WHY DOES WATER FREEZE?

We will now apply these principles to explain why, under certain conditions, water will freeze.

Ice has lower entropy than liquid water, so ΔS_{system} is *negative*.

The process is exothermic, so $\Delta S_{surroundings}$ is *positive*.

If the magnitude of $\Delta S_{surroundings}$ > the magnitude of ΔS_{system}, then ΔS_{total} is *positive* and the water will freeze.

We will now calculate ΔS_{total} for the change of water into ice at +5 °C and −5 °C using the following data.

$H_2O(l) \rightarrow H_2O(s) \qquad \Delta H = -6010\ J\ mol^{-1}$

- $S^{\ominus}$(water) = 69.9 J K^{-1} mol^{-1}
- $S^{\ominus}$(ice) = 47.9 J K^{-1} mol^{-1}

At +5 °C (278 K):

$$\Delta S_{system} = (47.9 - 69.9) = -22.0\ J\ K^{-1}\ mol^{-1}$$

$$\Delta S_{surroundings} = -\frac{-6010}{278}$$

$$= +\ 21.6\ J\ K^{-1}\ mol^{-1}$$

$$\Delta S_{total} = -0.4\ J\ K^{-1}\ mol^{-1}$$

The total entropy change is negative, meaning that the change is not thermodynamically spontaneous. The water will not freeze.

At −5 °C (268 K):

$$\Delta S_{system} = (47.9 - 69.9) = -22.0\ J\ K^{-1}\ mol^{-1}$$

$$\Delta S_{surroundings} = -\frac{-6010}{268}$$

$$= +22.4\ J\ K^{-1}\ mol^{-1}$$

$$\Delta S_{total} = +0.4\ J\ K^{-1}\ mol^{-1}$$

The total entropy change is positive, meaning that the change is thermodynamically spontaneous. The water will freeze.

It is interesting to note that what has changed between +5 °C and −5 °C is the entropy of the surroundings. The reason that $\Delta S_{surroundings}$ has changed in magnitude is simply because of the temperature change. The same amount of heat energy has been transferred to the surroundings but, because the entropy of the surroundings has a higher temperature when at +5 °C than when at −5 °C, the change in entropy is smaller, as explained earlier.

So, the reason we put water into a freezer when we want to make ice is because the entropy change of the surroundings (i.e. essentially the air inside the freezer) is large enough to compensate for the decrease in entropy when the water freezes.

CHECKPOINT

1. Hydrated cobalt(II) chloride dehydrates on heating according to the following equation:

$CoCl_2.6H_2O(s) \rightarrow CoCl_2(s) + 6H_2O(l) \quad \Delta H^{\ominus}_{298\ K} = +\ 88.1\ kJ\ mol^{-1}$

SKILLS CREATIVITY

(a) Calculate the standard entropy change of the system, $\Delta S^{\ominus}_{system}$.

$S^{\ominus}[CoCl_2.6H_2O(s)] = 343.0\ J\ K^{-1}\ mol^{-1}$

$S^{\ominus}[CoCl_2(s)] = 109.2\ J\ K^{-1}\ mol^{-1}$

$S^{\ominus}[H_2O(l)] = 69.9\ J\ K^{-1}\ mol^{-1}$

(b) Calculate the standard entropy change of the surroundings, $\Delta S^{\ominus}_{surroundings}$, at 298 K.

(c) Calculate the standard total entropy change, $\Delta S^{\ominus}_{total}$, at 298 K for this reaction.

(d) Explain whether hydrated cobalt(II) chloride can be stored at 298 K without it dehydrating?

2. Use the data provided to calculate the total standard entropy change, $\Delta S^{\ominus}_{total}$, at 298 K, for the following reaction:

$2Fe(s) + 1\frac{1}{2}O_2(g) \rightarrow Fe_2O_3(s)$

$\Delta H^{\ominus} = -822\ kJ\ mol^{-1}$

$S^{\ominus}[Fe(s)] = 27.2\ J\ K^{-1}\ mol^{-1}$

$S^{\ominus}[O_2(g)] = 205.0\ J\ K^{-1}\ mol^{-1}$

$S^{\ominus}[Fe_2O_3(s)] = 90.0\ J\ K^{-1}\ mol^{-1}$

12A 3 UNDERSTANDING ENTROPY CHANGES

SPECIFICATION REFERENCE

12.5(i) 12.5(ii) 12.5(iii) 12.19

LEARNING OBJECTIVES

- Understand why entropy changes occur during:
  (i) changes of state
  (ii) dissolving of a solid ionic lattice
  (iii) reactions in which there is a change in the number of moles from reactants to products
- Be able to use entropy and enthalpy changes of solution values to predict the solubility of ionic compounds and discuss trends in the solubility of ionic compounds covered in Unit 2

REACTIONS INVOLVING A CHANGE OF STATE

We have already mentioned that, in general, entropy increases in the order:

solid < liquid < gas

So, we would expect an increase in the entropy of the system if a gas is produced from a reaction involving a solid and/or a liquid.

EXAMPLE 1

When solid ammonium carbonate is added to pure ethanoic acid, bubbles of gas are rapidly produced. Despite its violent appearance, this is an endothermic reaction, as can be shown by placing a thermometer in the acid before the ammonium carbonate is added. The temperature falls considerably as the reaction takes place.

$$2CH_3COOH(l) + (NH_4)_2CO_3(s) \rightarrow 2CH_3COONH_4(aq) + H_2O(l) + CO_2(g)$$

Since the reaction is endothermic, $\Delta S_{surroundings}$ will be negative. However, there is a large increase in the entropy of the system, ΔS_{system}, because a gas is produced from a liquid and a solid. The magnitude of ΔS_{system} is greater than that of $\Delta S_{surroundings}$, and this makes ΔS_{total} positive, so the reaction is thermodynamically spontaneous.

EXAMPLE 2

Hydrated barium hydroxide reacts with solid ammonium chloride in a rapid endothermic reaction at room temperature:

$$Ba(OH)_2.8H_2O(s) + 2NH_4Cl(s) \rightarrow BaCl_2(s) + 10H_2O(l) + 2NH_3(g)$$

As in the previous example, the driving force of the reaction is ΔS_{system}, which overcomes the negative value of $\Delta S_{surroundings}$ caused by the endothermic nature of the reaction. The reactants are solids and the products are a solid, a liquid and a gas.

EXAMPLE 3

Magnesium burns in oxygen to form solid magnesium oxide:

$$Mg(s) + \tfrac{1}{2}O_2(g) \rightarrow MgO(s)$$

The reaction is highly exothermic, so $\Delta S_{surroundings}$ will be positive. ΔS_{system} is negative since a solid and a gas are changing into a solid. However, the magnitude of the value of $\Delta S_{surroundings}$ is greater than that of ΔS_{system}, making ΔS_{total} positive. So, the reaction is thermodynamically spontaneous.

DISSOLVING IONIC SOLIDS IN WATER

When an ionic solid dissolves in water, two changes take place:

- the lattice structure is broken down, and
- the ions become hydrated.

The breaking down of the lattice structure is an endothermic process, equivalent to the reverse of the lattice energy. It also results in an increased number of moles of particles present, so increases the entropy of the salt.

The hydration of the ions is an exothermic process, but results in the water molecules becoming more ordered as they arrange themselves in an orderly manner around the positive and negative ions. The increase in ordering of the water molecules produces a decrease in entropy of the water. This ordering of the water molecules is particularly significant when dissolving anhydrous solids in water.

To explain why some ionic solids are soluble in water, while others are insoluble, we need to look at both the enthalpy and the entropy changes involved.

The solubility of an ionic solid is determined by the total entropy change for the solid.

$$\Delta S_{total} = \Delta S_{system} + \Delta S_{surroundings}$$

Since $\quad \Delta S_{surroundings} = -\dfrac{\Delta_{sol}H}{T}$

this expression becomes:

$$\Delta S_{total} = \Delta S_{system} - \frac{\Delta_{sol}H}{T}$$

The value of ΔS_{total}, and hence the solubility of the solid, depends on the values of three factors:

- the entropy change of the system, ΔS_{system}
- the enthalpy change of solution, $\Delta_{sol}H$
- the temperature, in kelvin, of the water, T.

Let us have a close look at the dissolving of ammonium nitrate crystals in water at a temperature of 298 K.

$NH_4NO_3(s) \xrightarrow{aq} NH_4^+(aq) + NO_3^-(aq)$

$\Delta_{sol}H = +25.8\ kJ\ mol^{-1}$

$\Delta S_{system} = S[NH_4^+(aq)] + S[NO_3^-(aq)] - S[NH_4NO_3(s)]$

$= +113.4 + 146.4 - 151.1\ J\ K^{-1}\ mol^{-1}$

$= +108.7\ J\ K^{-1}\ mol^{-1}$

$$\Delta S_{surroundings} = -\frac{\Delta H_{sol}}{T} = -\frac{+25800}{298} = -86.6\ J\ K^{-1}\ mol^{-1}$$

$\Delta S_{total} = (+108.7 - 86.6) = +22.1\ J\ K^{-1}\ mol^{-1}$

Since ΔS_{total} is positive, the dissolving of ammonium nitrate in water at 298 K is thermodynamically spontaneous. Since the activation energy for this process is very low, ammonium chloride is soluble in water at 298 K.

We are now in a position to think about the solubility in water of some other ionic solids.

Table A shows the relevant thermodynamic data for some solids at a temperature of 298 K. The values have been quoted to the nearest whole number.

IONIC SOLID	$\Delta_{sol}H$/kJ mol^{-1}	$\Delta S_{surroundings}$/J K^{-1} mol^{-1}	ΔS_{system}/J K^{-1} mol^{-1}	ΔS_{total}/J K^{-1} mol^{-1}	SOLUBILITY
NaCl	+4	−13	+43	+56	soluble
NH_4Cl	+15	−50	+167	+117	soluble
AgCl	+66	−221	+33	−188	insoluble
$MgSO_4$	−91	+305	−213	+92	soluble
$CuSO_4$	−73	+245	−192	+53	soluble
$CaSO_4$	−18	+60	−145	−85	insoluble

table A

REACTIONS INVOLVING A CHANGE IN NUMBER OF MOLES FROM REACTANTS TO PRODUCTS

If you increase the number of moles that you have, you automatically increase the number of particles (i.e. atoms, molecules or ions) present. This will result in an increase in the number of ways that the particles can be arranged, and this increases the entropy of the system, making ΔS_{system} positive.

We will now look again at the reactions shown above, but this time we will consider the change in the number of moles from reactants to products.

EXAMPLE 1

$2CH_3COOH(l) + (NH_4)_2CO_3(s)$
$\rightarrow 2CH_3COONH_4(s) + H_2O(l) + CO_2(g)$

Number of moles of reactants = 3

Number of moles of products = 4

So, ΔS_{system} is positive.

EXAMPLE 2

$Ba(OH)_2.8H_2O(s) + 2NH_4Cl(s)$
$\rightarrow BaCl_2(s) + 10H_2O(l) + 2NH_3(g)$

Number of moles of reactants = 3

Number of moles of products = 13

So, ΔS_{system} is positive.

DID YOU KNOW?

When we state that the solids AgCl and $CaSO_4$ are insoluble in water, this is a simplification. Both solids have a measurable solubility in water, but the solubility is very small. What we are really saying is that the position of the equilibrium in both cases is very far over to the solid side of the equation. We will come back to this point in **Topic 13**, when we discuss the relationship between the total entropy change for a reaction and the equilibrium constant, *K*, for the reaction.

EXAMPLE 3

$$Mg(s) + \tfrac{1}{2}O_2 \rightarrow MgO(s)$$

Number of moles of reactants = 1.5

Number of moles of products = 1

So, ΔS_{system} is negative.

SOLUBILITY TRENDS IN GROUP 2

In **Topic 8 (Book 1: IAS)**, we mentioned that for Group 2 the solubility of the metal hydroxides increases down the group and the solubility of the metal sulfates decreases down the group. We can now explain the reason for these trends in solubility.

As we have mentioned earlier, the two factors that affect the solubility of ionic compounds are the standard enthalpy change of solution of the compound, $\Delta_{sol}H^{\ominus}$, and the standard entropy change of the system, $\Delta S^{\ominus}_{system}$.

It is difficult to work out $\Delta S^{\ominus}_{system}$ accurately for many ionic compounds. We can use the standard entropy values of the ions as a guide.

SOLUBILITY OF GROUP 2 METAL HYDROXIDES

Table B shows the standard enthalpy and standard entropy changes involved in dissolving the Group 2 metal hydroxides. The values are given here to the nearest whole number.

METAL HYDROXIDE	$\Delta H^{\ominus}_{sol}$	$\Delta S^{\ominus}_{surroundings}$ /J K^{-1} mol^{-1}	HYDRATED CATION	STANDARD ENTROPY OF HYDRATED CATION / J K^{-1} mol^{-1}
$Mg(OH)_2$	+3	−10	$Mg^{2+}(aq)$	−138
$Ca(OH)_2$	−16	+54	$Ca^{2+}(aq)$	−53
$Sr(OH)_2$	−46	+154	$Sr^{2+}(aq)$	−33
$Ba(OH)_2$	−52	+174	$Ba^{2+}(aq)$	+10

table B

The values for the standard entropy of the hydroxide ion have been left out here, because the ion is common to all of the metal hydroxides. The trend in the standard entropies of the cations mirrors the trend in the standard entropy change of the system, $\Delta S^{\ominus}_{system}$.

The standard enthalpy change of solution becomes more negative down the group. This favours solubility, as $\Delta S^{\ominus}_{surroundings}$ becomes more positive.

The values of the entropies of the hydrated cations become less negative (i.e. more positive), and this also favours solubility.

As both factors favour an increase in solubility, the metal hydroxides become more soluble as you go down the group.

SOLUBILITY OF GROUP 2 METAL SULFATES

Table C shows the standard enthalpy and standard entropy changes involved in dissolving the Group 2 metal sulfates. The values are given here to the nearest whole number.

METAL SULFATE	$\Delta H^{\ominus}_{sol}$	$\Delta S^{\ominus}_{surroundings}$ /J K^{-1} mol^{-1}	HYDRATED CATION	STANDARD ENTROPY OF HYDRATED CATION / J K^{-1} mol^{-1}
$MgSO_4$	−91	+305	$Mg^{2+}(aq)$	−138
$CaSO_4$	−18	+60	$Ca^{2+}(aq)$	−53
$SrSO_4$	−9	+30	$Sr^{2+}(aq)$	−33
$BaSO_4$	+19	−63	$Ba^{2+}(aq)$	+10

table C

Again, the values for the standard entropy of the sulfate ion have been left out here, because the ion is common to all of the metal sulfates. The trend in the standard entropies of the cations mirrors the trend in the standard entropy change of the system, $\Delta S^{\ominus}_{system}$.

The standard enthalpy change of solution becomes less negative down the group. This favours insolubility, as $\Delta S^{\ominus}_{surroundings}$ becomes less positive.

The values of the entropy of the hydrated cation become less negative (i.e. more positive), and this also favours solubility.

The overall decrease in $\Delta S^{\ominus}_{surroundings}$ down the group is 368 $J K^{-1} mol^{-1}$ (+ 305 to −63).

The overall increase in the standard entropy change of the hydrated cations is 148 $J K^{-1} mol^{-1}$ (−138 to +10).

As the decrease in the standard entropy of the hydrated cations is much less than the increase in the entropy of the surroundings, the Group 2 metal sulfates become less soluble as you go down the group.

LEARNING TIP

At this level, it is assumed that ΔS_{system} does not change with temperature, since the entropies of both the reactants and the products change by similar amounts.

The magnitude of $\Delta S_{surroundings}$ will always decrease with increasing temperature. If $\Delta S_{surroundings}$ is negative, it will become less negative. If $\Delta S_{surroundings}$ is positive, it will become less positive.

Evidence from thermodynamics does not indicate the rate at which a reaction will occur. A reaction that is thermodynamically spontaneous at a given temperature may have a high activation energy. This could result in the reaction not taking place at all.

CHECKPOINT

SKILLS CREATIVITY, PROBLEM-SOLVING

1. Predict whether there is likely to be an increase, a decrease or no change in the entropy of the system in each of the following reactions. In each case, give your reasons.

   (a) $CuSO_4.5H_2O(s) \rightarrow CuSO_4(s) + 5H_2O(l)$

   (b) $HCl(g) + NH_3(g) \rightarrow NH_4Cl(s)$

   (c) $SO_2(g) + \frac{1}{2}O_2(g) \rightarrow SO_3(g)$

   (d) $Co(H_2O)_6^{2+}(aq) + EDTA^{2-}(aq) \rightarrow Co(EDTA)(aq) + 6H_2O(l)$

2. This question is about the following reaction:

   $$H_2(g) + I_2(g) \rightarrow 2HI(g)$$

   (a) State why you might predict that the entropy change in the system for this reaction is zero.

   (b) The actual ΔS_{system} for this reaction is +22 $J K^{-1} mol^{-1}$. Explain why the total entropy change for this reaction is not zero.

12B 1 LATTICE ENERGY, $\Delta_{LE}H$ AND BORN–HABER CYCLES

SPECIFICATION REFERENCE
12.12(i) 12.12(ii) 12.12(iii)
12.13 12.18 PART

LEARNING OBJECTIVES

- Be able to define the terms:
  (i) standard enthalpy change of atomisation, $\Delta_{at}H$
  (ii) electron affinity
  (iii) lattice energy (as the exothermic process for the formation of one mole of an ionic solid from its gaseous ions).
- Understand that lattice energy provides a measure of the strength of ionic bonding.
- Construct Born–Haber cycles and carry out related calculations.
- Understand the effect of ionic charge and ionic radius on the value of the lattice energy of an ionic compound.

LATTICE ENERGY

In **Book 1**, we saw that bond enthalpies can be used as a measure of the strength of the covalent bonding in molecules. The equivalent energy change for ionic bonding in ionic compounds is lattice energy, $\Delta_{LE}H$ (or lattice enthalpy).

The lattice energy of a compound is the energy change when one mole of the ionic solid is formed from its gaseous ions. If standard conditions of 100 kPa and a stated temperature (usually 298 K) are applied (as shown by $^{\ominus}$), then the energy change is called the **standard lattice energy**.

The equation that represents the standard lattice energy of sodium chloride is:

$$Na^+(g) + Cl^-(g) \rightarrow NaCl(s) \qquad \Delta_{LE}H^{\ominus} = -780\,kJ\,mol^{-1}$$

For magnesium chloride it is:

$$Mg^{2+}(g) + 2Cl^-(g) \rightarrow MgCl_2(s) \qquad \Delta_{LE}H^{\ominus} = -2526\,kJ\,mol^{-1}$$

LEARNING TIP

The terms 'lattice energy' and 'lattice enthalpy' are commonly used as if they mean exactly the same thing. You will often find both terms used within the same textbook, article or website, including some university websites.

In fact, there is a difference between them. It relates to the conditions under which they are calculated. However, the difference is negligible when compared with the different values for lattice energies that you will find in different sources.

Unless you go on to study chemistry at degree level, you do not need to worry about the difference between the two terms.

DID YOU KNOW?

There are two ways of defining lattice energy. The one mentioned in the specification is the one we have already defined. This is sometimes called lattice energy of *formation*, since the compound is being formed from its ions. Using this definition, the energy change will always be negative.

However, some books define it as lattice energy of *dissociation*, in which it is the energy change when one mole of the compound is broken down (or dissociated) into its ions. In this case, the energy change will be positive.

FACTORS AFFECTING THE MAGNITUDE OF LATTICE ENERGY

The lattice energy of magnesium chloride, $Mg^{2+}(Cl^-)_2$, is much larger (i.e. more negative) than that of sodium chloride, Na^+Cl^-. A number of factors are responsible for this difference:

- The first thing to realise is that a magnesium ion carries twice the charge of a sodium ion.
- Second, there are more cation-to-anion interactions in magnesium chloride because there are twice as many chloride ions per cation than in sodium chloride.
- A third factor is the distance between the centres of the cations and their neighbouring anions, which is equal to the sum of their ionic radii. This distance is determined in part by the relative sizes of the ions involved (the Mg^{2+} ion is smaller than Na^+, so reducing the sum of the ionic radii) and also by the type of lattice structure the compound has. In fact, the relative ion sizes determine the type of lattice structure.

Table A shows the effects that both the inter-ionic distance and the charges on the ions have on the lattice energy.

COMPOUND	INTER-IONIC DISTANCE/nm	CHARGES ON THE IONS	LATTICE ENERGY/ kJ mol^{-1}
LiF	0.207	+1, −1	−1031
NaF	0.235	+1, −1	−918
CaF_2	0.233	+2, −1	−2630
Li_2O	0.214	+1, −2	−2814
MgO	0.212	+2, −2	−3791
Al_2O_3	0.193	+3, −2	−15 504

table A

Comparison of LiF and NaF, where the charges on the ions are the same, shows that a decrease in the distance between the centres of the two ions (the inter-ionic distance) results in a more negative value for the lattice energy.

Comparison of NaF and CaF_2, where the inter-ionic distances are almost the same, shows that an increase in charge of even one of the ions (Ca^{2+} as opposed to Na^+) results in a more negative value for the ionisation energy. Similar comparisons can be made between Li_2O and MgO and also between Li_2O and Al_2O_3.

Finally, as we will see later in this topic, there are also covalent interactions between the ions, and these affect the magnitude of the lattice energy. The values quoted in **table A** take these interactions into consideration since they are calculated values using a Born–Haber cycle (see later in this section).

LEARNING TIP

Lattice energies are determined by:

- the magnitudes of the charges on the ions
- the sum of the ionic radii
- the type of lattice structure
- the extent of covalent interactions between the ions.

At International A Level, you are not required to know the different types of lattice structures. In addition, the extent of the covalent interactions is already taken into consideration in the quoted values for lattice energy. Therefore, you need only consider the magnitude of the charges on the ions and the sizes of the ions.

STANDARD ENTHALPY CHANGE OF ATOMISATION, $\Delta_{at}H^{\ominus}$

The enthalpy change measured at a stated temperature (usually 298 K) and 100 kPa when one mole of gaseous atoms is formed from an element in its standard state is called the **standard enthalpy change of atomisation** of the element (see **Topic 6 (Book 1: IAS)**). It is given the symbol $\Delta_{at}H^{\ominus}$.

Equations representing some standard enthalpy changes of atomisation at 298 K are given below.

$C(s) \rightarrow C(g)$ $\Delta_{at}H^{\ominus} = +717\,kJ\,mol^{-1}$

$Na(s) \rightarrow Na(g)$ $\Delta_{at}H^{\ominus} = +107\,kJ\,mol^{-1}$

$\frac{1}{2}H_2(g) \rightarrow H(g)$ $\Delta_{at}H^{\ominus} = +218\,kJ\,mol^{-1}$

$\frac{1}{2}Cl_2(g) \rightarrow Cl(g)$ $\Delta_{at}H^{\ominus} = +122\,kJ\,mol^{-1}$

LEARNING TIP

Note that the standard enthalpy change of atomisation is the enthalpy change when one mole of gaseous *atoms* is formed. In the case of elements that exist as polyatomic molecules, it is *not* the enthalpy change when one mole of gaseous *molecules* is atomised.

ELECTRON AFFINITY

The first **electron affinity** of an element, 1st *EA*, is the energy change when each atom in one mole of atoms in the gaseous state gains an electron to form a −1 ion.

The equations below represent the first electron affinities of some elements.

$Cl(g) + e^- \rightarrow Cl^-(g)$ 1st $EA = -349\,kJ\,mol^{-1}$

$Br(g) + e^- \rightarrow Br^-(g)$ 1st $EA = -325\,kJ\,mol^{-1}$

$O(g) + e^- \rightarrow O^-(g)$ 1st $EA = -141\,kJ\,mol^{-1}$

The first electron affinity has a negative value for many elements, including the alkali metals. Examples include:

$Li(g) + e^- \rightarrow Li^-(g)$ 1st $EA = -60\,kJ\,mol^{-1}$

and

$Na(g) + e^- \rightarrow Na^-(g)$ 1st $EA = -53\,kJ\,mol^{-1}$

There is a notable exception with the noble gases. For these, repulsion caused by the electrons already present in the valence shell results in a positive value for the first electron affinity as the additional electron would have to occupy a new valence shell.

In contrast, second electron affinities, 2nd *EA*, tend to be positive, as energy is required to overcome the repulsion between the negative ion and the electron being added. For example:

$O^-(g) + e^- \rightarrow O^{2-}(g)$ 2nd $EA = +798\,kJ\,mol^{-1}$

Therefore, the formation of the oxide ion, O^{2-}, in the gaseous state from its atom in the gaseous state is an endothermic process overall:

$O(g) + 2e^- \rightarrow O^{2-}(g)$ 1st EA + 2nd $EA = +657\,kJ\,mol^{-1}$

This raises an interesting question: why does oxygen form O^{2-} ions in its binary ionic compounds rather than O^- ions, as the O^- ion would appear to be the more energetically favourable state? We will attempt to answer this question later in this topic.

BORN–HABER CYCLES

We are now in a position to consider the overall energy changes that take place when an ionic compound is made from its elements. These energy changes are summarised in an energy level diagram called the Born–Haber cycle. **Fig A** shows the Born–Haber cycle for sodium chloride.

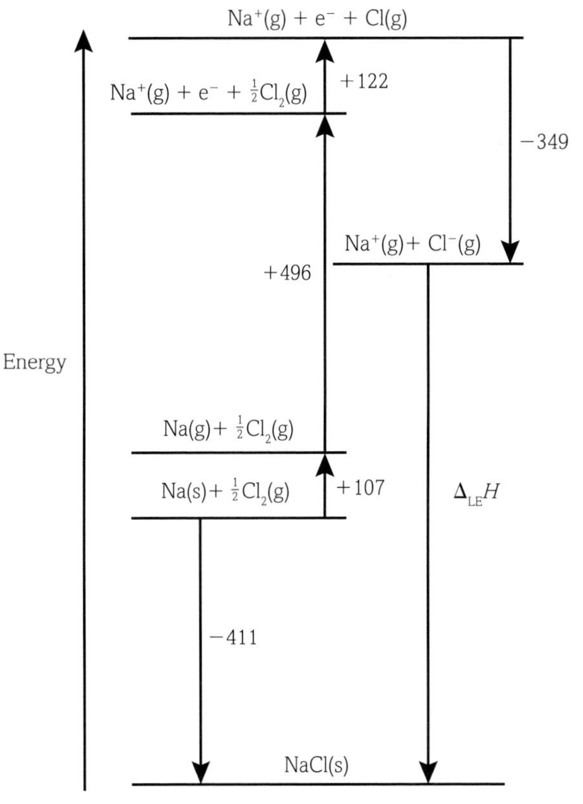

fig A Born-Haber cycle for sodium chloride.

EXAM HINT

Examiners may ask you to label enthalpy change of atomisation, enthalpy change of ionisation, electron affinity and enthalpy change of formation on Born-Haber cycle diagrams like **fig A**. Make sure that you can recognise each process and label it correctly.

The cycle includes the following energy changes, all of which can be determined experimentally:

- The enthalpy change of formation of NaCl(s), $\Delta_f H[NaCl(s)] = -411\,kJ\,mol^{-1}$
- $\Delta_{at}H[Na(s)] = +107\,kJ\,mol^{-1}$
- the first ionisation energy, $1^{st}\,IE[Na(g)] = +496\,kJ\,mol^{-1}$
- $\Delta_{at}H[Cl_2(g)] = +122\,kJ\,mol^{-1}$
- $EA[Cl(g)] = -349\,kJ\,mol^{-1}$

Applying Hess's Law to the cycle gives us:

$$+107 + 496 + 122 + (-349) + \Delta_{LE}H = -411$$

Hence:

$$\Delta_{LE}H[NaCl(s)] = -411 - 107 - 496 - 122 + 349\,kJ\,mol^{-1} = -787\,kJ\,mol^{-1}$$

LEARNING TIP

Note: see **Topic 2 (Book 1: IAS)** for the definition of ionisation energy and **Topic 6 (Book 1: IAS)** for an explanation of Hess's Law.

WHY IS CALCIUM OXIDE $Ca^{2+}O^{2-}$ AND NOT Ca^+O^-?

You may think that the answer to this question is obvious. You might say that calcium oxide is $Ca^{2+}O^{2-}$ because both ions would then have a stable noble gas electronic configuration.

If that is what you believe, then the following may come as a surprise. Acquiring a noble gas electronic configuration is not a reason for electronic changes to take place when atoms and molecules react together.

After all, many ions exist in stable compounds, in particular, cations of the transition metals, in which they do not have the electronic configuration of a noble gas. In fact, compounds in which the ions do have a noble gas electronic configuration are in the minority.

So, why does calcium oxide form as $Ca^{2+}O^{2-}$ and not as Ca^+O^-, and why are we asking the question?

You may remember that we have mentioned that the formation of $O^-(g)$ from O(g) is exothermic:

$$O(g) + e^- \rightarrow O^-(g) \qquad 1^{st}\,EA = -141\,kJ\,mol^{-1}$$

However, the formation of $O^{2-}(g)$ from O(g) is endothermic:

$$O(g) + 2e^- \rightarrow O^{2-}(g) \qquad 1^{st}\,EA + 1^{st}\,EA = +657\,kJ\,mol^{-1}$$

Since more energy is required, why is the formation of the O^{2-} ion preferred over the formation of O^-?

A similar pattern is observed with the formation of $Ca^+(g)$ and $Ca^{2+}(g)$ from Ca(g):

$$Ca(g) \rightarrow Ca^+(g) + e^- \qquad 1^{st}\,IE = +590\,kJ\,mol^{-1}$$

$$Ca(g) \rightarrow Ca^{2+}(g) + 2e^- \qquad 1^{st}\,IE + 2^{nd}\,IE = +1735\,kJ\,mol^{-1}$$

Similarly, since more energy is required, why is the formation of the Ca^{2+} ion preferred over the formation of the Ca^+ ion?

To answer this question it is necessary to consider *all* of the energy changes involved in the formation of an ionic compound from its elements – not just those involved in the formation of the gaseous ions from the gaseous atoms. In other words, we need to look at the information supplied by the Born–Haber cycle for each compound (see **fig B** and **fig C**).

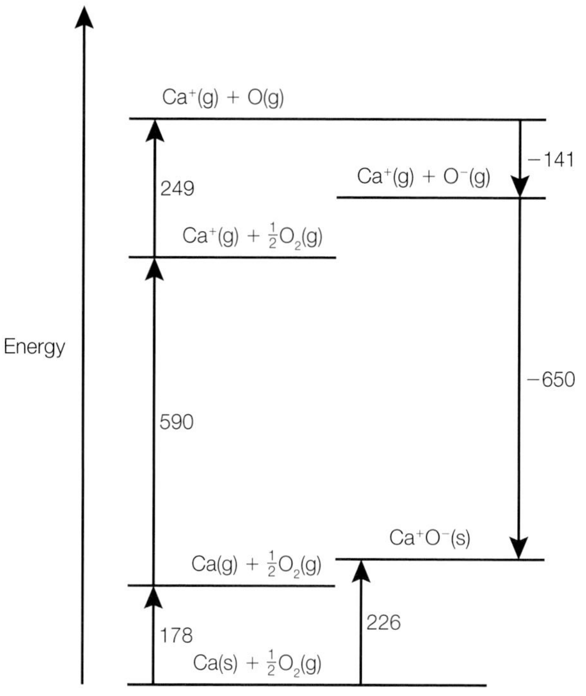

fig B Born–Haber cycle for $Ca^+O^-(s)$.

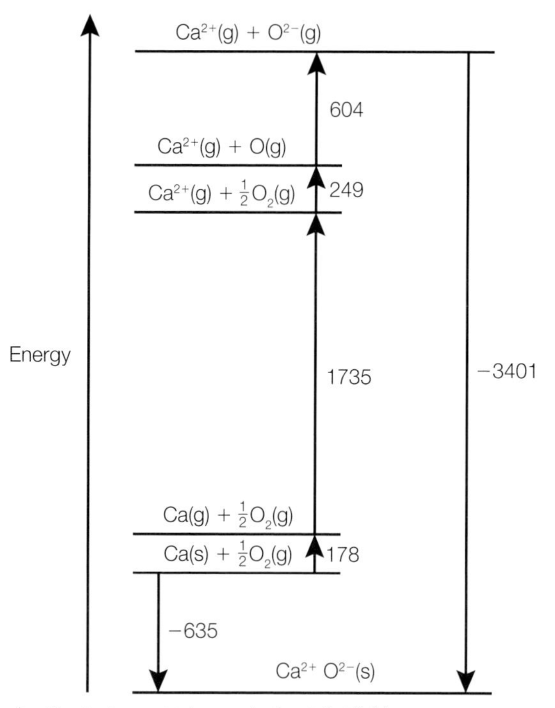

fig C Born–Haber cycle for $Ca^{2+}O^{2-}(s)$.

Using the equation:

$$\Delta_f H[Ca^+O^-(s)] = \Delta_{at}H[Ca(s)] + 1^{st}\,IE[Ca(g)] + \Delta_{at}H[\tfrac{1}{2}O_2(g)] + 1^{st}\,EA[O(g)] + \Delta_{LE}H[Ca^+O^-(s)]$$

we can calculate a value for $\Delta_f H[Ca^+O^-(s)]$ using a theoretical value for $\Delta_{LE}H[Ca^+O^-(s)]$ of $-650\,kJ\,mol^{-1}$.

In this way, the value calculated for $\Delta_f H[Ca^+O^-(s)]$ is:

$$178 + 590 + 249 + (-141) + (-650) = +226\,kJ\,mol^{-1}.$$

Similarly, the value calculated for $\Delta_f H[Ca^{2+}O^{2-}(s)]$ is -635 kJ mol^{-1}.

From these values, it is clear that the formation of $Ca^{2+}O^{2-}$ is energetically more favourable than the formation of $Ca^{+}O^{-}$.

The extra energy required to form the 2+ and 2− ions is more than compensated for by the much larger (i.e. more negative) lattice energy of $Ca^{2+}O^{2-}$ (-3401 kJ mol^{-1}) compared to that of $Ca^{+}O^{-}$ (-650 kJ mol^{-1}).

It is the *overall* energy change involved, and not the incorrect notion of desirability to obtain a noble gas electronic configuration, that determines how atoms and molecules will interact with one another to form a compound.

However, it is important to recognise that the conditions under which a reaction takes place will influence the overall energy change involved, and hence the nature of the compound formed.

For example, if iron is heated in chlorine gas, iron(III) chloride is the product. However, if hydrogen chloride gas is used in place of chlorine, iron(II) chloride is formed.

Therefore, changing the conditions of temperature and/or pressure can also have an effect on the exact compound formed.

CHECKPOINT

1. (a) Explain why the lattice energy of sodium fluoride, $Na^{+}F^{-}$, is larger (i.e. more negative) than the lattice energy of potassium chloride, $K^{+}Cl^{-}$.
   (b) Explain why the lattice energy of calcium oxide, $Ca^{2+}O^{2-}$, is approximately four times larger (i.e. more negative) than the lattice energy of potassium fluoride, $K^{+}F^{-}$.
   [Ionic radii: Ca^{2+} 0.100 nm; K^{+} 0.138 nm; O^{2-} 0.140 nm; F^{-} 0.133 nm.]

2. The Born–Haber cycle can be used to calculate the lattice energy, $\Delta_{LE}H$, for magnesium oxide.

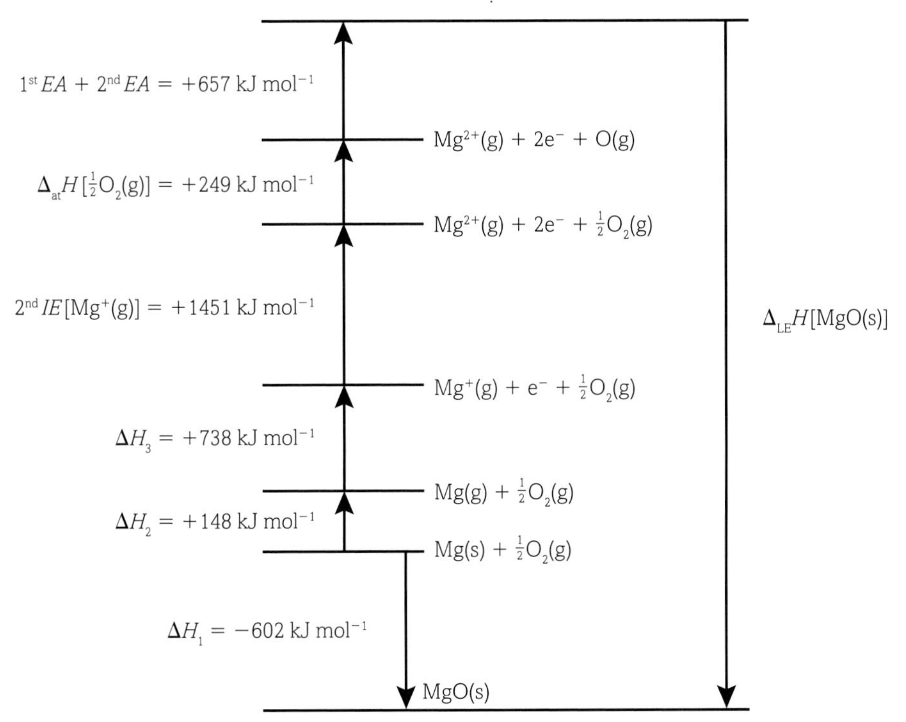

   (a) (i) State the names for each of the enthalpy changes labelled ΔH_1, ΔH_2 and ΔH_3.
   (ii) Give the formula missing at the top of the cycle, indicated by a question mark (?). Give the state symbols.
   (b) The equations representing the first and second electron affinities of oxygen are:
   $O(g) + e^- \rightarrow O^-(g)$ 1st *EA* = −141 kJ mol^{-1}
   $O^-(g) + e^- \rightarrow O^{2-}(g)$ 2nd *EA* = +798 kJ mol^{-1}
   Suggest why the second of these processes is endothermic.
   (c) Use the information in the Born-Haber cycle shown above to calculate the lattice energy of magnesium oxide, $\Delta_{LE}H$[MgO(s)].
   (d) Explain how the lattice energy of barium oxide differs from that of magnesium oxide.

SKILLS INTERPRETATION

SUBJECT VOCABULARY

standard lattice energy (of an ionic compound) the energy change measured at a stated temperature, usually 298 K, and 100 kPa when one mole of an ionic compound is formed from its ions in the gaseous state

standard enthalpy change of atomisation (of an element) the enthalpy change measured at a stated temperature, usually 298 K, and 100 kPa when one mole of gaseous atoms is formed from an element in its standard state

(first) electron affinity (of an element) the energy change when each atom in a mole of atoms in the gaseous state gains an electron to form a -1 ion

12B 2 EXPERIMENTAL AND THEORETICAL LATTICE ENERGIES

SPECIFICATION REFERENCE 12.14 12.15

LEARNING OBJECTIVES

- Understand that a comparison of the experimental lattice energy value (obtained from a Born–Haber cycle) with the theoretical value (obtained from electrostatic theory) in a particular compound indicates the degree of covalent bonding.
- Understand that polarisation of anions by cations leads to some covalency in an ionic bond, based on evidence from the Born–Haber cycle.

EXPERIMENTAL LATTICE ENERGY

The Born–Haber cycle allows us to calculate a value for the lattice energy of an ionic compound from knowledge of the other energy changes, all of which can be determined experimentally. The value of the lattice energy calculated this way is called the **experimental lattice energy**.

LEARNING TIP

Be very careful not to confuse experimental and theoretical lattice energies.

The experimental lattice energy is calculated from a Born-Haber cycle.

The theoretical lattice energy is calculated using the principles of electrostatics.

THEORETICAL LATTICE ENERGY

The type of lattice structure and the inter-ionic distance can be found by X-ray crystallography. Using this information, it is possible for us to calculate a value for the lattice energy of an ionic compound. However, we first need to make the following assumptions.

- The ions are in contact with one another.
- The ions are perfectly spherical.
- The charge on each ion is evenly distributed around the centre so that each ion can be considered as point charges.

A value for the **theoretical lattice energy** can be calculated using the principles of electrostatics. There are three main methods for performing such calculations. If you are interested, you can research them under the headings of the 'Born–Landé equation', the 'Born–Mayer equation' and the 'Kapustinskii equation'.

Table A shows a comparison of the experimental lattice energies (obtained by using a Born–Haber cycle) with the theoretical lattice energies (calculated using the principles of electrostatics) for various compounds.

EXAM HINT

The large differences between the theoretical and experimental lattice energies for AgCl and AgBr help to explain why these compounds are insoluble in water. The difference is smaller for AgF, and AgF is soluble in water.

COMPOUND	EXPERIMENTAL LATTICE ENERGY/kJ mol^{-1}	THEORETICAL LATTICE ENERGY/kJ mol^{-1}
NaF	−918	−912
NaCl	−780	−770
NaBr	−742	−735
AgF	−958	−920
AgCl	−905	−833
AgBr	−891	−816

table A

You will notice that there is good agreement between the experimental values and the theoretical values for the halides of sodium. However, the agreement is not so good for the halides of silver.

Agreement between the experimental and theoretical values of the lattice energy of a compound indicates that the ionic model is a good one for that compound. A significant difference suggests that the ionic model needs to be modified. In such compounds, the bonding in the lattice has a considerable covalent character, which makes the experimental value for the lattice energy more negative than the theoretical value.

The covalency in bonding is caused by **polarisation** of the anion by the cation. Polarisation results in distortion of the electron density within the anion, resulting in a higher electron density near the cation. This means that there is some electron density existing between the two ions. In other words, there will be a degree of covalent bonding in the compound.

EXTENT OF COVALENT CHARACTER: POLARISATION OF THE ANION

In an ionic lattice, the positive ion (cation) will attract the electrons of the anion. If the electrons are pulled towards the cation, the anion is said to be *polarised* because the even distribution of its electron density has been distorted (**fig A**).

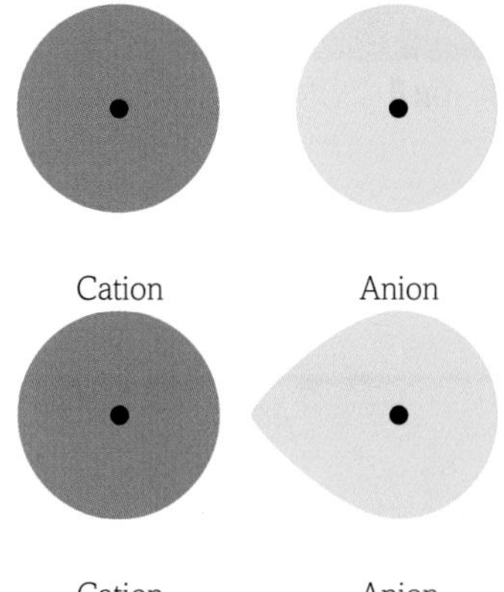

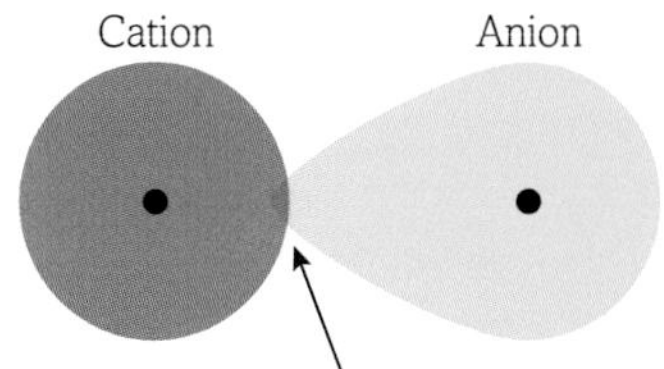

fig A A representation of a cation attracting the electrons of an anion in an ionic lattice.

The extent to which an anion is polarised by a cation depends on several factors. The two main factors are summarised below. These are known as Fajan's Rules.

Polarisation will be increased by:

- a high charge and small size of the cation (i.e. a high charge density of the cation)
- a high charge and large size of the anion.

HIGH CHARGE AND SMALL SIZE OF CATION

The ability of a cation to attract electrons from the anion towards itself is called its 'polarising power'. A cation with a high charge and a small radius has a large polarising power. An approximate value for the polarising power of a cation can be obtained by calculating its charge density (sometimes called surface charge density). The charge density of a cation is the charge divided by the surface area of the ion. If the ion is assumed to be a sphere, its surface area is equal to $4\pi r^2$, where r is the ionic radius.

A rough approximation of the charge density can be determined by dividing the charge by the square of its ionic radius. This calculation is beyond the scope of your International A Level course.

$$\text{charge density} \sim \frac{\text{charge}}{r^2}$$

Table B compares the extent of covalent bonding in sodium chloride and magnesium chloride.

COMPOUND	CHARGE DENSITY OF CATION	EXPERIMENTAL LATTICE ENERGY/kJ mol^{-1}	THEORETICAL LATTICE ENERGY/kJ mol^{-1}	PERCENTAGE DIFFERENCE	EXTENT OF COVALENT BONDING
NaCl	$\frac{1}{(0.095)^2} = 111$	−780	−770	1.28	very little
$MgCl_2$	$\frac{2}{(0.060)^2} = 556$	−2526	−2326	7.92	more than in NaCl

table B

The charge density of the magnesium ion is larger than that of the sodium ion, resulting in greater polarisation of the chloride ion and greater covalent bonding in $MgCl_2$ than in NaCl.

HIGH CHARGE AND LARGE SIZE OF THE ANION

The ease with which an anion is polarised depends on its charge and its size. Anions with a large charge and a large size are polarised most easily.

Table C compares the extent of covalent bonding in silver fluoride and silver iodide.

COMPOUND	CHARGE OF ANION	RADIUS OF ANION/nm	EXPERIMENTAL LATTICE ENERGY/kJ mol^{-1}	THEORETICAL LATTICE ENERGY/kJ mol^{-1}	PERCENTAGE DIFFERENCE	EXTENT OF COVALENT BONDING
AgF	−1	0.133	−958	−920	3.97	fairly large
AgI	−1	0.215	−889	−778	12.49	greater than in AgF

table C

The larger iodide ion is more easily polarised, which leads to a greater degree of covalent bonding in silver iodide.

▲ **fig B** Sodium chloride.

▲ **fig C** Silver chloride.

DID YOU KNOW?

The interesting case of silver compounds

Table D compares the extent of covalent bonding in sodium chloride and silver chloride.

COMPOUND	CHARGE DENSITY OF CATION	EXPERIMENTAL LATTICE ENERGY/ kJ mol^{-1}	THEORETICAL LATTICE ENERGY/ kJ mol^{-1}	PERCENTAGE DIFFERENCE	EXTENT OF COVALENT BONDING
NaCl	$\frac{1}{(0.095)^2} = 111$	−780	−770	0.13	very little
AgCl	$\frac{1}{(0.126)^2} = 63$	−905	−833	7.96	much greater than in NaCl

table D

Despite the polarising power of the Ag^+ ion being less than that of the Na^+ ion, there is considerably more covalent bonding in AgCl.

This is explained by considering the valence shell electronic configurations of the two ions concerned.

Na^+ $1s^2\ 2s^2\ 2p^6$

Ag^+ [Kr] $4d^{10}$

A d^{10} configuration offers less shielding than a p^6 electronic configuration, so compounds that have a d^{10} configuration show a greater tendency towards a covalent character.

For the same reason, zinc compounds also contain a significant degree of covalent bonding.

CHECKPOINT

SKILLS REASONING

1. The experimental and theoretical lattice energies, in kJ mol^{-1}, of calcium fluoride, CaF_2, and silver fluoride, AgF, are given below.

	experimental value	theoretical value
CaF_2	−2630	−2609
AgF	−958	−920

Suggest why there is good agreement between the two values for CaF_2, but there is a significant difference between the two values for AgF.

2. Anions in an ionic lattice can be polarised by the cations adjacent to them. The extent of polarisation depends on the nature of both the anion and the cation involved.
   (a) Explain what is meant by the term 'polarised' in this context.
   (b) State whether the oxide ion, O^{2-}, or the sulfide ion, S^{2-}, is more easily polarised.
   (c) Place the following cations (with ionic radii shown in parenthesis) in order of increasing polarising power:

   Mg^{2+} (0.072 nm) Al^{3+} (0.053 nm) Li^+ (0.074 nm)

   Na^+ (0.102 nm) Ca^{2+} (0.100 nm) K^+ (0.138 nm)

   Support your conclusion with suitable calculations.

3. Suggest why the oxide $Na^{2+}O^{2-}$ does not exist. State what further energy change, other than those quoted in the data books, you would require in order to confirm your suggestion.

SUBJECT VOCABULARY

experimental lattice energy the lattice energy calculated from a Born–Haber cycle
theoretical lattice energy the lattice energy calculated using the principles of electrostatics
polarisation the distortion of the electron density of a negative ion (anion)

12B 3 ENTHALPY CHANGES OF SOLUTION AND HYDRATION

SPECIFICATION REFERENCE 12.16 12.17 12.18

LEARNING OBJECTIVES

- Define the terms 'enthalpy change of solution, $\Delta_{sol}H$', and 'enthalpy change of hydration, $\Delta_{hyd}H$ of an ion'.
- Use energy cycles and energy level diagrams to calculate the enthalpy change of solution of an ionic compound using the enthalpy change of hydration and the lattice energy
- Understand the effect of ionic charge and ionic radius on the values of the enthalpy change of hydration and the lattice energy of an ionic compound.

ENTHALPY CHANGE OF SOLUTION, $\Delta_{sol}H$

The solubilities of ionic solids in water show a very wide variation, and there is no obvious pattern.

One of the factors that determines solubility is the value of the **enthalpy change of solution, $\Delta_{sol}H$**, the enthalpy change when one mole of an ionic solid dissolves in water to form an infinitely dilute solution.

The enthalpy change of solution for sodium chloride is the energy change associated with the following process:

$$NaCl(s) \xrightarrow{aq} Na^+(aq) + Cl^-(aq)$$

It is important to specify the extent of dilution of the final solution when quoting a value for the enthalpy change. Upon dilution, the ions in the solution move further apart (an endothermic process) and also become more hydrated (an exothermic process).

The relative importance of these two processes (endothermic and exothermic) changes with dilution, and they affect the value of $\Delta_{sol}H$ in a complicated way. For this reason, the quoted values for $\Delta_{sol}H$ refer to an **infinitely dilute solution**. This value cannot be determined experimentally and is found by a process of extrapolation.

In practice, there comes a point when further dilution has no measurable effect on $\Delta_{sol}H$. This is known as the point of infinite dilution.

Enthalpy changes of solution can be either negative or positive, as is shown in **table A**.

IONIC SOLID	EQUATION	$\Delta_{sol}H$/kJ mol^{-1}
NaCl	$NaCl(s) \xrightarrow{aq} Na^+(aq) + Cl^-(aq)$	+11.0
NaOH	$NaOH(s) \xrightarrow{aq} Na^+(aq) + OH^-(aq)$	−44.5
NH_4NO_3	$NH_4NO_3(s) \xrightarrow{aq} NH_4^+(aq) + NO_3^-(aq)$	+25.7
$MgSO_4$	$MgSO_4(s) \xrightarrow{aq} Mg^{2+}(aq) + SO_4^{2-}(aq)$	−91.3

table A

ENTHALPY CHANGE OF HYDRATION, $\Delta_{hyd}H$

The **enthalpy change of hydration**, $\Delta_{hyd}H$, is the enthalpy change when one mole of an ion in its gaseous state is completely hydrated by water. In practice, complete hydration is said to have occurred when the solution formed is at infinite dilution.

For the sodium and chloride ions, the enthalpy change of hydration is the enthalpy change for the following processes:

$$Na^+(g) \xrightarrow{aq} Na^+(aq)$$

and $$Cl^-(g) \xrightarrow{aq} Cl^-(aq)$$

When an ion is placed in water it immediately interacts with the water molecules. Water molecules are polar (see **Topic 3 (Book 1: IAS)**) and are attracted to both positive and negative ions. **Fig A** shows the hydration of sodium and chloride ions.

▲ **fig A** Hydration of sodium and chloride ions.

In the case of the sodium ion, the interaction is the result of the attraction between the δ^- oxygen atom of the water molecule and the cation. Such an interaction is often referred to as an **ion–dipole interaction**. With some other positive ions, notably those of the transition metals, a dative covalent bond is formed between the water molecule and the cation using one of the lone pairs of electrons on the oxygen atom.

Ion–dipole interactions exist in the hydrated chloride ion, but some hydrogen bonds are also formed between the $\delta+$ hydrogen atoms of the water molecules and the chloride ion, making use of the lone pairs of electrons on the chloride ion (see **Topic 7 (Book 1: IAS)**).

Enthalpy changes of hydration are always negative. Some examples are given in **table B**.

EXAM HINT

When you **write** an equation for the enthalpy change of hydration of an ion, make sure that you put the gaseous ion on the left-hand side.

ION	IONIC RADIUS/nm	EQUATION	$\Delta_{hyd}H$/kJ mol^{-1}
Na^+	0.102	$Na^+(g) \xrightarrow{aq} Na^+(aq)$	−406
K^+	0.138	$K^+(g) \xrightarrow{aq} K^+(aq)$	−322
Rb^+	0.149	$Rb^+(g) \xrightarrow{aq} Rb^+(aq)$	−301
Mg^{2+}	0.072	$Mg^{2+}(g) \xrightarrow{aq} Mg^{2+}(aq)$	−1920
Ca^{2+}	0.100	$Ca^{2+}(g) \xrightarrow{aq} Ca^{2+}(aq)$	−1650
Sr^{2+}	0.113	$Sr^{2+}(g) \xrightarrow{aq} Sr^{2+}(aq)$	−1480
Cl^-	0.180	$Cl^-(g) \xrightarrow{aq} Cl^-(aq)$	−363
Br^-	0.195	$Br^-(g) \xrightarrow{aq} Br^-(aq)$	−335
I^-	0.215	$I^-(g) \xrightarrow{aq} I^-(aq)$	−293

table B

LEARNING TIP

It is important to understand what is meant by 'magnitude' here. A value for a hydration enthalpy of −322 is smaller than a value of −406. The negative sign refers to the type of enthalpy change; in this case, it is an exothermic change. Less energy is given out if the change is −322 than if the change is −406. So −322 has a smaller magnitude than −406.

In mathematics, the negative sign could indicate a negative number, in which case −322 would be greater than −406.

We can also describe the trend in hydration enthalpies by saying that the values become less negative as we go down a group.

FACTORS AFFECTING THE MAGNITUDE OF THE HYDRATION ENTHALPY

The first thing you will notice from the values in **table B** is that $\Delta_{hyd}H$ is much more negative for 2+ ions than for 1+ ions. This makes sense because we would expect a doubly charged ion to have a stronger interaction with the water molecules compared to a singly charged ion. That is to say, the electrostatic force of attraction between a doubly charged ion and water molecules will be greater than that between a singly charged ion and water molecules.

As we go down a group (e.g. Na^+ to Rb^+ or Cl^- to I^-) the magnitude of $\Delta_{hyd}H$ decreases. This seems to correlate with an increase in ionic radius. This trend can be explained using a simple electrostatic model for hydration, similar to that used to explain the variation in lattice energies for ionic solids (see **Section 12B.1**). As the ions become larger, the electrostatic force of attraction between them and the water molecules decreases, and hence the energy released upon hydration decreases.

As with lattice energy, there is a strong correlation between $\Delta_{hyd}H$ and the charge densities of the ions. **Table C** shows this for four cations.

CATION	CHARGE DENSITY OF CATION	$\Delta_{hyd}H$/kJ mol^{-1}
Na^+	$\frac{1}{(0.095)^2} = 111$	-406
K^+	$\frac{1}{(0.133)^2} = 57$	-322
Mg^{2+}	$\frac{2}{(0.060)^2} = 556$	-1920
Ca^{2+}	$\frac{2}{(0.099)^2} = 204$	-1650

table C

The greater the charge density of the cation, the more negative the value of $\Delta_{hyd}H$.

RELATIONSHIP BETWEEN $\Delta_{sol}H$, $\Delta_{hyd}H$ AND $\Delta_{LE}H$

The relationship between $\Delta_{sol}H$, $\Delta_{hyd}H$ and $\Delta_{LE}H$ is best shown by an energy level diagram similar to that of a Born–Haber cycle as shown in **fig B**.

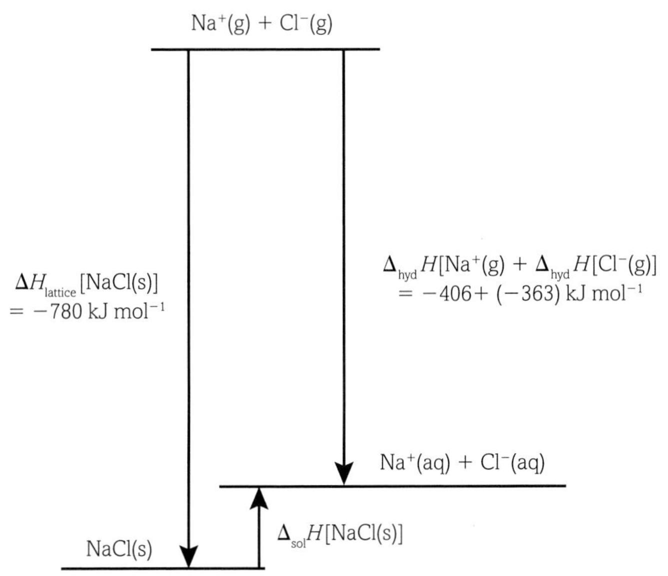

▲ **fig B** Energy level diagram for the dissolution of sodium chloride.

Applying Hess's Law:

$$\Delta_{LE}H[NaCl(s)] + \Delta_{sol}H[NaCl(s)] = \Delta_{hyd}H[Na^+(g)] + \Delta_{hyd}H[Cl^-(g)]$$

$$\Delta_{sol}H[NaCl(s)] = -406 + (-363)\text{ kJ mol}^{-1} - (-780)\text{ kJ mol}^{-1}$$

$$= +11\text{ kJ mol}^{-1}$$

The relationship can also be shown in the form of a Hess cycle (shown in question 3 below).

CHECKPOINT

SKILLS INTERPRETATION

1. When potassium fluoride dissolves in water, the lattice breaks up and the potassium and fluoride ions become hydrated.
   (a) Draw diagrams to represent (i) a hydrated potassium ion and (ii) a hydrated fluoride ion.
   (b) Name the type of interaction that occurs between the water molecules and the ion for both the hydrated potassium ion and the fluoride ion. Describe how each interaction occurs.
2. These data refer to some of the energy changes involved when magnesium chloride dissolves in water.

   $Mg^{2+}(Cl^-)_2(s) \xrightarrow{aq} Mg^{2+}(aq) + 2Cl^-(aq)$

   $\Delta_{sol}H[Mg^{2+}(Cl^-)_2(s)] = -155\text{ kJ mol}^{-1}$

   $Mg^{2+}(g) + 2Cl^-(g) \rightarrow Mg^{2+}(Cl^-)_2(s)$

   $\Delta_{LE}H[Mg^{2+}(Cl^-)_2(s)] = -2526\text{ kJ mol}^{-1}$

   $Mg^{2+}(g) \xrightarrow{aq} Mg^{2+}(aq)$

   $\Delta_{hyd}H[Mg^{2+}(g)] = -1920\text{ kJ mol}^{-1}$

   Use this information to calculate the enthalpy change of hydration of the chloride ion, $\Delta_{hyd}H[Cl^-(g)]$.
3. The diagram below is a Hess cycle for the dissolving of lithium fluoride in water.

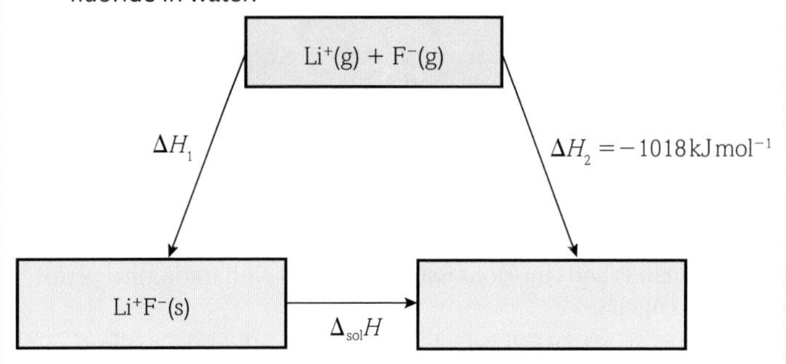

   (a) Complete the cycle by filling in the empty box.
   (b) State the name of the energy change represented by ΔH_1.
   (c) Apply Hess's Law to obtain an expression for $\Delta_{sol}H$ in terms of ΔH_1 and ΔH_2.
   (d) Calculate $\Delta_{sol}H[Li^+F^-(s)]$, given that $\Delta H_1 = -1031\text{ kJ mol}^{-1}$.
4. The standard enthalpy change of solution of sodium fluoride is +0.3 kJ mol^{-1}.

   A sample of sodium fluoride of mass 1 g is added to 250 cm^3 of water in a beaker and stirred with a thermometer graduated in intervals of 1 °C.

   Explain what is likely to happen to the reading on the thermometer as the sodium fluoride dissolves. No calculation is necessary.

SUBJECT VOCABULARY

enthalpy change of solution, $\Delta_{sol}H$ the enthalpy change when one mole of an ionic solid dissolves in water to form an infinitely dilute solution
infinitely dilute solution a solution in which there is so much water that adding any more does not cause a further enthalpy change
enthalpy change of hydration, $\Delta_{hyd}H$ is the enthalpy change when one mole of an ion in its gaseous state is completely hydrated by water
ion–dipole interaction the attraction in aqueous solution between ions and polar water molecules

HYDROGEN REVOLUTION

SKILLS CREATIVITY, REASONING

Like many car manufacturers, Toyota hopes to become a leader in the next generation of hydrogen-powered cars. Toyota calls its Mirai fuel cell car '… *the nearest thing yet to an ultimate eco-car*'. The following extract from the car manufacturer's blog explains how they work.

HOW DOES TOYOTA'S FUEL CELL VEHICLE WORK?

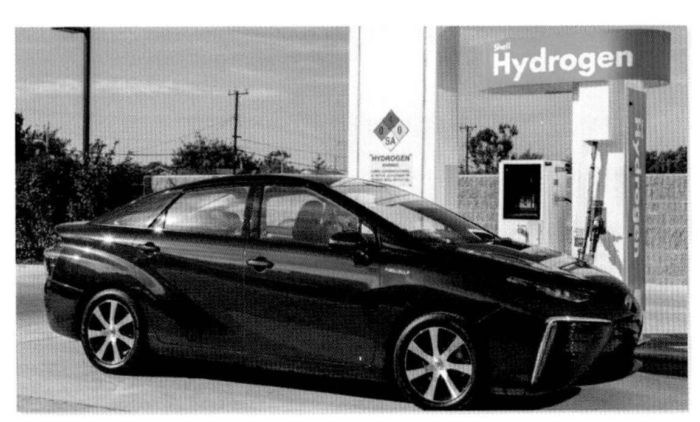

fig A The Toyota Mirai fuel cell car won the World Green Car Award in 2016.

On sale since 2015, the Toyota Mirai has been described as the nearest thing yet to the ultimate eco-car, and a key step in finding a solution to energy demands and emissions issues associated with traditional petrol and diesel engines.

But Toyota's new fuel cell vehicle is much more than the realisation of cutting-edge science theory.

How do fuel cell vehicles actually work?

A fuel cell converts fuel into electricity by forcing it to react with oxygen.

Hydrogen is the most common fuel used in today's fuel cells, but almost any hydrocarbon, including gas and alcohol, can be used. Fuel cells require a constant supply of fuel and oxygen to sustain the electricity-generating reaction.

It's worth pointing out that the idea of fuel cells is nothing new; in fact, the first examples were designed in the mid-1800s. However, it took more than 100 years for the idea to get off the ground – literally, as NASA adapted their use for the Apollo Moon project.

This environmentally friendly and highly energy-efficient process of generating electricity in a fuel cell produces no tailpipe emissions, but lots of pure water instead – great news if you are running one inside a spaceship. Back on Earth the same qualities make fuel cell vehicles ideal for achieving sustainable mobility, which is why Toyota has been trying to make this technology widely available as soon as possible.

That understood, it is now necessary to explain the functions of the two primary components used in a fuel cell vehicle.

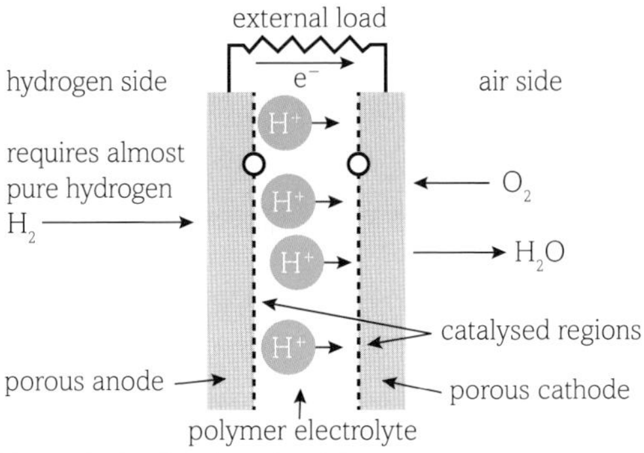

fig B How a fuel cell makes electricity.

Fuel cell

The fuel cell generates electricity through a chemical reaction between hydrogen and oxygen. This is achieved by supplying hydrogen to the negative anode of the fuel cell and ambient air to the positive cathode.

A fuel cell consists of individual cells within a membrane electrode assembly (MEA) sandwiched between separators. The MEA consists of a polymer electrolyte membrane with positive and negative catalyst layers on either side. Each cell yields less than one volt of electricity, so several hundred cells are connected in series to increase the voltage. This combined body of cells is called a stack, which is commonly referred to as a fuel cell unit.

Though it is possible to use other elements in a fuel cell, the advantage of hydrogen is its high energy efficiency. Since electricity can be produced directly from hydrogen without combustion, it is possible to convert 83% of the energy within a hydrogen molecule into electricity. This is more than double the energy efficiency of a petrol-powered engine.

High-pressure hydrogen tank

Hydrogen is stored in two high-pressure (70 MPa) tanks. The innermost layer features a polyamide resin liner that has high strength and superb resistance to hydrogen permeation. This is necessary because the diameter of the hydrogen molecule is the smallest known to science, and they tend to escape through inferior materials.

Further use of optimal materials has increased tank capacity (historically, most FCVs have needed four separate tanks to improve capacity, and therefore cruising range) and reduced weight. This can be seen in the winding angle, tension, volume and wall thickness of the carbon fibre used in the outer shell.

From 'Toyota The Official Blog of Toyota GB' by Joe Clifford blog.toyota.co.uk/how-does-toyotas-fuel-cell-vehicle-work

SCIENCE COMMUNICATION

1 Write a list of any advantages and disadvantages of fuel cell vehicles that are mentioned in the blog post. Do you think that the post provides a balanced viewpoint of the tecÚology? Justify your answer.

INTERPRETATION NOTE

Always remember to consider the source when you read scientific writing. Does the author have an interest in presenting a particular viewpoint?

CHEMISTRY IN DETAIL

2 (a) Write a balanced chemical equation for the reaction between hydrogen and oxygen to produce water.

(b) Explain why the volumes of hydrogen to oxygen must be in the ratio of 2 : 1 to ensure complete combustion.

(c) The enthalpy of combustion of one mole of hydrogen to produce water is $-286\ \text{kJ mol}^{-1}$ under standard conditions. The extract says that "83% of this energy can be converted into electricity" in a fuel cell. Calculate the energy converted to electricity by $50\ \text{dm}^3$ of hydrogen (measured at standard temperature and pressure) using the enthalpy of combustion value given.

3 Use the data in the table below to answer this question.

HALF-EQUATION	STANDARD ELECTRODE POTENTIAL/V
$H^+ + e^- \rightleftharpoons \frac{1}{2}H_2$	0.00
$\frac{1}{2}O_2 + 2H^+ + 2e^- \rightleftharpoons H_2O$	+1.23

(a) Write down the balanced equation for the reaction of hydrogen with oxygen in a fuel cell and calculate the cell emf.

The total entropy change, ΔS_{total}, can be calculated from the expression:

$$\Delta S_{total} = \frac{nFE}{T}$$

where T is the temperature in K.

(b) Calculate the total entropy change for the reaction in the fuel cell. You can assume that $F = 96500$ C and $T = 298$ K.

(c) What does the sign for the total entropy change tell you about the reaction?

4 State two challenges that manufacturers of fuel cell vehicles may face.

WRITING SCIENTIFICALLY

If the word 'explain' is used in a question, your answer needs to include a justification. This could involve reasoning or a mathematical explanation.

ACTIVITY

The concept of total entropy, ΔS_{total}, is a central one in chemistry and links a number of topics in the syllabus:

- $\Delta S_{total} = \frac{nFE}{T}$
- $\Delta S_{total} = R\ln K$
- $\Delta S_{total} = \Delta S_{system} - \frac{\Delta H}{T}$

Choose one of the three equations above and give a short (5–8 slides) presentation on the application of the equation to an aspect of the chemistry you have covered to date. You should also point out the limitations of using the equation in making predictions about chemical reactions.

DID YOU KNOW?

Although predicted as early as 1935, the creation of the metallic state of hydrogen was finally reported in 2011. Scientists at the Max Planck Institute in Germany reported that this form of hydrogen had finally been made at a pressure of 260 billion Pa. Metallic hydrogen helps explain, among other things, the strong magnetic field generated by the 'gas giant' Jupiter, which, unlike Earth, has no iron core.

12 EXAM PRACTICE

1 The table shows the ionic radii and charges for six different ions.

Ion	J^+	L^+	M^{2+}	X^-	Y^-	Z^{2-}
Ionic radius / nm	0.14	0.18	0.15	0.14	0.18	0.15

The ionic solids JX, LY and MZ have the same lattice structure. What is the order of magnitude of their lattice energies, giving the most exothermic first?

A JX > LY > MZ

B JX > MZ > LY

C MZ > JX > LY

D MZ > LY > JX [1]

(Total for Question 1 = 1 mark)

2 The table shows the lattice energies of rubidium fluoride, RbF, and caesium chloride, CsCl.

COMPOUND	LATTICE ENERGY / kJ mol^{-1}
rubidium fluoride, RbF	−783
caesium chloride, CsCl	−661

Which is most likely to be the lattice energy of caesium fluoride?

A −647 kJ mol^{-1} **B** −747 kJ mol^{-1}

C −847 kJ mol^{-1} **D** −947 kJ mol^{-1} [1]

(Total for Question 2 = 1 mark)

3 Why do calcium and chlorine react together to form $CaCl_2$(s) rather than CaCl(s)?

A Less energy is required to remove one electron from the calcium atom than to remove two electrons.

B More energy is released in forming chloride ions from chlorine molecules in the formation of $CaCl_2$(s) than in the formation of CaCl(s).

C The lattice energy of CaCl(s) is less exothermic than the lattice energy of $CaCl_2$(s).

D When CaCl(s) is formed from its elements, more energy is released than when $CaCl_2$(s) is formed from its elements. [1]

(Total for Question 3 = 1 mark)

4 How does the entropy of the system change when ammonium nitrate crystals dissolve in water?

A It remains the same.

B It decreases, because the hydrated ions are more ordered in the solution than they are in the crystal.

C It increases, because the ions in the crystal become hydrated in the solution.

D It increases, because the ions are arranged more randomly in the solution than they are in the crystal. [1]

(Total for Question 4 = 1 mark)

5 Which statement is true for the exothermic reaction

$$Zn(s) + 2HCl(aq) \rightarrow ZnCl_2(aq) + H_2(g)$$

A ΔH is positive

B $\Delta S_{surroundings}$ is positive

C ΔS_{system} is negative

D ΔS_{total} is negative [1]

(Total for Question 5 = 1 mark)

6 Calcium carbonate decomposes in an endothermic reaction when heated to 1113 K.

$$CaCO_3(s) \rightarrow CaO(s) + CO_2(g)$$

What are the signs of the entropy changes at 1113 K?

	ΔS_{system}	$\Delta S_{surroundings}$
A	+	+
B	+	−
C	−	+
D	−	−

[1]

(Total for Question 6 = 1 mark)

7 The Born–Haber cycle shown below can be used to calculate the lattice energy for magnesium oxide.

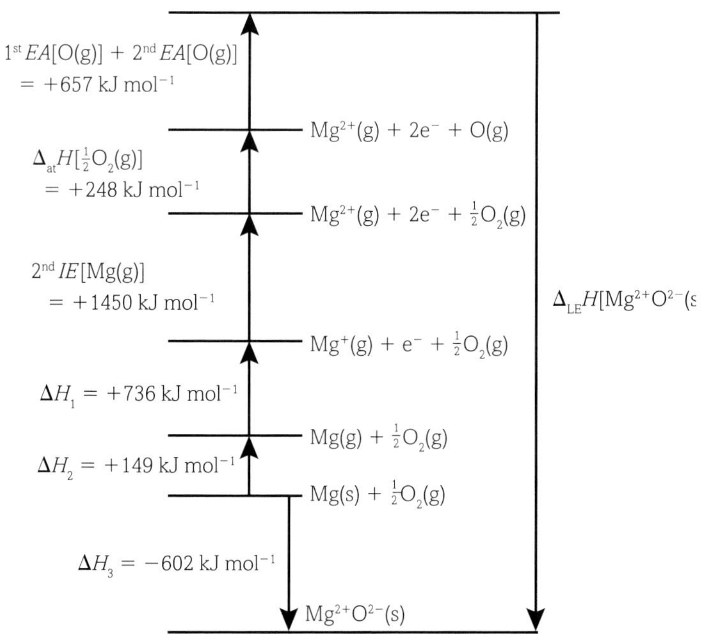

(a) Give the name of each of the enthalpy changes labelled ΔH_1, ΔH_2 and ΔH_3. [3]

(b) Write the missing formulae from the top of the Born–Haber cycle. Include state symbols. [2]

(c) The equations for the first and second electron affinities of oxygen are:

$O(g) + e^- \rightleftharpoons O^-(g)$ 1st $EA = -141\,kJ\,mol^{-1}$

$O^-(g) + e^- \rightleftharpoons O^{2-}(g)$ 2nd $EA = +798\,kJ\,mol^{-1}$

Suggest why the second electron affinity of oxygen is endothermic. [2]

(d) Use the data in the Born–Haber cycle to calculate the lattice energy of magnesium oxide. [2]

(Total for Question 7 = 9 marks)

8 Lattice energies can be found *either* via experimentally determined energy changes, using the Born–Haber cycle, *or* by theoretical calculation using the ionic model for the crystalline compound.

(a) State the energy changes that need to be determined in order to find the lattice energy of sodium fluoride, using the Born–Haber cycle. [4]

(b) The table lists the experimentally determined and theoretically calculated lattice energies for some compounds.

Compound	NaF	NaI	AgF	AgI
Experimental value/kJ mol^{-1}	−918	−705	−958	−889
Theoretical value/kJ mol^{-1}	−912	−687	−920	−778

Comment on the discrepancies between the experimental and theoretical values of the lattice energies. [4]

(c) The enthalpy changes of hydration, in kJ mol^{-1}, of the ions in the compounds listed in the table in part (b) are:

Na^+ −406 Ag^+ −464 F^- −506 I^- −293

(i) Using the experimental lattice energy values, show by calculation which of the four compounds is the most soluble. [2]

(ii) Explain what other information is necessary in order to make a valid comparison of the solubilities of the four compounds. [2]

(Total for Question 8 = 12 marks)

9 Entropy changes are important in determining the feasibility of a process.

(a) For each of the four processes below, predict the likely sign of the entropy change of the system, ΔS_{system}. In each case, justify your answer. [4]

(i) $N_2(g) + 3H_2(g) \rightarrow 2NH_3(g)$

(ii) $H_2O(s) \rightarrow H_2O(l)$

(iii) $2Na(s) + O_2(g) \rightarrow Na_2O_2(s)$

(iv) $S(s) + O_2(g) \rightarrow 2SO_2(g)$

(b) Magnesium oxide is used as a refractory lining for furnaces. It can be made by heating magnesium carbonate.

$MgCO_3(s) \rightarrow MgO(s) + CO_2(g)$ $\Delta H^\ominus = +117\,kJ\,mol^{-1}$

The table shows the standard entropies at 298 K of $MgCO_3(s)$, $MgO(s)$ and $CO_2(g)$.

Substance	$MgCO_3(s)$	$MgO(s)$	$CO_2(g)$
$S^\ominus$/J K^{-1} mol^{-1}	65.7	27.0	214.0

(i) Use the data to show that magnesium carbonate is stable at 298 K. [4]

(ii) Calculate the minimum temperature at which magnesium carbonate will decompose. [2]

(Total for Question 9 = 10 marks)

10 (a) The equation for the combustion of hydrogen is:

$H_2(g) + \frac{1}{2}O_2(g) \rightarrow H_2O(l)$ $\Delta H^\ominus = -285.5\,kJ\,mol^{-1}$

The table gives the standard entropies at 298 K for $H_2(g)$, $O_2(g)$ and $H_2O(l)$.

Substance	$H_2(g)$	$O_2(g)$	$H_2O(l)$
$S^\ominus$/J K^{-1} mol^{-1}	131.0	205.0	69.9

(i) Calculate the standard entropy change of the system at 298 K, $\Delta S^\ominus_{system}$, for the combustion of hydrogen. [2]

(ii) Calculate the standard entropy change of the surroundings at 298 K, $\Delta S^\ominus_{surroundings}$, for the combustion of hydrogen. [2]

(iii) Calculate the total standard entropy change at 298 K, $\Delta S^\ominus_{total}$, for the combustion of one mole of hydrogen. Give your answer to an appropriate number of significant figures. [3]

(b) Explain why hydrogen does not react with oxygen unless the mixture is ignited. [1]

(Total for Question 10 = 8 marks)

11 Ammonia can be made by reacting together nitrogen and hydrogen. The reaction is reversible.

$N_2(g) + 3H_2(g) \rightleftharpoons 2NH_3(g)$ $\Delta_r H^\ominus (700\,K) = -110.2\,kJ\,mol^{-1}$

$\Delta S^\ominus_{total} (700\,K) = -78.7\,J\,K^{-1}\,mol^{-1}$

(a) The $^\ominus$ symbol represents standard conditions of a specified temperature and standard pressure. In this instance, the temperature specified is 700 K. State the standard pressure for thermodynamic measurements. [1]

(b) Calculate $\Delta S^\ominus_{surroundings}$ (700 K). [2]

(c) Calculate $\Delta S^\ominus_{system}$ (700 K). [2]

(d) If the reaction mixture is left for long enough, in contact with a suitable catalyst, it is possible to establish an equilibrium between the nitrogen, hydrogen and the ammonia.

Comment on what the value of $\Delta S^\ominus_{total}$ (700 K) indicates about the relative proportions of nitrogen, hydrogen and ammonia in the equilibrium mixture at 700 K. [1]

(Total for Question 11 = 6 marks)

TOPIC 13 CHEMICAL EQUILIBRIA

A CHEMICAL EQUILIBRIA

Whenever chemical engineers design a chemical plant to manufacture a new chemical they must overcome a problem. They want the best possible yield from the reactants, but they find that most chemical reactions do not go to completion because they are reversible and can reach a state of dynamic equilibrium.

All reactions have a tendency to approach equilibrium, whether they seem to go to completion or whether they seem to form no products at all. In many cases, such as the combustion of a fuel, the equilibrium composition corresponds to the complete conversion of reactants into products.

Reactions that don't seem to take place at all, like the corrosion of gold in air, also tend to approach equilibrium, but in these cases the equilibrium composition corresponds to virtually unchanged reactants.

Many reactions approach an equilibrium in which there are significant amounts of both reactants and products present. Finding and controlling the position of equilibrium in such reactions is the focus of this topic.

MATHS SKILLS FOR THIS TOPIC

- Recognise and make use of appropriate units in calculations
- Recognise and use expressions in decimal and ordinary form
- Use ratios, fractions and percentages
- Use an appropriate number of significant figures
- Change the subject of an equation
- Substitute numerical values into algebraic equations using appropriate units for physical quantities
- Solve algebraic expressions
- Use logarithms in relation to quantities that range over several orders of magnitude

What prior knowledge do I need?

- Deducing expressions for K_c for both homogeneous and heterogeneous systems

Topic 9 (Book 1: IAS)

- The concept of reversible reactions and dynamic equilibrium
- The qualitative effect of change in concentration, temperature and pressure on the position of equilibrium

What will I study in this topic?

- How to deduce an expression for K_c, for homogeneous and heterogeneous systems, in terms of equilibrium concentrations
- How to deduce an expression for the equilibrium constant, K_p, in terms of partial pressures
- The quantitative effect of change in concentration
- How to predict the effect of change in temperature on the values of K_c and K_p
- How to predict the effect of a change in temperature on the position of equilibrium in terms of changes to K_c and K_p
- Why the value of an equilibrium constant is not altered by the addition of a catalyst

What will I study later?

- The application of equilibrium constants to acid–base equilibria

Topic 14

- The relationship between the equilibrium constant, K, and the total entropy change, ΔS_{total}

13A 1 EQUILIBRIUM CONSTANT, K_c

SPECIFICATION REFERENCE 13.1 13.3 PART

LEARNING OBJECTIVES

- Be able to deduce an expression for K_c, for homogeneous and heterogeneous systems, in terms of equilibrium concentrations.
- Be able to calculate a value, with units where appropriate, for the equilibrium constant K_c for homogeneous and heterogeneous reactions, from experimental data.

HOMOGENEOUS REACTIONS

A **homogeneous reaction** is one in which all reactants and products are in the same phase. For example, all are gases or all are in aqueous solution.

In **Topic 9 (Book 1: IAS)** we saw how the position of equilibrium of a reversible reaction can change when the conditions are altered.

In this section, we will study *quantitatively* how changes in concentration affect the equilibrium position.

For the reversible reaction:

$$N_2O_4(g) \rightleftharpoons 2NO_2(g)$$

it can be shown experimentally that at equilibrium and at a given temperature:

$$\frac{[NO_2(g)]^2_{eq}}{[N_2O_4(g)]_{eq}} = \text{a constant}$$

This value is called the **equilibrium constant** and is given the symbol K_c.

The subscript 'c' indicates that this equilibrium constant is calculated using concentrations. This is to distinguish it from another equilibrium constant, K_p, which uses partial pressures.

The subscript 'eq' indicates that the concentrations are those *at equilibrium*.

At a given temperature, the value of the equilibrium constant does not change with concentration, unlike the equilibrium position. The only factor that affects the value of the equilibrium constant is temperature.

There are two ways of obtaining a value for the equilibrium constant:

- the first is by experimentation, and this is the method we will look at in this section
- the second is by calculation using thermodynamic equations, and this will be covered in **Section 13A.5**.

For a general reaction:

$$wA + xB \rightleftharpoons yC + zD$$

$$K_c = \frac{[C]^y_{eq}[D]^z_{eq}}{[A]^w_{eq}[B]^x_{eq}}$$

For convenience, the 'eq' sign is often not included, but is taken for granted.

DETERMINATION OF THE EQUILIBRIUM CONSTANT

To determine the value of the equilibrium constant for a reaction, known amounts of reactants are allowed to reach equilibrium with their products. At equilibrium, the amount of one of the substances is measured and the amounts of the others can be calculated using the stoichiometry of the equation.

WORKED EXAMPLE 1

2.00 mol of ethanoic acid and 2.00 mol of ethanol are mixed and allowed to reach equilibrium with ethyl ethanoate and water at 298 K. The amount of ethanoic acid at equilibrium is found to be 0.67 mol. Calculate K_c for the reaction at 298 K.

Answer

Work out the amounts, and hence the concentrations, of each substance at equilibrium:

$$CH_3COOH(l) + CH_3CH_2OH(l) \rightleftharpoons CH_3COOCH_2CH_3(l) + H_2O(l)$$

amount at start/mol

$CH_3COOH(l)$	$CH_3CH_2OH(l)$	$CH_3COOCH_2CH_3(l)$	$H_2O(l)$
2.00	2.00	0	0

amount at equilibrium/mol

$CH_3COOH(l)$	$CH_3CH_2OH(l)$	$CH_3COOCH_2CH_3(l)$	$H_2O(l)$
0.67	0.67	(2.00 – 0.67) = 1.33	(2.00 – 0.67) = 1.33

concentration at equilibrium/mol dm^{-3}

$CH_3COOH(l)$	$CH_3CH_2OH(l)$	$CH_3COOCH_2CH_3(l)$	$H_2O(l)$
$\frac{0.67}{V}$	$\frac{0.67}{V}$	$\frac{1.33}{V}$	$\frac{1.33}{V}$

where V is the total volume in dm^3 of the reaction mixture.

$$K_c = \frac{[CH_3COOCH_2CH_3(l)][H_2O(l)]}{[CH_3COOH(l)][CH_3CH_2OH(l)]} = \frac{(1.33/V)(1.33/V)}{(0.67/V)(0.67/V)}$$

$= 3.94$ (no units)

K_c has no units because the same number of moles are present on each side of the equation. For the same reason, V also cancels.

WORKED EXAMPLE 2

0.100 mol of N_2O_4 is placed into a 0.100 dm^3 flask and allowed to reach equilibrium with NO_2 at 398 K. At equilibrium, there is 0.071 mol of N_2O_4. Calculate K_c at this temperature.

Answer

$N_2O_4(g)$	$\rightleftharpoons$ $2NO_2(g)$	
0.100	0	amount at start/mol
0.071	(0.100 – 0.071) × 2 = 0.058	amount at equilibrium/mol
0.071/0.100 = 0.71	0.058/0.100 = 0.58	concentration at equilibrium/mol dm^{-3}

$$K_c = \frac{[NO_2(g)]^2}{[N_2O_4(g)]} = \frac{0.58^2}{0.71} = 0.474 \text{ mol dm}^{-3}$$

Note: this time, K_c has units of mol dm^{-3}

$$\frac{\text{mol dm}^{-3} \times \text{mol dm}^{-3}}{\text{mol dm}^{-3}} = \text{mol dm}^{-3}$$

HETEROGENEOUS REACTIONS

A **heterogeneous reaction** is one in which at least one of the reactants and/or products is in a different phase to the others.

Examples include:

$$CaCO_3(s) \rightleftharpoons CaO(s) + CO_2(g)$$

$$3Fe(s) + 4H_2O(g) \rightleftharpoons Fe_3O_4(s) + 4H_2(g)$$

$$H_2O(l) \rightleftharpoons H^+(aq) + OH^-(aq)$$

WRITING EQUILIBRIUM CONSTANTS FOR HETEROGENEOUS REACTIONS

If we apply the equilibrium law to the reaction:

$$CaCO_3(s) \rightleftharpoons CaO(s) + CO_2(g)$$

we obtain:

$$\frac{[CaO(s)][CO_2(g)]}{[CaCO_3(s)]} = \text{a constant}$$

DID YOU KNOW?

The concentration term of a solid in a heterogeneous equilibrium is constant. It can therefore be absorbed into the equilibrium constant.

However, the concentration of a solid at a given temperature is determined by its density, which has a constant value. Hence, the expression can be simplified to:

$[CO_2(g)]$ = a constant

This constant is given the symbol K_c.

So, $K_c = [CO_2(g)]$ for the above equilibrium.

Similarly, for the equilibrium:

$$3Fe(s) + 4H_2O(g) \rightleftharpoons Fe_3O_4(s) + 4H_2(g)$$

$$K_c = \frac{[H_2(g)]^4}{[H_2O(g)]^4}$$

CHECKPOINT

1. A dynamic equilibrium is set up between carbon monoxide, hydrogen and methanol. The equation for the reaction is:

$$CO(g) + 2H_2(g) \rightleftharpoons CH_3OH(g)$$

The equilibrium concentrations, in mol dm^{-3}, for the three components are shown below:

$CO(g)$, 3.1×10^{-3}; $H_2(g)$, 2.4×10^{-2}; $CH_3OH(g)$, 2.6×10^{-5}

(a) Write an expression for K_c for this reaction.

(b) Calculate a value for K_c. State the units, if any.

SKILLS ADAPTIVE LEARNING

2. The reaction between ethanoic acid and ethanol is reversible. The equation for the reaction is:

$$CH_3COOH(l) + CH_3CH_2OH(l) \rightleftharpoons CH_3COOCH_2CH_3(l) + H_2O(l)$$

6.0 mol of ethanoic acid and 12.5 mol of ethanol are mixed together with some sulfuric acid to act as a catalyst. The total volume of the reaction mixture is V dm^3. The mixture is left for several days to reach equilibrium, after which only 1.0 mol of ethanoic acid remains.

(a) Write an expression for Kc for the reaction.

(b) Calculate the amounts of ethanol, ethyl ethanoate and water present in the equilibrium mixture.

(c) Use your answers to part (b) to calculate a value for Kc. Give your answer to an appropriate number of significant figures and state the units, if any.

EXAM HINT

In question 2, the amount of acid at equilibrium can be quantified by titration. You might expect the equilibrium to shift to the left as the acid is neutralised. However, the rate of this reaction is very slow (note that the reaction is left for several days). This means that the amount of acid can be quantified before the equilibrium has had time to respond.

3. The reaction between hydrogen and iodine to form hydrogen iodide is reversible.

$$H_2(g) + I_2(g) \rightleftharpoons 2HI(g)$$

0.30 mol of $H_2(g)$ is mixed with 0.20 mol of $I_2(g)$ and the mixture is allowed to reach equilibrium. At equilibrium, 0.14 mol of $H_2(g)$ is present.

Calculate the value of K_c for the reaction. Give your answer to an appropriate number of significant figures and state the units, if any.

4. Nitrogen and oxygen can exist in equilibrium with nitrogen(II) oxide. The equation for the reaction is:

$$N_2(g) + O_2(g) \rightleftharpoons 2NO(g) \qquad \Delta H \text{ is positive}$$

The equilibrium constant at 289 K for this reaction is 4.8×10^{-31}.

(a) What does the value of the equilibrium constant tell you about the position of equilibrium of this reaction at 298 K?

(b) An equilibrium mixture of volume 1.2 dm^3 of the three gases contains 1.1 mol of $N_2(g)$ and 4.0×10^{-16} mol of $NO(g)$. Calculate the equilibrium concentration of $O_2(g)$ in this mixture.

5. The equilibrium constant, K_c, for the reaction:

$$H_2O(g) + C(s) \rightleftharpoons H_2(g) + CO(g)$$

has a value of 4.92×10^{-5} mol dm^{-3} at 700 K.

A mixture of water vapour and carbon is heated to 700 K in a closed vessel and allowed to reach equilibrium with hydrogen and carbon monoxide.

Write an expression for K_c and use it to calculate the equilibrium concentrations of H_2 and CO in the above equilibrium mixture, at 700 K, when the equilibrium concentration of H_2O is 2.00×10^{-2} mol dm^{-3}.

SUBJECT VOCABULARY

homogeneous reaction a reaction in which all the reactants and products are in the same phase

equilibrium constant a number that expresses the relationship between the amounts of products and reactants present at equilibrium in a reversible chemical reaction at a given temperature

heterogeneous reaction a reaction in which at least one of the reactants and/or products is in a different phase to the others

13A 2 EQUILIBRIUM CONSTANT, K_p

SPECIFICATION REFERENCE

13.2 13.3 PART

LEARNING OBJECTIVES

- Be able to deduce an expression for K_p for homogeneous and heterogeneous systems in terms of equilibrium partial pressures in atm.
- Be able to calculate a value, with units where appropriate, for the equilibrium constant K_p for homogeneous and heterogeneous reactions from experimental data.

HOMOGENEOUS REACTIONS

For reversible reactions involving gases we can express the concentrations of the reactants and products in terms of their partial pressures. The **partial pressure** of an individual gas in a mixture of gases is defined as the pressure that the gas would exert if it alone occupied the volume of the mixture. The total pressure of the mixture is then equal to the sum of the partial pressures of all the gases present in the mixture.

If we have a mixture of two gases, A and B, then the total pressure, p, of the mixture is related to the partial pressures of A and B by the following equation:

$$p = p_A + p_B$$

where p_A = partial pressure of A and p_B = partial pressure of B.

For the general reaction:

$$wA(g) + xB(g) \rightleftharpoons yC(g) + zD(g)$$

we can define another equilibrium constant, K_p, in terms of partial pressures:

$$K_p = \frac{(p_C)^y(p_D)^z}{(p_A)^w(p_B)^x}$$

CALCULATING PARTIAL PRESSURES

The partial pressure of an individual gas in a mixture of gases is calculated by multiplying its mole fraction by the total pressure, for example:

$$p_A = x_A p$$

where x_A = the mole fraction of gas A in the mixture.

$$\text{mole fraction of A} = \frac{\text{number of moles of A}}{\text{total number of moles of gas}}$$

WORKED EXAMPLE

1.00 mol of PCl_5 vapour is heated to 500 K in a sealed vessel. The equilibrium mixture, at a pressure of 6 atm, contains 0.60 mol of chlorine. Calculate K_p for the reaction:

$$PCl_5(g) \rightleftharpoons PCl_3(g) + Cl_2(g)$$

Answer

$PCl_5(g)$	$\rightleftharpoons$ $PCl_3(g)$	+ $Cl_2(g)$	Total	
1.00	0	0	1.00	amount at start/mol
1.00 – 0.60	0.60	0.60	1.60	amount at equilibrium/mol
$\frac{0.40}{1.60}$	$\frac{0.60}{1.60}$	$\frac{0.60}{1.60}$		mole fraction at equilibrium
0.25 × 6	0.375 × 6	0.375 × 6		partial pressure/atm

$$K_p = \frac{(p_{PCl_3})(p_{Cl_2})}{(p_{PCl_5})} = \frac{(0.375 \times 6)(0.375 \times 6)}{(0.25 \times 6)} = 3.38 \text{ atm}$$

EXAM HINT

Note that you must use (p) for partial pressures. Do *not* use square brackets, which are for substances measured in mol dm^{-3}.

HETEROGENEOUS REACTIONS

SOLIDS AND GASES

For solids that are in equilibrium with gases, the partial pressure terms of any solids are not included in the expression for K_p. You will recall that K_c has a similar rule.

Consider the reaction:

$$CaCO_3(s) \rightleftharpoons CaO(s) + CO_2(g)$$

The equilibrium constant in terms of partial pressures is given by:

$$K_p = p_{CO_2(g)}$$

This equation suggests that for a particular temperature, the pressure of carbon dioxide in the equilibrium mixture is constant regardless of the masses of calcium carbonate and calcium oxide present. This prediction can be confirmed experimentally.

At 1073 K, the pressure of carbon dioxide in equilibrium with $CaCO_3(s)$ and $CaO(s)$ is 0.25 atm.

Hence, at 1073 K, K_p = 0.25 atm.

For this reaction, K_p increases with increasing temperature. It only exceeds 1 atm at temperatures above 1173 K. So, calcium carbonate must be heated to this temperature for it to decompose at 1 atm of pressure.

DID YOU KNOW?

A similar situation exists for liquids that are in equilibrium with gases. For the equilibrium between liquid water and water vapour:

$$H_2O(l) \rightleftharpoons H_2O(g)$$

$$K_p = p_{H_2O(g)}$$

The partial pressure of water (p_{H_2O}) in equilibrium with its liquid is also known as the 'saturated vapour pressure' of water.

At 298 K, K_p, and hence p_{H_2O}, is equal to 0.03 atm; whereas at 373 K, K_p, and hence p_{H_2O}, is equal to 1.00 atm.

Since the boiling temperature of a liquid is defined as the temperature at which the saturated vapour pressure equals the external pressure, the boiling temperature of water at 1 atm is 373 K (100 °C).

CHECKPOINT

SKILLS PROBLEM-SOLVING

1. One stage in the manufacture of sulfuric acid by the contact process involves the reaction between sulfur dioxide and oxygen to form sulfur trioxide:

   $$2SO_2(g) + O_2(g) \rightleftharpoons 2SO_3(g)$$

   (a) Write an expression for K_p for this reaction.

   (b) An equilibrium is set up for this reaction at 700 K.

   At this temperature:
   - the partial pressure of SO_2 is 0.100 atm
   - the partial pressure of O_2 is 0.500 atm
   - K_p for the reaction is 3.00×10^4 atm^{-1}

   Calculate the partial pressure of SO_3 in this equilibrium mixture and hence determine the percentage of SO_3 present.

2. The reaction between carbon monoxide and chlorine to form phosgene, $COCl_2$, is reversible. The equation for the reaction is:

   $$CO(g) + Cl_2(g) \rightleftharpoons COCl_2(g)$$

   Some carbon monoxide and chlorine are allowed to react and reach a position of equilibrium. The equilibrium partial pressures of the mixture are:

   CO(g), 2.47×10^{-8} atm; Cl_2(g), 2.47×10^{-8} atm; $COCl_2$(g), 4.08×10^{-10} atm

   (a) What is meant by the term 'partial pressure'?

   (b) Write an expression for K_p for this reaction and calculate its value, giving the units, if any.

3. When chlorine gas is heated to a high temperature, the molecules dissociate into chlorine atoms.

   $$Cl_2(g) \rightleftharpoons 2Cl(g)$$

   Some chlorine gas is placed in a closed container and heated to 1400 K until equilibrium is established. The partial pressure of Cl_2(g) is 0.84 atm and that of Cl(g) is 0.030 atm.

   (a) Determine the mole fraction of Cl in the equilibrium mixture.

   (b) (i) Write an expression for K_p for this reaction.

   (ii) Calculate the value of K_p for this reaction at 700 K. Give the units, if any.

SUBJECT VOCABULARY

partial pressure (of a gas in a mixture of gases) the pressure that the gas would exert if it alone occupied the volume of the mixture

13A 3 FACTORS AFFECTING EQUILIBRIUM CONSTANTS 1

SPECIFICATION REFERENCE 13.6 13.7

LEARNING OBJECTIVES

- Know the effect of changing the temperature on the equilibrium constant (K_c and K_p) for both exothermic and endothermic reactions.
- Understand that the effect of temperature on the position of equilibrium is explained using a change in the value of the equilibrium constant.

EFFECT OF TEMPERATURE ON K_p AND K_c

In this section we will discuss the effects of temperature on equilibrium constants. **Table A** gives the values of the equilibrium constant, K_p, at different temperatures for two gaseous reactions.

$N_2(g) + 3H_2(g) \rightleftharpoons 2NH_3(g)$; $\Delta H^\ominus = -92.2\,kJ\,mol^{-1}$		$N_2O_4(g) \rightleftharpoons 2NO_2(g)$; $\Delta H^\ominus = +57.2\,kJ\,mol^{-1}$	
T/K	K_p/atm^{-2}	T/K	K_p /atm
298	6.76×10^5	298	1.15×10^{-1}
400	4.07×10^1	400	4.79×10^1
500	3.55×10^{-2}	500	1.70×10^3
600	1.66×10^{-3}	600	1.78×10^4

table A

For the *exothermic* reaction between $N_2(g)$, $H_2(g)$ and $NH_3(g)$, the value of K_p *decreases* with increasing temperature.

The opposite is true for the *endothermic* reaction between $N_2O_4(g)$ and $NO_2(g)$.

Exactly the same trend is observed with the values of K_c.

Table B summarises the effect of changing the temperature on the value of the equilibrium constant.

THERMICITY OF REACTION	INCREASE IN TEMPERATURE	DECREASE IN TEMPERATURE
exothermic	*K* decreases	*K* increases
endothermic	*K* increases	*K* decreases

table B

CHANGE IN TEMPERATURE ALTERS THE EQUILIBRIUM POSITION

If the equilibrium constant for a reaction changes with a change in temperature, then it follows that the equilibrium position also changes. This is illustrated by the data in **table C**, which shows the value of K_c and the amounts of reactants and products at equilibrium for the reaction between hydrogen and iodine to form hydrogen iodide:

$$H_2(g) + I_2(g) \rightleftharpoons 2HI(g) \qquad \Delta H^\ominus = -9\,kJ\,mol^{-1}$$

In each case, one mole of hydrogen is mixed with one mole of iodine and equilibrium is allowed to establish, at the given temperature, in a vessel of volume 1 dm^3.

TEMPERATURE	K_p	AMOUNT/mol		
		$H_2(g)$	$I_2(g)$	$HI(g)$
500	160	0.14	0.14	1.72
700	54	0.21	0.21	1.58

table C

As the value of K_p decreases, the percentages of $H_2(g)$ and $I_2(g)$ increase, and the percentage of HI(g) decreases. The equilibrium shifts to the left, in the endothermic direction, as the temperature is increased.

CHECKPOINT

1. A dynamic equilibrium is set up between carbon monoxide, hydrogen and methanol. The equation for the reaction is:

   $$CO(g) + 2H_2(g) \rightleftharpoons CH_3OH(g)$$

   When the reaction is carried out at a higher temperature, the value of K_c decreases.

   Explain whether the forward reaction is exothermic or endothermic.

SKILLS ADAPTIVE LEARNING

2. The reaction between hydrogen and iodine to form hydrogen iodide is reversible:

   $$H_2(g) + I_2(g) \rightleftharpoons 2HI(g)$$

   Some $H_2(g)$ is mixed with $I_2(g)$ and the mixture is allowed to reach equilibrium. At equilibrium, 0.14 mol of $H_2(g)$ is present.

   The experiment is repeated with the same amounts of $H_2(g)$ and $I_2(g)$, but at a higher temperature. This time the amount of $H_2(g)$ at equilibrium is greater than 0.14 mol.

   Explain what this information tells you about the reaction.

3. Nitrogen and oxygen can exist in equilibrium with nitrogen(II) oxide. The equation for the reaction is:

   $$N_2(g) + O_2(g) \rightleftharpoons 2NO(g) \qquad \Delta H \text{ is positive}$$

   The equilibrium constant at 298 K for this reaction is 4.8×10^{-31}.

   Explain the change, if any, in the proportion of NO(g) at a temperature higher than 298 K.

13A 4 FACTORS AFFECTING EQUILIBRIUM CONSTANTS 2

SPECIFICATION REFERENCE 13.4 13.5

LEARNING OBJECTIVES

- Understand how, if at all, a change in temperature, pressure or the presence of a catalyst affects the equilibrium composition in a homogeneous or heterogeneous system.
- Understand that the value of the equilibrium constant is not affected by changes in concentration or pressure or by the addition of a catalyst.

EFFECT OF CONCENTRATION

In **Topic 9 (Book 1: IAS)** we saw that the equilibrium position changes when the concentration of a component of the equilibrium mixture is altered. In this section we will discuss the effects of concentration, pressure and catalysts on equilibrium constants and provide an explanation for this phenomenon in terms of K_c.

The most important thing to appreciate is that the only factor that can change the value of K_c is a change in temperature. That is, K_c *is constant at a given temperature.*

When the concentration of one of the components of an equilibrium mixture is altered, there is an immediate change in the **reaction quotient**, Q_c, which is the mathematical relationship between the concentrations of the components.

At this point, the reaction quotient is no longer equal to the equilibrium constant, K_c. Therefore, the equilibrium composition changes until the reaction quotient and equilibrium constant become equal.

The following examples illustrate this process.

REACTION QUOTIENT, Q_c

For a general reaction:

$$wA + xB \rightleftharpoons yC + zD$$

the reaction quotient, Q_c, is given by the expression:

$$Q_c = \frac{[C]^y[D]^z}{[A]^w[B]^x}$$

where the concentration terms are not necessarily those at equilibrium.

At equilibrium $Q_c = K_c$.

WORKED EXAMPLE 1

The equation for the reaction between hydrogen and iodine to form hydrogen iodide is:

$$H_2(g) + I_2(g) \rightleftharpoons 2HI(g)$$

Explain the effect on the position of equilibrium of increasing the concentration of hydrogen at a constant temperature and constant volume.

$$K_c = \frac{[HI(g)]^2_{eq}}{[H_2(g)]_{eq}[I_2(g)]_{eq}} \qquad Q_c = \frac{[HI(g)]^2}{[H_2(g)][I_2(g)]}$$

- If $[H_2(g)]$ is suddenly increased, then Q_c decreases and is no longer equal to K_c.
- In order for equilibrium to be re-established, the magnitude of the denominator in the expression for Q_c has to decrease in value.
- This will produce a subsequent increase in the value of the numerator.
- The two values will adjust until the value of Q_c becomes equal to that of K_c and a new equilibrium system is established.

The net result of an increase in $[H_2(g)]$ is that the equilibrium shifts to the right. This is in agreement with the qualitative prediction that we made in **Topic 9 (Book 1: IAS)**.

QUALITATIVE PREDICTIONS

In **Topic 9 (Book 1: IAS)** we said that qualitative predictions do not always give us the correct result. The worked example below shows this.

WORKED EXAMPLE 2

The amounts of N_2, H_2 and NH_3 in an equilibrium mixture are 0.510 mol, 0.197 mol and 0.204 mol, respectively:

$$N_2(g) + 3H_2(g) \rightleftharpoons 2NH_3(g)$$

The total volume of the gaseous mixture is 1.00 dm^3.

Work out the direction of the shift in equilibrium position if 0.140 mol of N_2 gas is suddenly added to the equilibrium system at constant temperature and pressure.

Show your calculations.

$$Q_c = \frac{[NH_3]^2}{[N_2][H_2]^3}$$

$$= \frac{(n_{NH_3}/V)^2}{(n_{N_2}/V)(n_{H_2}/V)^3}$$

where n is the amount (in mol) and V is the total volume (in dm^3) of the mixture.

(Remember that concentration = amount ÷ volume.)

Step 1: Calculation of K_c.

$$K_c = \frac{(0.204)^2/1^2}{(0.510)/1 \times (0.197)^3/1^3} = 10.7$$

Step 2: Calculation of new total number of moles of gas.

Original total moles of gas = (0.510 + 0.197 + 0.204) = 0.911

New total moles of gas = (0.911 + 0.140) = 1.051

Step 3: Calculation of new volume of gas mixture.

Since the pressure remains constant, the gas mixture will expand. If we assume that the gas mixture behaves like an ideal gas, then the volume of gas is proportional to the number of moles:

$$n_1/V_1 = n_2/V_2$$

So the new volume of the gas mixture =

$(1.051 \div 0.911) \times 1\ dm^3 = 1.154\ dm^3$

Step 4: Calculation of Q_c immediately after the addition of nitrogen.

$$Q_c = \frac{(0.204)^2/(1.154)^2}{(0.650)/1.154 \times (0.197)^3/(1.154)^3} = 11.15$$

Step 5: Compare Q_c with K_c.

11.15 > 10.7, therefore $Q_c > K_c$.

For equilibrium to be re-established, Q_c must decrease. This can only happen by the equilibrium shifting to the left, to increase the denominator in the equation.

The addition of more nitrogen, at constant pressure, results in a shift of the equilibrium position to the left. Our prediction told us that the equilibrium would shift to the right, away from the nitrogen added.

The problem with making predictions is that two conflicting changes are occurring here:

1. the increase in the number of moles of nitrogen should result in a shift of equilibrium to the right
2. the increase in volume should result in a shift to the left.

We have no way of knowing which effect is greater unless we perform the calculations shown above.

EFFECT OF PRESSURE

As with concentration, a change in pressure, at constant temperature, on an equilibrium system containing gases has no effect on the value of either K_c or K_p.

Once again, any change in the equilibrium position occurs in order to maintain the equilibrium constant at a constant value.

If the partial pressure of only *one* of the component gases is changed, then the overall effect on K_p can be predicted in the same way the effects of changes in concentration on K_c were described in the previous section.

However, if the *total* pressure of a gaseous system is suddenly increased or decreased, then the partial pressures of *all* the gases will either increase or decrease. The effect on the position of equilibrium can be explained using the equilibrium law. This is shown in the following worked example.

WORKED EXAMPLE 3

Consider the reaction:

$$N_2(g) + 3H_2(g) \rightleftharpoons 2NH_3(g)$$

for which:

$$K_p = \frac{(p_{NH_3})^2}{(p_{N_2})\,(p_{H_2})^3}$$

Suppose the equilibrium partial pressures of nitrogen, hydrogen and ammonia are *a*, *b* and *c* atm, respectively. Then:

$$K_p = \frac{c^2}{ab^3}$$

If the total pressure is suddenly doubled, the partial pressures of nitrogen, hydrogen and ammonia will be doubled to 2*a*, 2*b* and 2*c* respectively:

$$Q_p = \frac{(2c)^2}{2a \times (2b)^3} = \frac{4c^2}{16ab^3}$$

The value of Q_p is now one-quarter of the value of K_p. In order to re-establish equilibrium the numerator has to increase, with a subsequent decrease in the denominator, until $Q_p = K_p$. This is achieved by some nitrogen and hydrogen reacting to form more ammonia.

Hence, an increase in pressure increases the equilibrium yield of ammonia by shifting the equilibrium to the right.

EFFECT OF ADDING A CATALYST

The expression for an equilibrium constant of a given reaction contains only those substances included in the overall stoichiometric equation for the reaction. A catalyst does *not* appear in the overall equation, and so cannot influence the value of the equilibrium constant, and hence the equilibrium position.

We can think about this in another way. In **Topic 9 (Book 1: IAS)** you learned that a catalyst increases the rate of both the forward reaction and the backward reaction to the same extent. The catalyst therefore increases the rate at which equilibrium is established, but has no effect on the final concentrations of reactants and products at equilibrium. Therefore, a catalyst has no effect on the value of the equilibrium constant for a reaction.

CHECKPOINT

1. A dynamic equilibrium is set up between carbon monoxide, hydrogen and methanol. The equation for the reaction is:

$$CO(g) + 2H_2(g) \rightleftharpoons CH_3OH(g)$$

State the effect that an increase in pressure, at constant temperature, has on the value of K_c and also on the position of equilibrium. Justify your answers.

SKILLS CREATIVITY

2. The reaction between ethanoic acid and ethanol is reversible. The equation for the reaction is:

$$CH_3COOH(l) + CH_3CH_2OH(l) \rightleftharpoons CH_3COOCH_2CH_3(l) + H_2O(l)$$

Some ethanoic acid and ethanol are mixed together, and some sulfuric acid is added to act as a catalyst. The mixture is left for several days to reach equilibrium.

(a) Explain what happens to the composition of the equilibrium mixture if some more ethanol is added.

(b) Explain what happens to the composition of the equilibrium mixture if some more sulfuric acid is added.

3. The reaction between hydrogen and iodine to form hydrogen iodide is reversible:

$$H_2(g) + I_2(g) \rightleftharpoons 2HI(g)$$

Some $H_2(g)$ is mixed with $I_2(g)$ and the mixture is allowed to reach equilibrium. At equilibrium, 0.14 mol of $H_2(g)$ is present.

The mixture was compressed to reduce its volume and then left to reach equilibrium at the original temperature. Explain the change, if any, in the composition of the equilibrium mixture.

SUBJECT VOCABULARY

reaction quotient a measure of the relative amounts of products and reactants present during a reaction at any given time

13A 5 RELATING ENTROPY TO EQUILIBRIUM CONSTANTS

SPECIFICATION REFERENCE 13.8 13.9

LEARNING OBJECTIVES

- Understand the effect of a change in temperature on
  (i) the value of ΔS_{total}
  (ii) the magnitude of the equilibrium constant, since $\Delta S_{total} = R \ln K$
- Be able to apply knowledge of the value of equilibrium constants to predict the extent to which a reaction takes place.

EFFECT OF TEMPERATURE CHANGE ON THE VALUE OF ΔS_{total}

We are going to use ideas from **Topic 12** on entropy and apply them to equilibrium situations.

In **Topic 12**, we introduced the equation for calculating the total entropy change:

$$\Delta S_{total} = \Delta S_{system} + \Delta S_{surroundings}$$

We also stated that ΔS_{total} is positive for all spontaneous changes.

There is very little change in ΔS_{system} with a change in temperature, unless there is a change in state of one of the reactants or products. However, there are significant changes to $\Delta S_{surroundings}$.

The entropy change of the surroundings during a chemical reaction is given by:

$$\Delta S_{surroundings} = -\frac{\Delta H}{T}$$

where ΔH is the enthalpy change and T is the absolute temperature (i.e. the temperature measured in kelvin).

We can use this information to find out if a reaction is spontaneous at a particular temperature.

WORKED EXAMPLE 1

We will look at the decomposition of calcium carbonate into calcium oxide and carbon dioxide:

$$CaCO_3(s) \rightarrow CaO(s) + CO_2(g) \qquad \Delta H = +177.9\ kJ\ mol^{-1}$$

$$\Delta S_{system} = +160.4\ J\ K^{-1}\ mol^{-1}$$

At 293 K (20 °C), $\Delta S_{surroundings} = -\frac{+177\,900\ J\ mol^{-1}}{293\ K} = -607.2\ J\ K^{-1}\ mol^{-1}$.

$$\Delta S_{total}(293\ K) = (+160.4 - 607.2) = -446.8\ J\ K^{-1}\ mol^{-1}$$

At 1173 K (900 °C), $\Delta S_{surroundings} = +177\,900\ J\ mol^{-1}/1173\ K$
$= -151.7\ J\ K^{-1}\ mol^{-1}$

$$\Delta S_{total}(1173\ K) = (+160.4 - 151.7) = +8.7\ J\ K^{-1}\ mol^{-1}$$

So the decomposition of calcium carbonate is not spontaneous at 293 K, but it is when heated to 1173 K.

fig A Speed skaters can go fast because the pressure of the skates on the ice is changing the equilibrium position between water as a solid and water as a liquid.

RELATIONSHIP BETWEEN TOTAL ENTROPY CHANGE, ΔS_{total}, AND THE EQUILIBRIUM CONSTANT, K

For a reversible reaction that can reach equilibrium, the equilibrium position can be reached from either side of the reaction. This means that both the forward and the backward reactions are spontaneous, so ΔS_{total} must be positive in both directions.

You may think this is impossible. To explain it we can look at the equilibrium reaction

$$N_2O_4(g) \rightleftharpoons 2\ NO_2(g)$$

colourless brown

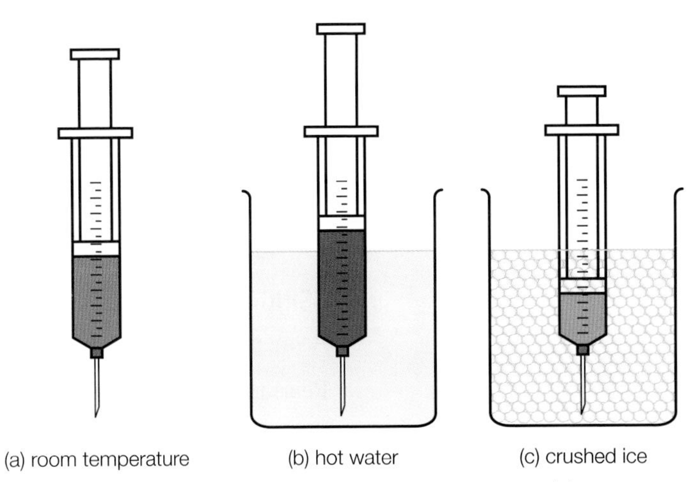

fig B A mixture of N_2O_4 gas and NO_2 gas in a gas syringe: (a) at room temperature; (b) immersed in hot water; (c) immersed in crushed ice.

The differences in colour in **fig B** show how the position of equilibrium changes as the temperature changes.

In hot water, the mixture becomes darker brown. The equilibrium position has shifted to the right to produce more NO_2. Note that the volume has become larger because the mixture expands as it gets hotter.

In iced water, the mixture becomes lighter brown. This shows that the equilibrium position has shifted to the left to produce more N_2O_4. Note that the volume has become smaller because the reaction mixture has cooled.

Fig C shows how it is possible for the reaction to be spontaneous in each direction.

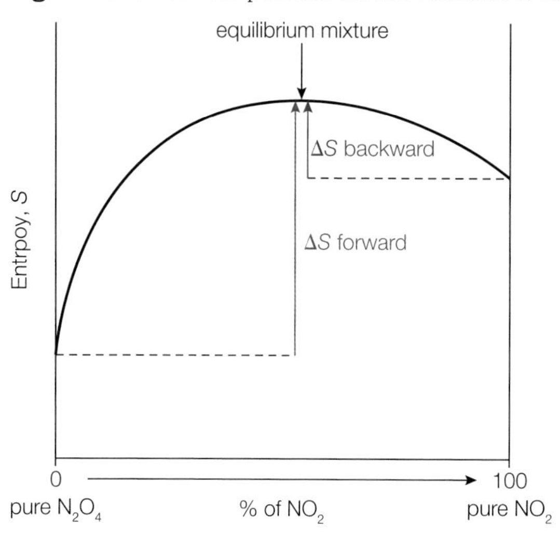

▲ **fig C** A graph of entropy against percentage NO_2 in a mixture of N_2O_4 and NO_2.

LEARNING TIP

The thermodynamic value of K calculated from $\Delta S^\ominus_{total} = R\ln K$ has no units. This is because you can take the logarithm of a number but not of a quantity with units. The value of K calculated from experimental data does have units.

To convert an experimental value of K_p into a thermodynamic value, you need to remove the units by dividing each partial pressure by the standard pressure of 1 atm. Since each partial pressure has been divided by a factor of 1, the numerical value does not alter.

The same is true for K_c. To convert an experimental value of K_c into a thermodynamic value, you need to remove the units by dividing each concentration by the standard concentration of 1 mol dm^{-3}.

The entropy of N_2O_4 is less than the entropy of the equilibrium mixture. The change in entropy from pure N_2O_4 to the equilibrium mixture is positive, so the change is spontaneous.

Similarly, the change in entropy from NO_2 to the equilibrium mixture is also positive.

The entropy change for a mixture of the two gases in any proportions moving to the equilibrium mixture is also positive, so such changes are spontaneous.

Fig C also shows that neither the forward reaction nor the backward reaction can go to completion, as the entropy change from the equilibrium mixture to either reactant is negative.

At equilibrium, the total entropy change is zero and

$$\Delta S_{total}\,[\text{forward reaction}] = \Delta S_{total}\,[\text{backward reaction}]$$

The relationship between the total entropy of the reaction and the equilibrium constant is given by

$$\Delta S_{total} = R\ln K$$

where K can be K_c or K_p, and R is the gas constant.

Note that this matches with the conclusion that a high value of K corresponds to a large positive total entropy change.

USING TOTAL ENTROPY CHANGE TO CALCULATE AN EQUILIBRIUM CONSTANT

As mentioned above, $\Delta S_{total} = R\ln K$.

Rearranging the equation gives us:

$$\ln K = \frac{\Delta S_{total}}{R}$$

So:

$$K = e^{\Delta S_{total}/R}$$

WORKED EXAMPLE 2

The equation for the reaction to form sulfur trioxide from sulfur dioxide and oxygen is:

$SO_2(g) + \frac{1}{2}O_2(g) \rightleftharpoons SO_3(g) \qquad \Delta S^{\ominus}_{total} = 238.3\,J\,K^{-1}\,mol^{-1}$

Calculate a value for the equilibrium constant, K, for this reaction ($R = 8.31\,J\,mol^{-1}\,K^{-1}$).

$K = e^{\Delta S_{total}/R}$

$= e^{238.3/8.31}$

$= 2.84 \times 10^{12}$

To solve this calculation, press the e^x or $e^{\square}$ button on your calculator. Then enter 238.3 ÷ 8.31 followed by the equals button. The answer will be displayed.

LEARNING TIP

Remember that a value of K calculated from thermodynamic data has no units.

RELATIONSHIP BETWEEN EQUILIBRIUM CONSTANT AND POSITION OF EQUILIBRIUM

There is no firm rule about the relationship between the equilibrium constant, K, and the position of equilibrium of the reaction.

However, as a general rule we can say that a very large value of K suggests that the position of equilibrium is well over to the products side (i.e. the right-hand side of the equation).

Similarly, a very small value of K suggests that the position of equilibrium is well over to the reactants side (i.e. the left-hand side of the equation).

CHECKPOINT

SKILLS ADAPTIVE LEARNING

1. Hydrogen can be manufactured by reacting methane with steam. The equation for the reaction is:

$CH_4(g) + H_2O(g) \rightleftharpoons CO(g) + 3H_2(g); \qquad \Delta H_r^{\ominus} = +206\,kJ\,mol^{-1}$

(a) The equilibrium constant, K_p, for this reaction, at a given temperature, has a numerical value of 8.54.

Calculate the total entropy change at the temperature of the reaction.

(b) Use your answer to (a) to calculate the temperature at which this reaction reaches equilibrium. ($\Delta S^{\ominus}_{system} = +225\,J\,K^{-1}\,mol^{-1}$; $R = 8.31\,J\,mol^{-1}\,K^{-1}$)

(c) Use the magnitude and signs of the entropy changes to explain the effect of a temperature increase on the equilibrium constant for this reaction.

2. This question is about the equilibrium reaction:

$Fe^{2+}(aq) + Ag^{+}(aq) \rightleftharpoons Fe^{3+}(aq) + Ag(s)$

The equilibrium is reached slowly.

(a) State two observations that are made if aqueous solutions containing iron(II) ions and silver ions are mixed and allowed to stand for several hours.

(b) (i) $\Delta S^{\ominus}_{total}$ for this reaction is + 47.64 J K^{-1} mol^{-1}.

Calculate the numerical value for the equilibrium constant, K_c.

($R = 8.31\,J\,mol^{-1}\,K^{-1}$)

(ii) $\Delta S^{\ominus}_{system}$ for this reaction is −208.3 J K^{-1} mol^{-1}.

Explain why this value is negative.

(iii) Calculate $\Delta S^{\ominus}_{surroundings}$ for this reaction.

(iv) Use your answer to (b)(iii) to calculate $\Delta H^{\ominus}$ for this reaction, and then explain the effect of increasing the temperature on the value of ΔS_{total}.

CATASTROPHE FOR CORAL?

You should know about the impact of rising carbon dioxide levels on the Earth's atmosphere. The following extract describes the impact on life in our oceans.

OCEANS ABSORB HUMAN CARBON POLLUTION

As humans burn fossil fuels and release greenhouse gases, those gases enter the atmosphere where they cause increases in global temperatures and climate consequences such as more frequent and severe heat waves, droughts, changes to rainfall patterns, and rising seas. But for many years scientists have known that not all of the carbon dioxide we emit ends up in the atmosphere. About 40% actually gets absorbed in the ocean waters.

I like to use an analogy from everyday experience: the ocean is a little like a soda. When we shake soda, it fizzes. That fizz is the carbon dioxide coming out of the liquid (that is why sodas are called "carbonated beverages"). We're doing the reverse process in the climate. Our carbon dioxide is actually going into the oceans.

The process of absorption is not simple – the amount of carbon dioxide that the ocean can hold depends on the ocean temperatures. Colder waters can absorb more carbon dioxide; warmer waters can absorb less. Scientists believe that as the oceans warm, they will become less and less capable of taking up carbon dioxide. As a result, more of our carbon pollution will stay in the atmosphere, worsening global warming. But it's clear that, at least for now, the oceans are doing us a great favour by absorbing large amounts of carbon pollution.

▲ **fig A** Coral bleaching near Christmas Island, Australia. This is an example of the effect of global warming in our oceans.

Knowing the rate at which the oceans absorb carbon dioxide is key to understanding how fast climate change will occur. This is an active area of research. In particular, scientists are closely watching the oceans to see if their ability to absorb is changing over time.

From an article in The Guardian, 'Scientists Study Ocean Absorption of Human Carbon Pollution' by John Abraham, 16 February 2017, https://www.theguardian.com/environment/climate-consensus-97-per-cent/2017/feb/16/scientists-study-ocean-absorption-of-human-carbon-pollution

SCIENCE COMMUNICATION

1 You may have heard people say that "a picture is worth a thousand words". Carry out an Internet search to find five images (e.g. photos or diagrams) that powerfully show the main issues in climate change. Add a short caption to each of the images you choose.

CHEMISTRY IN DETAIL

2 The equilibrium between gaseous carbon dioxide and dissolved carbon dioxide can be represented by the equation:

$$CO_2(g) \rightleftharpoons CO_2(aq)$$

Suggest the impact that an increasing concentration of atmospheric carbon dioxide will have on the concentration of dissolved carbon dioxide in the oceans.

3 Once dissolved, carbon dioxide reacts with water as follows:

$$CO_2(aq) + H_2O(l) \rightleftharpoons HCO_3^-(aq) + H^+(aq)$$

(a) Write an expression for the equilibrium constant (K_c) for this reaction (remember that H_2O is not included in the expression for K_c).

(b) A sample of seawater is shown to have 0.07 g of CO_2 per dm^3 dissolved in it. Calculate the concentration of this solution in mol dm^{-3}.

(c) Assuming that the value for K_c in (a) is 4.5×10^{-7} mol dm^{-3}, calculate the pH of this solution. State any assumptions you have made.

THINKING BIGGER TIP

The material covered in Topic 13 is developed further in Topic 14. You may prefer to wait until you have studied Topic 14 before answering questions 2 and 3.

CHEMISTRY TIP

The hydrogen carbonate ion (HCO_3^-) is sometimes referred to as the bicarbonate ion.

DID YOU KNOW

The UK's Met Office has predicted that CO_2 levels will rise to a record 415 ppm in 2019.

ACTIVITY

The evidence for anthropogenic climate change is now very strong. Using the Internet (including the useful website https://climate.nasa.gov), carry out research to answer the following questions:

(i) What is meant by anthropogenic climate change?

(ii) Are there any other gases responsible for climate change? If so, what are they and how are they produced?

(iii) Some scientists now call the current geological time period the Anthropocene. What does this term mean and why do you think scientists are using it?

You can present your answers in a PowerPoint presentation or as a poster.

13 EXAM PRACTICE

1 An equilibrium is represented by the following equation:

$N_2(g) + 3H_2(g) \rightleftharpoons 2NH_3(g)$; $\Delta H^\ominus = -92\ \text{kJ mol}^{-1}$

Which change would affect both the value of the equilibrium constant, K_p, and the proportion of ammonia present at equilibrium?

A adding a catalyst of finely divided iron

B reducing the temperature at constant pressure

C increasing the amount of nitrogen

D increasing the pressure at constant temperature [1]

(Total for Question 1 = 1 mark)

2 In a reversible reaction, what is the effect of adding a catalyst on the rate constant k_1 for the forward reaction, the rate constant k_{-1} for the backward reaction and on the equilibrium constant K?

	k_1	k_{-1}	K
A	increases	decreases	no effect
B	increases	decreases	increases
C	increases	increases	no effect
D	increases	increases	increases

[1]

(Total for Question 2 = 1 mark)

3 Hydrogen and iodine vapour exist in equilibrium with hydrogen iodide at a constant temperature in a gas syringe:

$H_2(g) + I_2(g) \rightleftharpoons 2HI(g)$

What will increase when the pressure is increased at constant temperature?

A the activation energy of the reaction

B the enthalpy change of the reaction

C K_p

D the partial pressure of hydrogen iodide [1]

(Total for Question 3 = 1 mark)

4 A nitrogen–hydrogen mixture, initially in the mole ratio of 1:3, reaches equilibrium with ammonia when 50% of the nitrogen has reacted.

$N_2(g) + 3H_2(g) \rightleftharpoons 2NH_3(g)$

The total pressure of the equilibrium mixture is P.

What is the partial pressure of ammonia in the equilibrium mixture?

A $P/6$ **B** $P/4$ **C** $P/3$ **D** $P/2$ [1]

(Total for Question 4 = 1 mark)

5 Nitrogen dioxide decomposes on heating according to the following equation:

$2NO_2(g) \rightleftharpoons 2NO(g) + O_2(g)$

When 4.0 mol of nitrogen dioxide is put into a 1 dm^3 container and heated, the equilibrium mixture obtained contains 0.8 mol of oxygen.

What is the numerical value of the equilibrium constant, K_c, at the temperature of the experiment?

A $\dfrac{0.8 \times 0.8}{2.4}$ **B** $\dfrac{(0.8)^2 \times 0.8}{2.4}$

C $\dfrac{1.6 \times 0.8}{(2.4)^2}$ **D** $\dfrac{(1.6)^2 \times 0.8}{(2.4)^2}$ [1]

(Total for Question 5 = 1 mark)

6 This question is about the equilibrium:

$N_2(g) + 3H_2(g) \rightleftharpoons 2NH_3(g)$ $\Delta H^\ominus = -92\ \text{kJ mol}^{-1}$

Which of the following statements is *not* correct?

A The units of K_p are atm^{-2}.

B K_p increases when the temperature decreases.

C K_p increases when the pressure increases.

D K_p increases when the total entropy change, ΔS_{total}, increases. [1]

(Total for Question 6 = 1 mark)

7 Nitrogen and hydrogen react together to form ammonia in a reversible reaction that can reach a position of equilibrium. The equation for the reaction is:

$N_2(g) + 3H_2(g) \rightleftharpoons 2NH_3(g)$; $\Delta H_r = -92\ \text{kJ mol}^{-1}$

At a fixed temperature and a total pressure of 5.00 atm, a sealed vessel of volume 20.0 dm^3 contained 1.00 mol of $N_2(g)$, 2.00 mol of $H_2(g)$ and 1.00 mol of $NH_3(g)$.

(i) Calculate the value, stating units, of the equilibrium constant, K_c, for this reaction. [5]

(ii) Calculate the value, stating units, of the equilibrium constant, K_p, for this reaction. [5]

(Total for Question 7 = 10 marks)

8 When phosphorus(V) chloride, PCl_5, is heated to a constant temperature in a sealed vessel, it forms a gaseous equilibrium mixture with phosphorus(III) chloride, PCl_3, and chlorine. The equation for the reaction is:

$PCl_5(g) \rightleftharpoons PCl_3(g) + Cl_2(g)$

(a) The equilibrium constant may be expressed in terms of either concentration, K_c, or partial pressure, K_p.

(i) Write an expression for K_p. [1]

(ii) If the units of partial pressure are atm, state the units of K_p for the above equilibrium. [1]

(b) Explain, in terms of the expression for K_p, the effect on the position of equilibrium of increasing the pressure at constant temperature. [4]

(c) The relationship between K_c and K_p is given by the expression:

$K_p = K_c\,(0.0821\,T)^{\Delta n}$

where T = the absolute temperature in kelvin (K) and

Δn = the change in number of moles from reactants to products.

Calculate a value for K_c if the value for K_p at 500 K is 0.810 atm. [3]

(Total for Question 8 = 9 marks)

9 A chemist has discovered a method for making a commercially important chemical, R. The reaction involves the following reversible reaction:

$2P(g) + Q(g) \rightleftharpoons R(g)$

The process is normally carried out in industry at a pressure of 500 atm and a temperature of 573 K.

(a) (i) Write an expression for K_p for this reaction. [1]

(ii) Give the units of K_p in this expression. [1]

(b) During their initial research, the chemist carried out several experiments, all at 500 atm pressure, using mixtures of P and Q, starting with 2.0 mol of P and 1.0 mol of Q. In each case they allowed the reaction mixture to reach equilibrium before determining the percentage of Q converted.

The table gives the results of two of his experiments.

	Experiment 1	**Experiment 2**
Temperature/K	423	573
Percentage of Q converted	50	70

(i) Calculate, using the results from experiment 1, the value of K_p. [4]

(ii) Deduce the conditions of temperature and pressure that would give the highest yield of R. [2]

(iii) Explain why the conditions you have suggested in part (ii) may be different from those used industrially. [2]

(Total for Question 9 = 10 marks)

10 Hydrogen and iodine react together, in the presence of a suitable catalyst, in a reversible reaction. The equation for the reaction is:

$H_2(g) + I_2(g) \rightleftharpoons 2HI(g); \quad \Delta H^\ominus = -9\,kJ\,mol^{-1}$

(a) The graph shows the changes in concentration when 2.0×10^{-3} mol of H_2 were mixed with 2.0×10^{-3} mol of I_2 in a sealed container of volume 1 dm^3 and left to reach equilibrium.

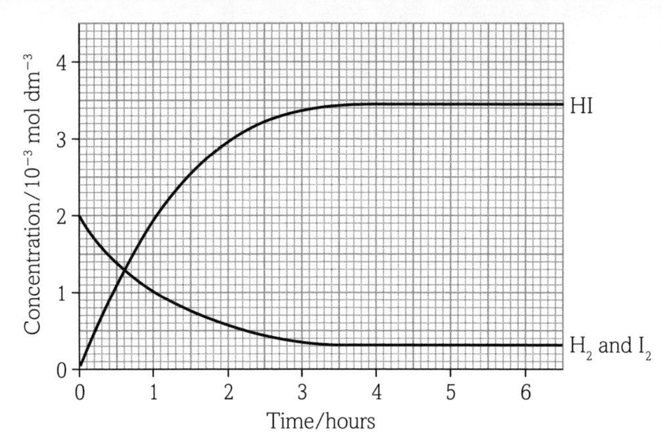

(i) Write the expression for the equilibrium constant, K_c, for this reaction. [1]

(ii) Use the graph to determine the time at which equilibrium was established. [1]

(iii) Use the graph to determine the equilibrium concentrations of HI(g), H_2(g) and I_2(g). [2]

(iv) Calculate a value for the equilibrium constant, K_c. State the units. [2]

(b) Explain how the value of K_c changes when:

(i) the total pressure at equilibrium is increased at constant temperature [2]

(ii) the temperature is increased at constant pressure. [2]

(Total for Question 10 = 10 marks)

11 The composition of an equilibrium mixture produced at 2 atm pressure is shown here:

$CH_4(g)$ +	$H_2O(g)$ ⇌	$CO(g)$ +	$3H_2(g)$	
0.80	0.80	1.20	3.60	Amount in equilibrium mixture/mol

(a) Give the expression for the equilibrium constant, K_p, for the reaction and calculate its value. Include units in your answer. [6]

(b) The total entropy change, in J mol^{-1} K^{-1}, is related to the equilibrium constant by the equation

$\Delta S^\ominus_{total} = R \ln K$

Calculate the total entropy change for this reaction.

(R = 8.31 J mol^{-1} K^{-1}) [1]

(Total for Question 11 = 7 marks)

TOPIC 14 ACID–BASE EQUILIBRIA

A STRONG AND WEAK ACIDS | B ACID–BASE TITRATIONS

Acids and bases play an important part in our lives. They are extremely important in industry. Life itself depends on the complex networks of acid–base reactions taking place inside living organisms. To understand life, and death, we need to understand acids and bases and the reactions they undergo. In this topic we will build on the knowledge you have already gained on the properties of acids and bases. In particular, we will look at their properties quantitatively using the concept of equilibrium constants. We will also see how the strengths of acids are related to their molecular structures.

The pH of your blood plasma is kept fairly constant at around 7.4 when your body is working normally. If it rises or falls by more than 0.4 from this normal value, you are likely to die. The pH of your blood plasma can fall as a result of disease or shock, both of which can generate acidic conditions in your body. You are also likely to die if the pH of your blood plasma rises; this could happen during recovery from severe burns. To survive, your body must control its own pH. If your natural control systems fail, then you must be saved with medicine such as intravenous electrolyte solutions.

In this topic we will look at how different ions affect pH, and how these ions can be used to control pH.

MATHS SKILLS FOR THIS TOPIC

- Recognise and make use of appropriate units in calculation
- Recognise and use expressions in decimal and ordinary form
- Use ratios, fractions and percentages
- Use calculators to find and use power, exponential and logarithmic functions
- Use an appropriate number of significant figures
- Change the subject of an equation
- Substitute numerical values into algebraic equations using appropriate units for physical quantities
- Solve algebraic expressions
- Use logarithms in relation to quantities that range over several orders of magnitude

What prior knowledge do I need?

- Reactions of acids and bases
- A qualitative appreciation of the significance of pH of aqueous solutions

Topic 11

- Acid-catalysed reactions such as the iodination of propanone

Topic 13

- Calculation of equilibrium constants based on concentrations (i.e. K_c)
- An understanding of the effect of changes in temperature on the value of equilibrium constants

What will I study in this topic?

- Acid–base reactions in terms of proton transfer
- The relationship between hydrogen ion concentration and pH
- How to calculate the pH of aqueous solutions
- The difference between strong and weak acids
- How to draw and interpret titration curves
- How to select a suitable indicator for an acid–base titration
- The concept of buffer solutions

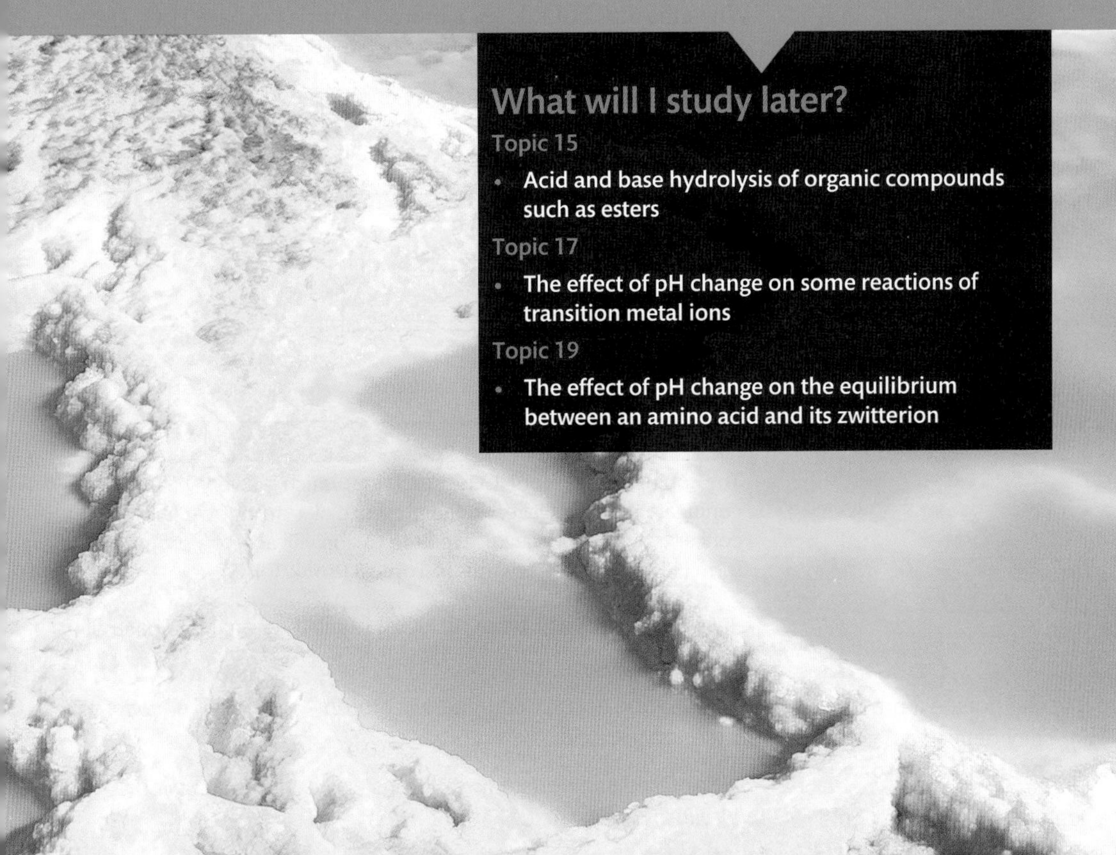

What will I study later?

Topic 15

- Acid and base hydrolysis of organic compounds such as esters

Topic 17

- The effect of pH change on some reactions of transition metal ions

Topic 19

- The effect of pH change on the equilibrium between an amino acid and its zwitterion

14A 1 THE BRØNSTED–LOWRY THEORY

SPECIFICATION REFERENCE 14.1 14.2 14.6

LEARNING OBJECTIVES

- Understand that a Brønsted–Lowry acid is a proton donor and a Brønsted–Lowry base is a proton acceptor and that acid–base reactions involve proton transfer.
- Be able to identify Brønsted–Lowry conjugate acid–base pairs.
- Understand the difference between a strong and a weak acid in terms of the degree of dissociation.

BRØNSTED–LOWRY ACIDS AND BASES

In 1923, physical chemists Johannes Nicolaus Brønsted in Denmark and Thomas Martin Lowry in England independently proposed the theory that carries their names. In the Brønsted–Lowry theory, acids and bases are defined by the way they react with each other.

▲ **fig A** Johannes Nicolaus Brønsted (1879 -1947).

▲ **fig B** Thomas Martin Lowry (1874-1936).

They defined an acid as a substance that can donate a proton, i.e. a **proton donor** (hydrogen ion, H^+). They defined a base as a substance that can accept a proton, i.e. a **proton acceptor**.

We might expect that any substance containing hydrogen could act as an acid. In practice, a substance behaves as an acid only if the hydrogen carries a slight positive charge. As an example, this is the case when it is bonded to a highly electronegative atom to the right of the Periodic Table, e.g. oxygen or a halogen.

In order to accept a proton, a base has to contain a lone pair of electrons that it can use to form a dative covalent bond with the proton. So, a base must contain an atom to the right-hand side of the Periodic Table, and this is often oxygen.

CONJUGATE ACID–BASE PAIRS

An equilibrium is established when hydrogen chloride dissolves in water. The following equation represents this equilibrium:

$$HCl(aq) + H_2O(l) \rightleftharpoons H_3O^+(aq) + Cl^-(aq)$$

In the forward reaction:

- HCl is acting as an acid because it is donating a proton to H_2O.
- H_2O is behaving as a base as it is accepting a proton from HCl.

In the reverse reaction:

- H_3O^+ is behaving as an acid because it is donating a proton to Cl^-.
- Cl^- is behaving as a base because it is accepting a proton from H_3O^+.

When the acid HCl loses a proton it forms a base, Cl^-. These two species are called a **conjugate acid–base pair**.

When the acid H_3O^+ loses a proton it forms the base H_2O. These two species also form a conjugate acid–base pair.

So the equilibrium mixture above contains two conjugate acid–base pairs:

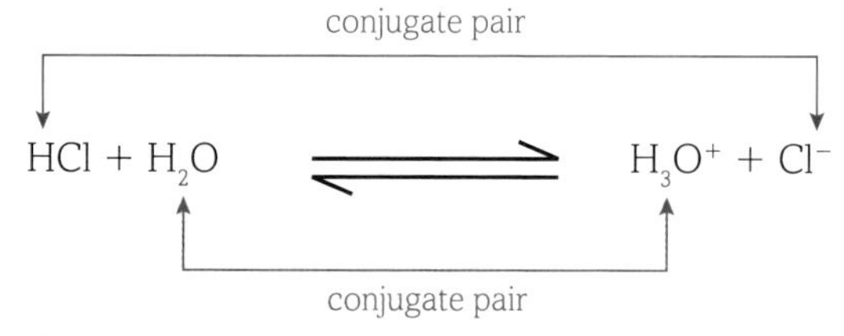

▲ **fig C** Conjugate acid-base pairs formed from hydrochloric acid and water.

EXAM HINT

The conjugate base for HCl is the Cl^- ion, which is a very weak conjugate base. However, in the presence of a strong acid (such as concentrated sulfuric acid), even the Cl^- ion will accept a proton to form HCl fumes. You saw this in **Topic 8 (Book 1: IAS)**.

Cl^- is the **conjugate base** of HCl; H_2O is the conjugate base of H_3O^+.

Hydrochloric acid can donate *one* proton – it is called a *monoprotic* or monobasic acid.

Sulfuric acid can donate *two* protons and is therefore called a *diprotic* or dibasic acid.

Step 1: $H_2SO_4 \rightarrow H^+ + HSO_4^-$

Step 2: $HSO_4^- \rightarrow H^+ + SO_4^{2-}$

Similarly, some bases such as the carbonate ion can accept more than one proton:

Step 1: $CO_3^{2-} + H^+ \rightarrow HCO_3^-$

Step 2: $HCO_3^- + H^+ \rightarrow H_2CO_3$

The carbonate ion is therefore described as a *diprotic* or diacidic base.

DID YOU KNOW?

- Acids that donate a maximum of one, two or three protons are called monoprotic, diprotic or triprotic, respectively. They are also called monobasic, dibasic or tribasic.
- Bases that can accept one, two or three protons are called monoprotic, diprotic or triprotic, respectively. They are also called monoacidic, diacidic or triacidic.

We will use the terms monobasic, dibasic, etc. for acids throughout this book.

ANOTHER EXAMPLE OF A CONJUGATE ACID–BASE PAIR

An equilibrium is established when ammonia dissolves in water. The following equation represents this equilibrium:

$$NH_3(aq) + H_2O(l) \rightleftharpoons NH_4^+(aq) + OH^-(aq)$$

In the forward reaction:

- NH_3 is acting as a base because it is accepting a proton from H_2O.
- H_2O is acting as an acid because it is donating a proton to NH_3.

In the reverse reaction:

- NH_4^+ is behaving as an acid because it is donating a proton to OH^-.
- OH^- is behaving as a base because it is accepting a proton from NH_4^+.

So the equilibrium contains two acid–base conjugate pairs:

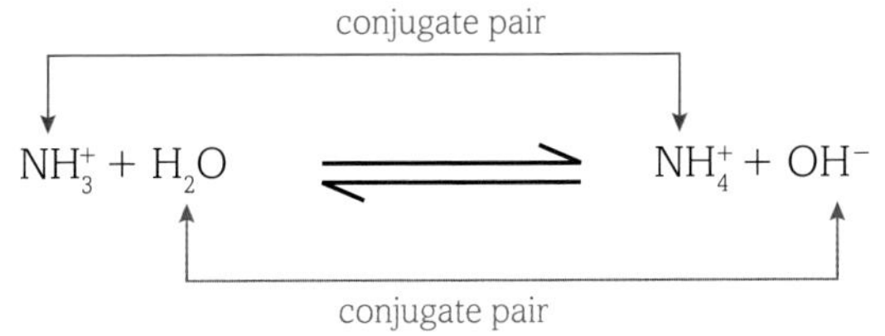

▲ **fig D** Conjugate acid-base pairs formed from ammonia and water.

EXAM HINT

The equilibrium that ammonia forms with water helps us to understand the test for the ammonium ion that you saw in **Topic 8 (Book 1: IAS)**. When we add hydroxide ions to a solution containing ammonium ions, the equilibrium shifts to the left. Ammonia gas is liberated and we can test for it using damp red litmus paper.

AMPHOTERIC SUBSTANCES

In its reaction with HCl, H_2O behaves as a base by accepting a proton. However, in its reaction with NH_3, H_2O behaves as an acid by donating a proton.

A substance that can act as either an acid or a base is described as being **amphoteric**.

DID YOU KNOW?

You may also see the term amphiprotic. An amphiprotic substance is one that can both donate and accept protons. So water is described as being amphiprotic. Other examples of amphiprotic substances are amino acids and the hydrogen sulfate ion, HSO_4^-.

All amphiprotic substances are also amphoteric, but the opposite is not true. There are amphoteric substances, like aluminium oxide, that do not donate or accept protons when they act as acids or bases, respectively. These substances are in the category of acting as acids or bases according to the Lewis theory. The Lewis theory states that an acid is an electron pair acceptor and a base is an electron pair donor.

The advantage of the Lewis theory is that it shows the similarity between acid–base reactions that involve proton transfer and other acid–base reactions that do not. It is a useful way of explaining why some reactions in organic chemistry that are catalysed by acids can also be catalysed by some substances that can accept a pair of electrons. For example, the nitration of benzene is catalysed by sulfuric acid, but the reaction also occurs if nitric acid is used with boron trifluoride, BF_3.

Neither the term 'amphiprotic' nor the Lewis theory of acids and bases is required learning for your International A Level course.

NITRIC ACID AS A BASE

The most common method to nitrate benzene (see **Topic 18**) is to use the 'nitrating mixture' of concentrated nitric acid and concentrated sulfuric acid. This mixture forms the equilibrium:

$$H_2SO_4 + HNO_3 \rightleftharpoons HSO_4^- + H_2NO_3^+$$

Acid 1 Base 2 Base 1 Acid 2

In this reaction:

- H_2SO_4 is an acid; its **conjugate base** is HSO_4^-
- HNO_3 is a base; its **conjugate acid** is $H_2NO_3^+$.

It seems strange to refer to nitric acid as a base. However, in this reaction the nitric acid *is* behaving as a base.

STRONG AND WEAK ACIDS

A strong acid is defined as one that is almost completely **dissociated** in aqueous solution.

Virtually all the hydrogen chloride molecules are dissociated in a dilute solution of hydrochloric acid. We usually represent this by using a single direction arrow in the equation for the dissociation:

$$HCl(aq) \rightarrow H^+(aq) + Cl^-(aq)$$

LEARNING TIP

From now on, we are going to use $H^+(aq)$ to represent the acid protons in aqueous solution, rather than H_3O^+. It is not certain how many water molecules are bonded to a given hydrogen ion. The formula could just as easily be $H_5O_2^+$ or $H_7O_3^+$. For this reason, $H^+(aq)$ is possibly the best representation of acid protons in aqueous solution. The choice of formula should not affect your understanding of the concepts involved.

LEARNING TIP

Ionisation is also used to describe the conversion of acid molecules into ions when an acid dissolves in water. You may see this term used in other books.

DID YOU KNOW?

Although, strictly speaking, an equilibrium exists between HCl molecules, H^+ ions and Cl^- ions, the acid is considered to be completely dissociated in dilute solution. So, strong acids are often said to be 100% dissociated.

By contrast, a weak acid is defined as one that is only **partially dissociated** (often less than 10%) in aqueous solution. Organic acids such as ethanoic acid are typically weak acids.

We represent partial dissociation by using the reversible arrow sign ($\rightleftharpoons$) in the equation for the dissociation:

$$CH_3COOH(aq) \rightleftharpoons CH_3COO^-(aq) + H^+(aq)$$

CHECKPOINT

1. Give the formula of the conjugate acid of each of the following species:
   (a) CH_3COO^-
   (b) CH_3NH_2
   (c) HSO_4^-
2. Give the formula of the conjugate base of each of the following species:
   (a) $HClO_4$
   (b) H_3O^+
   (c) HSO_4^-
3. In each of the following examples, identify the two conjugate acid-base pairs. In each case, identify the species that are acting as a Brønsted–Lowry acid.
   (a) $H_2CO_3 + H_2O \rightleftharpoons HCO_3^- + H_3O^+$
   (b) $HCO_3^- + H_2O \rightleftharpoons CO_3^{2-} + H_3O^+$
   (c) $CH_3COOH + HNO_3 \rightleftharpoons CH_3COOH_2^+ + NO_3^-$
4. Explain why the following reaction may be described as an acid-base reaction:
   $NH_4^+ + NH_2^- \rightarrow 2NH_3$

SUBJECT VOCABULARY

proton donor an acid is a proton donor
proton acceptor a base is a proton acceptor
conjugate acid-base pair either a base and its conjugate acid or an acid and its conjugate base
conjugate base when an acid donates a proton, the species formed is the conjugate base of the acid
conjugate acid when a base accepts a proton, the species formed is the conjugate acid of the base
amphoteric (substance) a substance that can act both as an acid and as a base
dissociated acid molecules are said to be dissociated when they have split to form ions
partially dissociated only a small fraction of the acid molecules have dissociated

14A 2 HYDROGEN ION CONCENTRATION AND THE pH SCALE

SPECIFICATION REFERENCE

14.3 14.4 14.5 14.6 14.7 14.8 14.9 14.12 PART

LEARNING OBJECTIVES

- Be able to define the term 'pH'.
- Be able to calculate pH from the hydrogen ion concentration.
- Be able to calculate the concentration of hydrogen ions in a solution, in mol dm^{-3}, from its pH, using the expression $[H^+] = 10^{-pH}$.
- Understand the difference between a strong acid and a weak acid in terms of the degree of dissociation.
- Be able to calculate the pH of a strong acid.
- Be able to deduce the expression for the acid dissociation constant, K_a, for a weak acid.
- Be able to calculate the pH of a weak acid from K_a or pK_a values, making relevant assumptions.
- Be able to define the term 'pK_a'.

HYDROGEN ION CONCENTRATION AND pH

STRONG ACIDS

As already mentioned, we assume strong acids are dissociated completely when they are dissolved in water. This means that the hydrogen ion concentration is related directly to the concentration of the acid.

For example, a solution of HCl of concentration 0.100 mol dm^{-3} will produce a hydrogen ion concentration of 0.100 mol dm^{-3}.

The **pH** of an aqueous solution is related to the hydrogen ion concentration by the following equation:

$$pH = -\lg [H^+] \qquad \text{or} \qquad pH = \lg \frac{1}{[H^+]}$$

The hydrogen ion concentration, $[H^+]$, is measured in mol dm^{-3}.

Which equation you decide to use is a matter of personal preference.

DID YOU KNOW?

Logarithms can only be taken of a number, not a quantity with a unit. So, strictly speaking, the hydrogen ion concentration has to be divided by the standard concentration, $c^{\ominus}$, which has a value of 1 mol dm^{-3}.

Therefore, the correct expression for calculating pH is $-\lg [H^+]/c^{\ominus}$. This, however, is not something you need to consider at this level.

LEARNING TIP

Do not worry if you are unfamiliar with the use of logarithms. They are merely a way to convert a scale of numbers in powers of 10 to a linear scale. For example:

lg 100 = 2 (as 100 is 10^2)
lg 10 = 1 (as 10 is 10^1)
lg 1 = 0 (as 1 is 10^0)
lg 0.01 = −2 (as 0.01 is 10^{-2})

The lg of a number can easily be found by using the 'lg' or 'log' button on your calculator.

The accepted abbreviation for logarithm to the base 10 is lg, but you can also use $\log_{10}$ or even log.

WORKED EXAMPLE 1

1 Calculate the pH of aqueous solutions of the following monobasic strong acids. In each case, assume that the acid is completely dissociated. Give your answers to two decimal places.

(a) 0.00100 mol dm^{-3} HCl
(b) 0.0500 mol dm^{-3} HNO_3
(c) 0.150 mol dm^{-3} HBr

Answer

(a)
$$pH = -\lg [H^+] = -\lg (0.00100)$$
$$\lg (0.00100) = -3$$
$$\text{so, } -\lg (0.00100) = +3.00$$
$$pH = 3.00$$

(b)
$$pH = -\lg (0.0500)$$
$$\lg (0.0500) = -1.30$$
$$pH = 1.30$$

(c)
$$pH = -\lg (0.150)$$
$$\lg (0.150) = -0.82$$
$$pH = 0.82$$

2 Calculate the pH of an aqueous solution of 10.00 mol dm^{-3} of HCl. In this solution the HCl is 55% dissociated. Give your answer to two decimal places.

Answer

$$[H^+] = 0.55 \times 10.00 \text{ mol dm}^{-3} = 5.50 \text{ mol dm}^{-3}$$
$$pH = -\lg (5.50)$$
$$\lg (5.50) = 0.74$$
$$\text{so, } pH = -0.74$$

DID YOU KNOW?

pH and activity

The measured pH of a concentrated acid is never as low as the calculated value, even though the acid may be completely dissociated. This is because the ions that are close together in solution interact with one another. This makes their effective concentration less than the actual concentration. This effective concentration is called the activity. For example, a solution of HCl of 1.00 mol dm^{-3} has an effective hydrogen ion concentration of 0.81 mol dm^{-3}. This makes the pH of this solution 0.09, not 0.00, as calculated from a hydrogen ion concentration of 1.00 mol dm^{-3}.

Only at concentrations below 0.10 mol dm^{-3} do activities and concentrations have similar values.

This activity effect means that the minimum pH and maximum pH that concentrated solutions can have are 0.30 and 14.30, respectively. This activity effect can be ignored at International A Level, which explains the calculation of −0.74 above.

CALCULATING HYDROGEN ION CONCENTRATION FROM pH

It is a simple matter to calculate the hydrogen ion concentration from a given pH value.

You should use the following equation:

$$[H^+(aq)] = 10^{-pH}$$

WORKED EXAMPLE 2

Calculate the hydrogen ion concentration of a solution with a pH of 4.8:

$[H^+(aq)] = 10^{-4.8}$

$= 1.58 \times 10^{-5}$ mol dm^{-3}

To solve this calculation, press the 10^X or $10^{■}$ button on your calculator. Then enter the negative pH value, followed by the equals button. The answer will then be displayed.

WEAK ACIDS

Determining the hydrogen ion concentration of an aqueous solution of a weak acid is more complicated. This is because a significant amount of undissociated acid is present in solution. It is necessary to refer to the acid dissociation constant, K_a, for the acid.

ACID DISSOCIATION CONSTANT

If we use HA to represent a weak acid, then the equation for its dissociation in aqueous solution is:

$$HA(aq) \rightleftharpoons H^+(aq) + A^-(aq)$$

When we apply the equilibrium law to this reaction we obtain:

$$\frac{[H^+(aq)][A^-(aq)]}{[HA(aq)]} = \text{a constant}$$

This constant is called the acid dissociation constant and is given the symbol K_a.

We will now calculate the hydrogen ion concentration of an aqueous solution of ethanoic acid of concentration 0.0500 mol dm^{-3}. The value of K_a for ethanoic acid is 1.74×10^{-5} mol dm^{-3} at 298 K.

$$CH_3COOH(aq) \rightleftharpoons CH_3COO^-(aq) + H^+(aq)$$

$$K_a = \frac{[CH_3COO^-(aq)][H^+(aq)]}{[CH_3COOH(aq)]}$$

Every time a molecule of CH_3COOH dissociates, a CH_3COO^- ion and a H^+ ion are formed.

This means that $[CH_3COO^-(aq)] = [H^+(aq)]$.

So, the expression for K_a can be simplified to:

$$K_a = \frac{[H^+(aq)]^2}{[CH_3COOH(aq)]}$$

At this stage, it is important to recognise that the concentrations in the expression for K_a are the *equilibrium* concentrations. However, if the value of K_a is very small (as it is in this case) then the concentration of the undissociated acid at equilibrium is very similar to the initial concentration of the acid. It is therefore reasonable to say the concentration at equilibrium is the same as the initial concentration.

Therefore,

$$K_a = \frac{[H^+(aq)]^2}{0.0500} = 1.74 \times 10^{-5} \text{ mol dm}^{-3}$$

$$[H^+(aq)] = \sqrt{(0.0500 \times 1.74 \times 10^{-5})} = 9.33 \times 10^{-4} \text{ mol dm}^{-3}$$

This gives a pH of 3.03.

LEARNING TIP

If you are worried that the approximation we made in the calculation is not justified, then it is possible to solve the precise equation:

$$K_a = [H^+(aq)]^2/(0.0500 - [H^+(aq)]) = 1.74 \times 10^{-5} \text{ mol dm}^{-3}$$

The solution to this equation gives a value for $[H^+(aq)]$ of 9.327×10^{-4} mol dm^{-3}

The difference is significant only for very accurate work.

There is also one other approximation that we made in both the original calculation and this calculation. As we will see in **Section 14A.3**, water ionises very slightly to form hydrogen ions and hydroxide ions. We have ignored the contribution from the dissociation of water to the total hydrogen ion concentration. However, unless the acid is *very* dilute, this approximation is also justified.

The following equation can be used to calculate the hydrogen ion concentration of an aqueous solution of a weak monobasic acid:

$$[H^+(aq)] = \sqrt{(K_a \times [\text{acid}])}$$

where K_a is the dissociation constant for the weak acid.

K_a AND pK_a VALUES

Table A (next page) lists some weak monobasic organic acids together with their K_a and **pK_a** values at 298 K, where:

$$pK_a = -\lg K_a$$

You will notice that the *larger* the value of K_a the stronger the acid.

By contrast, the *smaller* the value of pK_a the stronger the acid.

NAME OF ACID	FORMULA OF ACID	K_a/mol dm^{-3}	pK_a
propanoic acid	CH_3CH_2COOH	1.35×10^{-5}	4.87
ethanoic acid	CH_3COOH	1.74×10^{-5}	4.76
benzoic acid	C_6H_5COOH	6.31×10^{-5}	4.20
methanoic acid	$HCOOH$	1.60×10^{-4}	3.80
chloroethanoic acid	$CH_2ClCOOH$	1.38×10^{-3}	2.86
dichloroethanoic acid	$CHCl_2COOH$	5.13×10^{-2}	1.29
trichloroethanoic acid	CCl_3COOH	2.24×10^{-1}	0.65

increasing acid strength ↓

table A

CALCULATING THE pH OF A DIBASIC ACID

Sulfuric acid is the most common dibasic acid. It dissociates in two stages:

$$H_2SO_4(aq) \rightarrow H^+(aq) + HSO_4^-(aq)$$

$$HSO_4^-(aq) \rightleftharpoons H^+(aq) + SO_4^{2-}(aq)$$

H_2SO_4 is a strong acid and is therefore fully dissociated. HSO_4^- is a weak acid ($K_a = 0.0100$ mol dm^{-3}.)

In a 0.500 mol dm^{-3} aqueous solution of H_2SO_4 the contribution to the [H^+(aq)] from the H_2SO_4 will be 0.500 mol dm^{-3}. If we assume that the contribution to the [H^+(aq)] from the HSO_4^- ion is x mol dm^{-3}, then we have the following relationship:

$$K_a\,(HSO_4^-) = 0.0100 = (0.500 - x)x/0.500$$

Solving this quadratic equation gives $x = 0.0098$.

This gives a total [H^+(aq)] of 0.05098 mol dm^{-3} with a subsequent pH of 0.293.

This calculation shows that the contribution to the [H^+(aq)] from the HSO_4^- ions is negligible. This is because its dissociation is significantly reduced because of the high [H^+(aq)] from the full first ionisation of the H_2SO_4.

A second interesting reason for performing this calculation is because many books state that 0.5 mol dm^{-3} H_2SO_4(aq) can be used as the acid solution in a standard hydrogen electrode (see **Topic 16** on standard electrode potentials). As the standard hydrogen electrode requires a [H^+(aq)] of 1.00 mol dm^{-3} (pH = 0.00), this is clearly incorrect.

CHECKPOINT

1. Calculate the pH of each of the following aqueous solutions of strong monobasic acids. Assume the acid is fully dissociated in each case. Give your answers to two decimal places.
   (a) 0.0100 mol dm^{-3} HI
   (b) 0.500 mol dm^{-3} HNO_3
   (c) 0.00405 mol dm^{-3} HCl
2. Calculate the pH of a mixture of 20.0 cm^3 of 1.00 mol dm^{-3} HCl(aq) and 5 cm^3 of 1.00 mol dm^{-3} NaOH(aq).
3. Calculate the pH of each of the following aqueous solutions of weak monobasic acids. Give your answers to two decimal places.
   (a) 0.100 mol dm^{-3} HCOOH [K_a(HCOOH) = 1.60×10^{-4} mol dm^{-3}]
   (b) 1.00 mol dm^{-3} HF [K_a(HF) = 5.62×10^{-4} mol dm^{-3}]
   (c) 0.505 mol dm^{-3} NH_4Cl [$K_a(NH_4^+)$ = 5.62×10^{-10} mol dm^{-3}]
4. The pH of an aqueous solution of a weak acid, HA, of concentration 0.0305 mol dm^{-3} is 4.97. Calculate the dissociation constant, K_a, for this weak acid.

SKILLS PROBLEM-SOLVING

SUBJECT VOCABULARY

pH (of an aqueous solution) the reciprocal of the logarithm to the base 10 of the hydrogen ion concentration, measured in moles per cubic decimetre, pH = −lg [H^+]. This definition is difficult to remember, so either of the two equations given on page 83 can be used to define pH

pK_a = −lg K_a

14A 3 IONIC PRODUCT OF WATER, K_w

SPECIFICATION REFERENCE

14.10 14.11 14.12 PART

LEARNING OBJECTIVES

- Be able to define the ionic product of water, K_w.
- Be able to define the term 'pK_w'.
- Be able to calculate the pH of a strong base from its concentration, using K_w or pK_w.

DISSOCIATION OF WATER

Pure water has a slight electrical conductivity, so it must contain some ions. It self-ionises according to the following equation:

$$H_2O(l) \rightleftharpoons H^+(aq) + OH^-(aq)$$

If we apply the equilibrium law to this reaction we obtain:

$$\frac{[H^+(aq)][OH^-(aq)]}{[H_2O(l)]} = \text{a constant}$$

As $[H_2O(l)]$ is constant at a given temperature, the expression may be simplified to:

$$[H^+(aq)][OH^-(aq)] = \text{a constant}$$

This constant is called the **ionic product of water** and is given the symbol K_w.

The value of K_w at 298 K is 1.00×10^{-14} mol² dm⁻⁶.

A neutral solution is defined as one in which the hydrogen ion concentration is equal to the hydroxide ion concentration. This is the case for pure water.

At 298 K, $[H^+(aq)] = 1.00 \times 10^{-7}$ mol dm^{-3} $[(1.00 \times 10^{-14})^{½}]$

So, the pH of pure water at 298 K is 7.00 $[-\lg(1.00 \times 10^{-7})]$

LEARNING TIP

The pH of a neutral solution is often said to be 7.00. However, this is true only for a solution that has a temperature of 298 K.

As with all equilibrium constants, K_w varies with temperature.

At 288 K it has a value of 4.52×10^{-15} mol² dm^{-6}.

At this temperature, $[H^+(aq)] = 6.72 \times 10^{-8}$ mol dm^{-3}, giving a pH of 7.17.

So, at 288 K a neutral solution has a pH of 7.17.

Similarly it can be shown that at 308 K, the pH of a neutral solution is 6.84.

LEARNING TIP

The equation should strictly be $pK_w = -\lg K_w/(c^\ominus)^2$, where $c^\ominus$ is the standard concentration of 1 mol dm^{-3}, in order to be able to take the logarithm of a dimensionless quantity. However we will use the simplified version of the equation.

K_w AND pK_w

The relationship between K_w and pK_w is given by the following equation:

$$pK_w = -\lg K_w$$

At 298 K, when $K_w = 1.00 \times 10^{-14}$ mol² dm^{-6}, $pK_w = 14.00$

Table A gives the values of K_w and pK_w at various temperatures.

TEMPERATURE/K	273	283	293	303	313
K_w/mol² dm^{-6}	1.14×10^{-15}	2.93×10^{-15}	6.81×10^{-15}	1.47×10^{-14}	2.92×10^{-14}
pK_w	14.94	14.53	14.17	13.83	13.53

table A

pH OF AQUEOUS SOLUTIONS OF STRONG BASES

An acid dissolved in water produces so many hydrogen ions that the small contribution from the water is insignificant, unless the acid concentration is very small.

However, even the most alkaline solutions contain some hydrogen ions because water ionises.

Sodium hydroxide is a strong base, so in dilute aqueous solutions we can consider its ions to be completely dissociated.

A sodium hydroxide solution of concentration 0.100 mol dm^{-3} therefore has a hydroxide ion concentration of 0.100 mol dm^{-3}.

If $[OH^-(aq)] = 0.100$ mol dm^{-3} and $[H^+(aq)][OH^-(aq)] = 1.00 \times 10^{-14}$ mol^2 dm^{-6}, then

$[H^+(aq)] = 1.00 \times 10^{-14}/0.100$ mol dm^{-3}

$= 1.00 \times 10^{-13}$ mol dm^{-3}

The pH of this solution is therefore 13.0.

WORKED EXAMPLE

Calculate the pH, at 298 K, of an aqueous solution of potassium hydroxide of concentration 0.0200 mol dm^{-3}.

K_w (298 K) $= 1.00 \times 10^{-14}$ mol^2 dm^{-6}

Answer

As potassium hydroxide is a strong base we may assume that its ions are completely dissociated.

$[OH^-(aq)] = 0.0200$ mol dm^{-3}

So, $[H^+(aq)] = 1.00 \times 10^{-14}/0.0200 = 5.00 \times 10^{-13}$ mol dm^{-3}

pH $= -\lg(5.00 \times 10^{-13}) = 12.3$

LEARNING TIP

You can use the logarithmic form of the equation $[H^+(aq)][OH^-(aq)] = 1.00 \times 10^{-14}$, which is pH + pOH = 14.0, where pOH $= -\lg[OH^-(aq)]$ if you do not like working with the logarithms of small numbers.

So, in the worked example above:

pH = 14.0 − pOH

= 14.0 − (−lg (0.0200))

= (14.0 − 1.70) = 12.3

CHECKPOINT

1. The ionic product of water, K_w, has a value of 1.00×10^{-14} mol^2 dm^{-6} at 298 K and a value of 6.81×10^{-15} mol^2 dm^{-6} at 293 K. Use this information, where relevant, to answer the following questions.

(a) Calculate the pH of water at (i) 298 K and (ii) 293 K.

SKILLS PROBLEM-SOLVING

(b) Even though pure water at 298 K and at 293 K has different pH values, both samples of water are said to be neutral. Explain why.

(c) Is the following reaction exothermic or endothermic?

$H_2O(l) \rightarrow H^+(aq) + OH^-(aq)$

Explain how you arrived at your answer.

2. Calculate the pH at 298 K of each of the following aqueous solutions of strong bases. Assume the base is fully dissociated in each case. Give your answers to two decimal places.

(a) 0.0100 mol dm^{-3} NaOH

(b) 0.0500 mol dm^{-3} $Ca(OH)_2$

(c) 0.0315 mol dm^{-3} KOH

[$K_w = 1.00 \times 10^{-14}$ mol^2 dm^{-6} at 298 K]

SUBJECT VOCABULARY

ionic product of water, K_w the product of the concentration of the hydrogen ions and the hydroxide ions, both measured in mol dm^{-3}

$K_w = [H^+(aq)][OH^-(aq)]$

14A 4 ANALYSING DATA FROM pH MEASUREMENTS

SPECIFICATION REFERENCE
14.13(i) 14.13(ii)
14.14 CP11

LEARNING OBJECTIVES

- Be able to analyse data from the following experiments:
  (i) measuring the pH of a variety of substances, including equimolar solutions of strong and weak acids, strong and weak bases, and salts
  (ii) comparing the pH of a strong and weak acid after dilution 10, 100 and 1000 times.
- Be able to calculate K_a for a weak acid from experimental data given the pH of an aqueous solution containing a known mass of acid.

COMPARING SOLUTIONS THROUGH pH MEASUREMENT

STRONG AND WEAK ACIDS

The relative strengths of different acids can be determined by measuring the pH of equimolar aqueous solutions of the acids, at the same temperature.

Table A shows the pH of 0.100 mol dm^{-3} aqueous solutions of various acids at 298 K.

FORMULA OF ACID	HCl	$CHCl_2COOH$	$CH_2ClCOOH$	$HCOOH$	CH_3COOH	CH_3CH_2COOH
pH OF 0.100 mol dm^{-3} AQUEOUS SOLUTION	1.00	1.14	1.93	2.38	2.87	2.93

decreasing acid strength →

table A

The higher the value of the pH, the weaker the acid.

STRONG AND WEAK BASES

The same method can be used to determine the relative strengths of bases.

Table B shows the pH of 0.100 mol dm^{-3} aqueous solutions of various bases at 298 K.

FORMULA OF BASE	NH_3	CH_3NH_2	$(CH_3)_2NH$	$CH_3CH_2NH_2$	$CH_3CH_2CH_2NH_2$	$NaOH$
pH OF 0.100 mol dm^{-3} AQUEOUS SOLUTION	11.13	11.82	11.83	11.84	11.86	13.00

increasing base strength →

table B

The higher the value of the pH, the stronger the base.

SALTS

Table C shows the pH of aqueous solutions of various salts of concentration 0.100 mol dm^{-3} at 298 K.

FORMULA OF SALT	$NaCl$	KNO_3	CH_3COONa	NH_4Cl	CH_3COONH_4
pH OF AQUEOUS SOLUTION	7.00	7.00	8.88	5.13	7.00

table C

The pH of NaCl is 7.00 because the salt is made from a strong acid (HCl) and a strong base (NaOH). The same is true for KNO_3, which is a product of the strong acid HNO_3 and the strong base KOH.

An aqueous solution of CH_3COONa is alkaline because it is the product of a weak acid (CH_3COOH) and a strong base (NaOH).

An aqueous solution of NH_4Cl is acidic because it is the product of a strong acid (HCl) and a weak base (NH_3).

EXAM HINT

You can explain why the pH of 0.100 mol dm^{-3} NH_4Cl is acidic by showing how the NH_4^+ ion behaves in aqueous solution. The NH_4^+ ion behaves as a weak acid in aqueous solution, and it dissociates as shown by the following equation:

$NH_4^+ \rightleftharpoons NH_3 + H^+$

An aqueous solution of CH_3COONH_4 is neutral (pH = 7.00) because it is the product of a weak acid (CH_3COOH) and a weak base (NH_3), and the relative strengths of the acid and base are the same. This is shown by their dissociation constants:

CH_3COOH: $K_a = 1.74 \times 10^{-5}\,mol\,dm^{-3}$

NH_3: $K_b = 1.74 \times 10^{-5}\,mol\,dm^{-3}$

LEARNING TIP

K_a is a measure of acid strength and K_b is a measure of base strength.

K_b for a base is calculated in a similar way to K_a for an acid. The equilibrium set up when ammonia dissolves in water is:

$$NH_3(aq) + H_2O(l) \rightleftharpoons NH_4^+(aq) + OH^-(aq)$$

$$K_b = \frac{[NH_4^+][OH^-]}{[NH_3]}$$

You will not be asked to calculate values for K_b in the exam.

DID YOU KNOW?

The reason why some salts form alkaline or acidic solutions is that they undergo hydrolysis.

To learn more about this, research 'salt hydrolysis'.

It is important to recognise that different salts have different pH values because this will explain why, for example, the pH of the solution formed when a strong acid reacts with an equivalent amount of a weak base is less than 7. We will meet this concept when we look at acid–base titration curves in **Section 14B.1**.

SUMMARY TABLE FOR AQUEOUS SOLUTIONS AT 298 K

SALT OF A STRONG ACID AND A STRONG BASE	pH = 7 (solution is neutral)
SALT OF A WEAK ACID AND A STRONG BASE	pH > 7 (solution is alkaline)
SALT OF A STRONG ACID AND A WEAK BASE	pH < 7 (solution is acidic)
SALT OF A WEAK ACID AND A WEAK BASE	pH depends on relative strengths of acid and base; if $K_a = K_b$, pH = 7 if $K_a > K_b$, pH < 7 if $K_a < K_b$, pH > 7

table D

EFFECT OF DILUTION ON THE pH OF AQUEOUS SOLUTIONS OF ACIDS

STRONG ACIDS

Table E shows the pH of five aqueous solutions of hydrochloric acid. In each case, the acid has been diluted by a factor of ten from 1.00 (1.00×10^0) to 0.000100 (1.00×10^{-4}) $mol\,dm^{-3}$. All solutions are at a temperature of 298 K.

CONCENTRATION/mol dm^{-3}	1.00 (1.00×10^0)	0.100 (1.00×10^{-1})	0.0100 (1.00×10^{-2})	0.00100 (1.00×10^{-3})	0.00100 (1.00×10^{-4})
pH	0.00	1.00	2.00	3.00	4.00

table E

You will notice that the pH increases by a factor of one unit for each 10-fold decrease in concentration.

If we follow this to its logical conclusion, then an aqueous solution of hydrochloric acid of concentration $1.00 \times 10^{-8}\,mol\,dm^{-3}$ should have a pH of 8.00. This is clearly nonsense because this would mean that a solution of an acid was alkaline.

Earlier in this topic we mentioned the fact that the contribution to the hydrogen ion concentration from the dissociation of water can usually be ignored. When solutions are as dilute as $10^{-8}\,mol\,dm^{-3}$ this is no longer the case. The pH of $10^{-8}\,mol\,dm^{-3}$ hydrochloric acid is very close to 7, as the contribution to the concentration of hydrogen ions from the water ($10^{-7}\,mol\,dm^{-3}$) is now greater than that of the acid.

WEAK ACIDS

Table F shows the pH of five aqueous solutions of ethanoic acid (CH_3COOH). In each case the acid has been diluted by a factor of ten from 1.00 (1.00×10^0) to 0.000100 (1.00×10^{-4}) $mol\,dm^{-3}$. Once again, all solutions are at a temperature of 298 K.

CONCENTRATION/mol dm^{-3}	1.00 (1.00×10^0)	0.100 (1.00×10^{-1})	0.0100 (1.00×10^{-2})	0.00100 (1.00×10^{-3})	0.000100 (1.00×10^{-4})
pH	2.38	2.88	3.38	3.88	4.38

table F

With a weak acid, the pH value increases by a factor of about 0.50 for each 10-fold decrease in concentration.

PRACTICAL SKILLS CP11

Determining K_a of a weak acid from experimental data

Note that this is a different way of doing **Core Practical 11: Finding the K_a value for a weak acid** to that mentioned in your Lab Book. The following experiment can be performed to determine K_a of benzoic acid (C_6H_5COOH), which is a weak monobasic acid.

- Accurately weigh between 0.40 and 0.50 g of benzoic acid. Then dissolve it in a small volume (say 50 cm^3) of deionised water contained in a beaker. (Benzoic acid is not very soluble, so it may be necessary to warm the water to get it to dissolve. Allow the solution to cool before performing the next stage.)
- Transfer the solution to a 250 cm^3 volumetric flask. Add several washings from the beaker using deionised water, and then make up to the mark with deionised water.
- Put a stopper in the flask and then turn the flask upside down several times to mix the solution.
- Withdraw a sample of the solution and place it in a small beaker.
- Measure the pH of the solution using a calibrated pH meter.

Sample results

Mass of benzoic acid = 0.49 g

pH of solution = 3.00

Analysis of results

$C_6H_5COOH(aq) \rightleftharpoons C_6H_5COO^-(aq) + H^+(aq)$

Molar mass of benzoic acid (C_6H_5COOH) = 122 g mol^{-1}

$n(C_6H_5COOH)$ in 250 cm^3 of solution = 0.49/122 mol

So, $[C_6H_5COOH(aq)] = (0.49/122) \times 4$ mol dm^{-3}

pH = 3.00

So, $[H^+(aq)] = 10^{-3.00} = [C_6H_5COO^-(aq)]$

$K_a = (10^{-3.00})^2 \div ((0.49/122) \times 4)$

$= 6.22 \times 10^{-5}$ mol dm^{-3}

The Data Book value is 6.32×10^{-5} mol dm^{-3} at 298 K.

Safety Note: Wear eye protection. Avoid skin contact with, and do not raise dust from, the benzoic acid.

Our experimental result for the K_a of benzoic acid is close to the accepted value. However, that does not necessarily mean that we have performed an accurate experiment. There may have been a number of errors that have, by chance, cancelled out one another.

To start with, we have assumed that the $[C_6H_5COOH(aq)]$ at equilibrium is identical to the original concentration of benzoic acid. Obviously, this is not correct because it must have dissociated slightly to produce a solution of pH = 3.00.

The major uncertainty in this experiment is the measurement of the pH value. Even a small error leads to a large discrepancy in the final answer. A pH of 3.10 would give a final answer of 3.92×10^{-5}, whereas a value of 2.90 would give 9.87×10^{-5} for the value of K_a.

Transfer errors may also be significant when weighing out small amounts.

CHECKPOINT

1. Predict whether aqueous solutions of the following salts will be neutral, acidic or alkaline. Justify your answers.
   (a) Ammonium nitrate, NH_4NO_3
   (b) Potassium propanoate, CH_3CH_2COOK
   (c) Sodium nitrate, $NaNO_3$
2. What information is required in order to make a prediction about the pH of an aqueous solution of ammonium methanoate, $HCOONH_4$?

SKILLS PROBLEM-SOLVING

3. Calculate K_a for chloroethanoic acid from the following data. 1.89 g of chloroethanoic acid was dissolved in 50 cm^3 of water and the solution was diluted to 250 cm^3 in a volumetric flask. The pH of this solution was 1.99.

14B 1 ACID–BASE TITRATIONS, pH CURVES AND INDICATORS

SPECIFICATION REFERENCE 14.15 14.16

LEARNING OBJECTIVES

- Be able to draw and interpret titration curves, using all combinations of strong and weak monoprotic and diprotic acids with bases, and apply these principles to diprotic acids and bases.
- Be able to select a suitable indicator for a titration, using a titration curve and appropriate data.

ACID–BASE TITRATIONS

END POINT AND EQUIVALENCE POINT

When you carry out a simple acid–base titration, you usually use an indicator to tell you when the acid and base are mixed in exactly the right proportions to react in equivalent amounts, as dictated by the stoichiometric equation. When the indicator changes colour, this is often described as the 'end point' of the titration.

The **equivalence point** is when the acid and base have reacted together in the exact proportions as dictated by the stoichiometric equation. When titrating an aqueous solution of a monobasic acid with an aqueous solution of a monoacidic base of the same concentration, 25 cm^3 of acid will react exactly with 25 cm^3 of base.

The pH at the equivalence point depends on the combination of acid and base used. For example, if you are titrating aqueous sodium hydroxide with dilute hydrochloric acid, then the pH at the equivalence point is 7.00 (at 298 K), as both the base and the acid are strong. The solution at the equivalence point will contain the salt sodium chloride.

If, however, ethanoic acid (a weak acid) is titrated against sodium hydroxide (a strong base), the solution at the equivalence point will contain the salt sodium ethanoate, and the pH will be greater than 7 (see **Section 14A.4**).

If hydrochloric acid (a strong acid) is titrated against aqueous ammonia (a weak base), the solution at the equivalence point will contain the salt ammonium chloride, and the pH will be less than 7.

LEARNING TIP

The term 'neutralisation point' should *not* be used to describe the point at which the acid and base have reacted in the exact proportions as dictated by the stoichiometric equation. As seen by the examples listed, the pH of the solution formed is not always 'neutral' (i.e. does not always have a pH of 7.00 at 298 K).

Remember that the term 'end point' refers to when the colour of the indicator just changes colour - this does not always occur at the equivalence point (see discussion of indicator choice later in this section).

TITRATION OF A STRONG ACID WITH A STRONG BASE

As you add an aqueous solution of an acid to an aqueous solution of a base, you might expect there to be a gradual change in the pH of the solution formed. This is not the case. When the pH of the solution is plotted against the volume of acid added, the shape of the curve depends on the nature of the acid and base used. The curves produced in this way are called 'pH titration curves'.

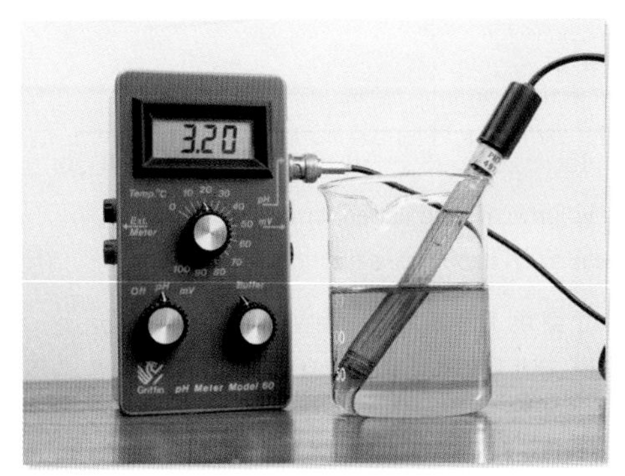

▲ **fig A** A pH meter.

We will start by looking at the pH titration curve when a strong acid is added to a strong base.

The following graph is produced when adding 1.00 mol dm^{-3} HCl(aq) to a 25 cm^3 sample of 1.00 mol dm^{-3} NaOH(aq). The pH is measured using a pH meter such as the one shown in **fig A**.

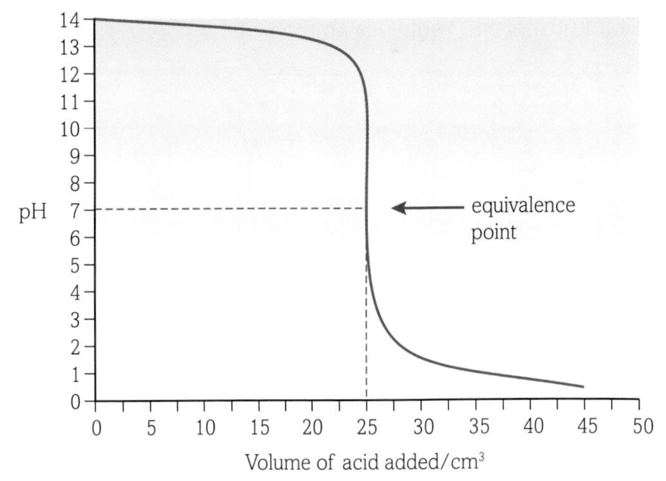

▲ **fig B** pH curve for a strong acid–strong base titration.

You can see in **fig B** that the pH falls only a very small amount until it is quite near the equivalence point. Then there is a really steep plunge. If you calculate the values, the pH falls all the way from 11.30 when you have added 24.90 cm^3 to 2.70 when you have added 25.10 cm^3.

There is a large 'steep section' to the curve. As we will see later in this section, this is an important point to consider when choosing an appropriate acid–base indicator to determine the end point of this titration.

TITRATION OF A WEAK ACID WITH A STRONG BASE

For this example we will add 1.00 mol dm^{-3} ethanoic acid to 25 cm^3 of 1.00 mol dm^{-3} sodium hydroxide.

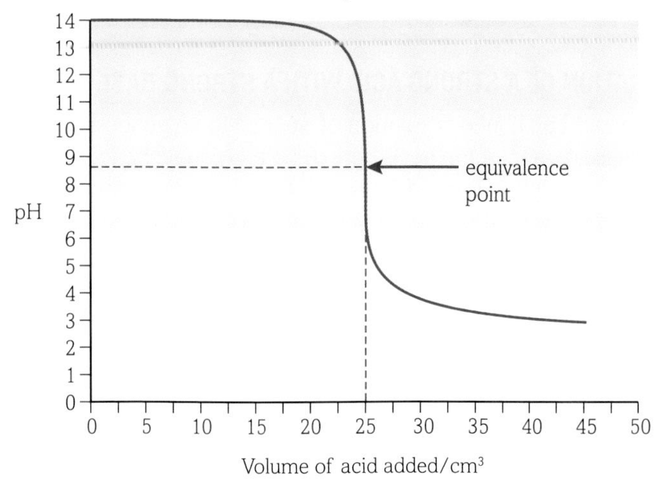

▲ **fig C** pH curve for a weak acid–strong base titration.

The curve (**fig C**) is the same as that for a strong acid–strong base up to the equivalence point, but there is a difference once the acid is present in excess. Past the equivalence point, the solution contains a mixture of ethanoic acid and sodium ethanoate. This mixture acts as a buffer solution and therefore resists any large change in pH upon addition of further acid. (See **Section 14B.2**).

Note that the pH at the equivalence point is between 8 and 9; it is *not* 7.

TITRATION OF A STRONG ACID WITH A WEAK BASE

For this example we will add 1.00 mol dm^{-3} hydrochloric acid to 25 cm^3 of 1.00 mol dm^{-3} aqueous ammonia.

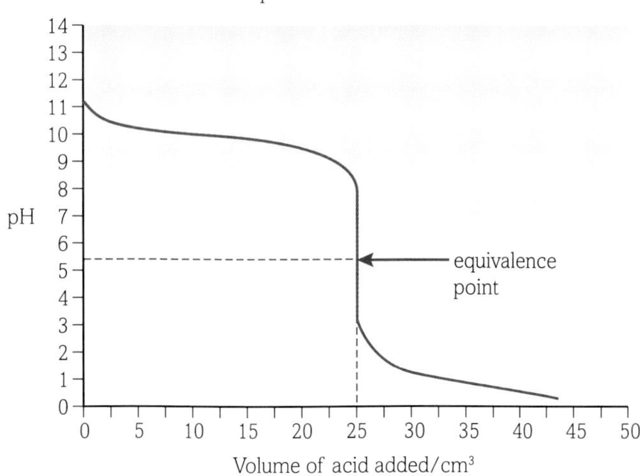

▲ **fig D** pH curve for a strong acid–weak base titration.

When the acid is first added the pH starts to fall quite sharply, but the curve quickly levels out (**fig D**). This is because a buffer solution has been formed, containing ammonia and ammonium chloride (again, see **Section 14B.2**).

Notice that the pH at the equivalence point is less than 7 because the salt formed, ammonium chloride, is composed of a strong acid and a weak base.

TITRATION OF A WEAK ACID WITH A WEAK BASE

For this example we will add 1.00 mol dm^{-3} ethanoic acid to 25 cm^3 of 1.00 mol dm^{-3} aqueous ammonia.

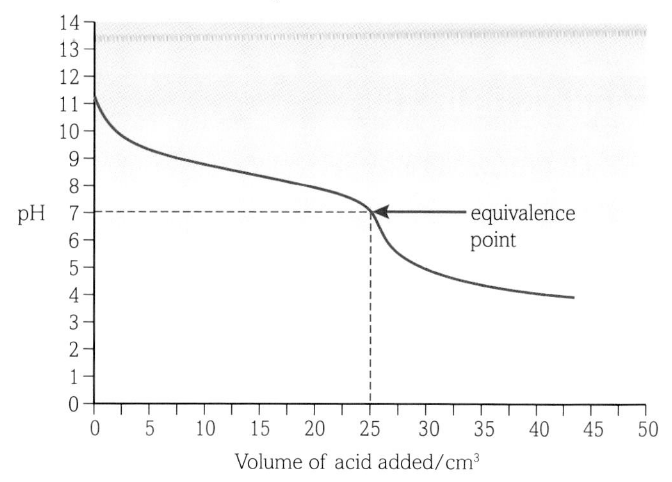

▲ **fig E** pH curve for a weak acid–weak base titration.

Notice that there is not any steep section to this graph (**fig E**). Instead, there is what is known as a 'point of inflection'. The lack of a steep section means that it is difficult to do a titration of a weak acid against a weak base using an indicator. The reason for this will be explained in **Section 14B.3**.

ACID–BASE INDICATORS

Earlier in this section we looked at the pH titration curves obtained when an aqueous solution of an acid is added to an aqueous solution of a base. The four cases we considered were:

- strong acid–strong base
- weak acid–strong base
- strong acid–weak base
- weak acid–weak base.

We are now going to use these four curves to help us understand why different indicators are required for different types of titrations.

An acid–base indicator is either a weak acid or a weak base. Most indicators are weak acids (HIn). For an indicator that is a weak acid, its dissociation in aqueous solution can be represented as:

$$HIn(aq) \rightleftharpoons H^+(aq) + In^-(aq)$$

The molecule, HIn, and its conjugate base, In^-, have different colours in aqueous solution. For methyl orange, these are red and yellow, respectively:

$$HIn(aq) \rightleftharpoons H^+(aq) + In^-(aq)$$

red yellow

When $[H^+(aq)]$ is sufficiently large the equilibrium will shift far enough to the left for the red colour to predominate. If $[H^+(aq)]$ is very low then the equilibrium will lie far over to the right and the yellow colour will predominate. Therefore, the indicator changes colour according to the pH of the solution.

There will be a stage at which [HIn(aq)] = [In⁻(aq)] and the indicator will appear orange. The exact pH at which this stage is reached can be determined using the equilibrium constant, K_{In}, for methyl orange.

$$K_{In} = \frac{[H^+(aq)][In^-(aq)]}{[HIn(aq)]} = 2.00 \times 10^{-4}\,mol\,dm^{-3}$$

When [HIn(aq)] = [In⁻(aq)], the expression becomes:

$$[H^+(aq)] = 2.00 \times 10^{-4}\,mol\,dm^{-3}$$

This gives a pH of 3.70 for the 'half-way' stage. So, methyl orange will change colour at a pH of 3.70.

Note that this pH value is also the same as the value of pK_{In} for methyl orange. Therefore, the pH at which different indicators will change colour can be determined from their pK_{In} values.

pH RANGE OF INDICATORS

As a 'rule of thumb', the red colour of methyl orange will first predominate when [HIn(aq)] is ten times [In⁻(aq)], and the yellow colour will predominate when [In⁻(aq)] is ten times [HIn(aq)].

The approximate pH at which each colour predominates can be calculated as follows.

When [HIn(aq)] = 10[In⁻(aq)]:

$$\frac{[H^+(aq)][In^-(aq)]}{10[In^-(aq)]} = 2.00 \times 10^{-4}\,mol\,dm^{-3}$$

$$[H^+(aq)] = 2.00 \times 10^{-3}\,mol\,dm^{-3}$$

So, the pH at which the red colour first predominates is 2.70.

A similar calculation will show that the pH at which the yellow colour first predominates is 4.70.

The 'pH range' of methyl orange is therefore approximately 2.70 to 4.70.

The *exact* pH range of methyl orange is 3.10 to 4.40. This means that at pH values below 3.10, methyl orange will appear red. At pH values above 4.40, methyl orange will appear yellow. Between 3.10 and 4.40, methyl orange will be a shade of orange.

Table A shows the pK_{In} values, pH ranges and colours of several common indicators.

INDICATOR	pK_{In}	pH RANGE	COLOUR	
			HIn(aq)	In⁻(aq)
methyl orange	3.70	3.10–4.40	red	yellow
bromophenol blue	4.00	2.80–4.60	yellow	blue
bromothymol blue	7.00	6.00–7.60	yellow	blue
phenol red	7.90	6.80–8.40	yellow	red
phenolphthalein	9.30	8.20–10.00	colourless	red

table A

CHOICE OF INDICATOR

A good indicator shows a complete colour change upon the addition of one drop of acid from the burette. This is necessary in order to accurately determine the end point of the titration.

Therefore, there has to be a minimum pH change equivalent to the pH range of the indicator in order for the indicator to successfully determine the end point.

STRONG ACID–STRONG BASE TITRATION

Let us first of all consider the use of methyl orange and phenolphthalein as indicators for a strong acid–strong base titration.

Fig F shows the pH titration curve for 25 cm³ of 1.00 mol dm⁻³ NaOH(aq) titrated with 1.00 mol dm⁻³ HCl(aq). The pH ranges of methyl orange and phenolphthalein have also been included.

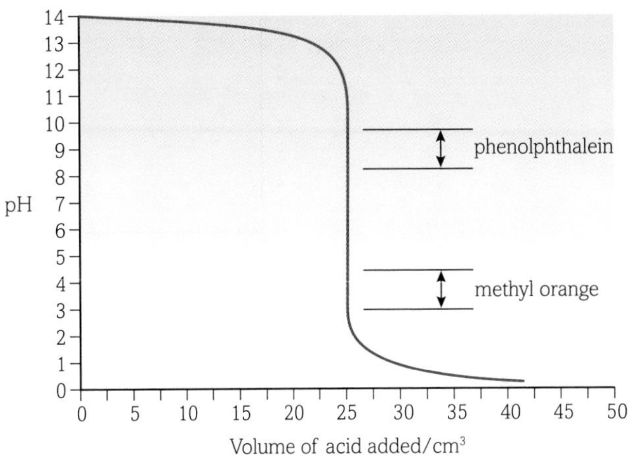

fig F Strong acid–strong base pH curve.

You will notice that the pH range of each indicator falls within the steep section of the curve, where a large pH change is occurring upon the addition of just one drop of acid. This means that both indicators will change colour at the end point and therefore both are suitable indicators to use in this titration.

WEAK ACID–STRONG BASE TITRATION

We will now consider the suitability of each indicator for a weak acid–strong base titration.

Fig G shows the pH titration curve for 25 cm³ of 1.00 mol dm⁻³ NaOH(aq) titrated with 1.00 mol dm⁻³ CH_3COOH(aq). Once again, the pH ranges of methyl orange and phenolphthalein have also been included.

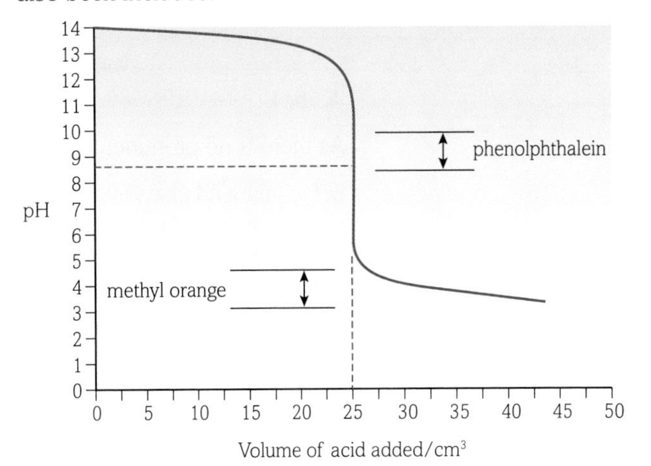

fig G Weak acid–strong base pH curve.

This time, only the pH range of phenolphthalein falls within the steep section of the curve. So, phenolphthalein is suitable, but methyl orange is not.

STRONG ACID–WEAK BASE TITRATION

Fig H shows the pH titration curve for 25 cm^3 of 1.00 mol dm^{-3} NH_3(aq) titrated with 1.00 mol dm^{-3} HCl(aq). Again, the pH ranges of methyl orange and phenolphthalein have also been included.

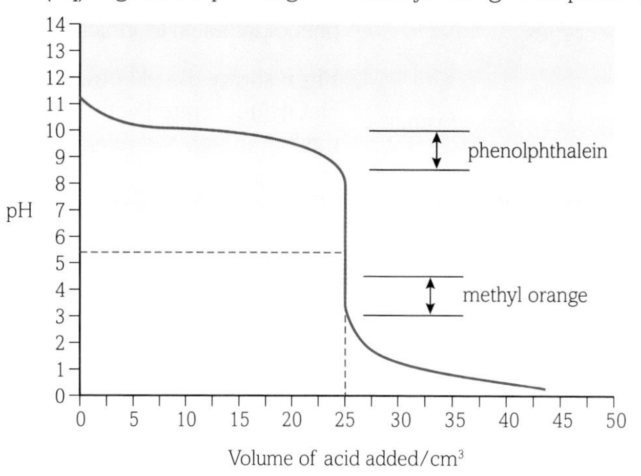

▲ **fig H** Strong acid–weak base pH curve.

This time, only the pH range of methyl orange falls within the steep section of the curve. So, methyl orange is suitable, but phenolphthalein is not.

WEAK ACID–WEAK BASE TITRATION

Fig I shows the pH titration curve for 25 cm^3 of 1.00 mol dm^{-3} NH_3(aq) titrated with 1.00 mol dm^{-3} CH_3COOH(aq). Again, the pH ranges of methyl orange and phenolphthalein have also been included.

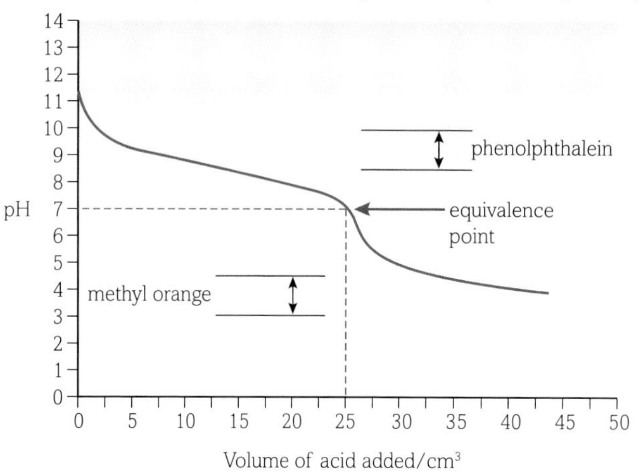

▲ **fig I** Weak acid–weak base pH curve.

As there is no prominent steep section to the curve, neither indicator is suitable.

In fact, the end point of a titration of a weak acid and a weak base cannot be determined using an acid–base indicator. The end point of such a titration is best determined by measuring the temperature changes (thermometric titration) or electrical conductivity changes (conductometric titration).

TITRATION CURVES WITH DIPROTIC ACIDS

A diprotic acid is an acid that produces two H^+ ions per acid molecule. Examples of diprotic acids are sulfuric acid, H_2SO_4, and carbonic acid, H_2CO_3.

A diprotic acid dissociates in water in two stages:

Stage 1: $H_2X(aq) \rightarrow H^+(aq) + HX^-(aq)$

Stage 2: $HX^-(aq) \rightarrow H^+(aq) + X^{2-}(aq)$

Because of the successive dissociations (one following immediately after the other), the titration curves of diprotic acids have two equivalence points, as shown in **fig J**.

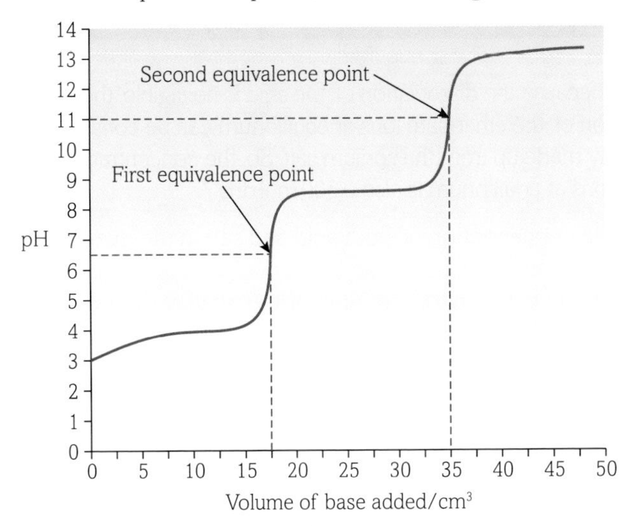

▲ **fig J** Diprotic acid–strong base pH curve.

The equations for the acid–base reactions occurring between a diprotic acid, H_2X, and the base sodium hydroxide, NaOH, are as follows.

From the beginning of the reaction to the first equivalence point:

$$H_2X + NaOH \rightarrow NaHX + H_2O$$

From the first equivalence point to the second equivalence point:

$$NaHX + NaOH \rightarrow Na_2X + H_2O$$

From the beginning of the reaction through to the second equivalence point (overall reaction):

$$H_2X + 2NaOH \rightarrow Na_2X + 2H_2O$$

At the first equivalence point, all H^+ ions from the first dissociation have reacted with NaOH.

At the second equivalence point, all H^+ ions from *both* reactions have reacted (twice as many as at the first equivalence point).

Therefore, the volume of NaOH added at the second equivalence point is exactly twice that at the first equivalence point.

CHOOSING THE BEST INDICATOR

The best indicator to choose for a particular titration is the one whose pK_{In} value is as close as possible to the pH at the equivalence point.

Bromothymol blue (pK_{In} = 7.00) is a particularly good indicator for a strong acid–strong base titration.

EXAM HINT

In an exam, you will be given a data book that contains the pK_{In} values, pH ranges and colours of 10 common indicators.

CHECKPOINT

SKILLS INTERPRETATION

1. The equation for the reaction between hydrochloric acid and ammonia is:

$$HCl(aq) + NH_3(aq) \rightarrow NH_4Cl(aq)$$

A 25.0 cm^3 sample of 0.0200 mol dm^{-3} HCl(aq) was placed in a conical flask. Aqueous ammonia was added gradually from a burette and the pH was measured after each addition, until the pH no longer changed.

The pH curve for this titration is shown below.

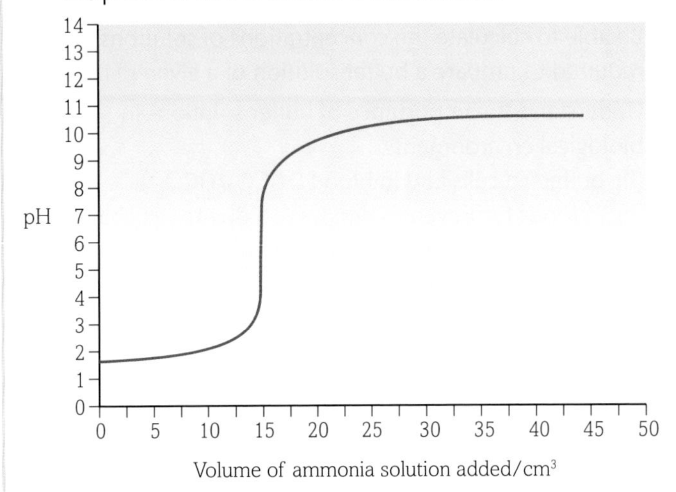

(a) State how the curve suggests that ammonia is a weak base.

(b) Use the information given to calculate the concentration of the ammonia solution.

(c) The pH ranges for three indicators are:

Thymol blue: 1.2 to 2.8

Methyl red: 4.2 to 6.3

Thymolphthalein: 9.3 to 10.5

Explain which of these three indicators is the most suitable for this titration.

SKILLS INTERPRETATION

2. (a) Calculate the pH at 298 K of an aqueous solution of CH_3COOH of concentration 0.100 mol dm^{-3}.
[K_a = 1.74 × 10^{-5} mol dm^{-3} at 298 K]

(b) Sketch the pH titration curve for the addition of 50.0 cm^3 of 0.100 mol dm^{-3} NaOH(aq) to 25.0 cm^3 of 0.100 mol dm^{-3} CH_3COOH(aq).

(c) State two differences in the pH curve that would be obtained if the titration were repeated using 25.0 cm^3 of 0.0500 mol dm^{-3} CH_3COOH(aq) instead of 25.0 cm^3 of 0.100 mol dm^{-3} CH_3COOH(aq).

SUBJECT VOCABULARY

equivalence point the point in a titration when the acid and base have reacted together in the exact proportions as dictated by the stoichiometric equation

14B 2 BUFFER SOLUTIONS

SPECIFICATION REFERENCE

14.17 14.18 14.19 14.20 14.22

LEARNING OBJECTIVES

- Know what is meant by the term 'buffer solution'.
- Understand the action of a buffer solution.
- Be able to calculate the pH of a buffer solution given appropriate data.
- Be able to calculate the concentrations of solutions required to prepare a buffer solution of a given pH.
- Understand the importance of buffer solutions in biological environments:
  (i) buffers in cells and in blood (H_2CO_3/HCO_3^-)
  (ii) in foods to prevent deterioration due to pH change (caused by bacterial or fungal activity).

WHAT IS A BUFFER SOLUTION?

Many experiments, particularly in biochemistry, have to be performed in aqueous solutions of fairly constant pH. Unfortunately, it is impossible to make a solution whose pH is totally unaffected by the addition of even small amounts of acid or base. However, it is possible to make a solution whose pH remains *almost* unchanged when small amounts of acid or base are added. Such a solution is called a **buffer solution**.

There are many ways of making a buffer solution, but two of the most common are:

- to mix a weak acid with its conjugate base
- to mix a weak base with its conjugate acid.

BUFFER MADE FROM A WEAK ACID AND ITS CONJUGATE BASE

The simplest example of this is ethanoic acid and sodium ethanoate. The salt of the weak acid has to be soluble in water, which is why sodium and potassium salts are commonly used to make buffer solutions.

In this mixture the acid is partially dissociated, whereas the salt is fully dissociated. The relevant equations are:

$$CH_3COOH(aq) \rightleftharpoons CH_3COO^-(aq) + H^+(aq)$$

$$CH_3COONa(aq) \rightarrow CH_3COO^-(aq) + Na^+(aq)$$

This mixture will produce a buffer solution with a pH less than 7. The exact pH depends on the concentration of both acid and its conjugate base, and can be calculated as follows.

CALCULATING THE pH OF A BUFFER SOLUTION

For our example we will use a solution that has been made by mixing equal volumes of $1.00\ mol\ dm^{-3}$ ethanoic acid and $1.00\ mol\ dm^{-3}$ sodium ethanoate at 298 K.

If we assume that the extent of dissociation of the acid is negligible, then the concentration of CH_3COOH at equilibrium will be $0.500\ mol\ dm^{-3}$.

Also, again because the dissociation of the acid is negligible, the concentration of the ethanoate ions at equilibrium can be considered to be entirely made up from the sodium salt. So, the concentration of ethanoate ions at equilibrium is also $0.500\ mol\ dm^{-3}$.

Note that the concentration of both acid and salt in the mixture is half of the concentrations used because equal volumes of each solution were mixed. The total volume of the mixture is twice that of the original volume of each solution used.

At 298 K,

$$K_a = \frac{[CH_3COO^-(aq)][H^+(aq)]}{[CH_3COOH(aq)]} = 1.74 \times 10^{-5}\ mol\ dm^{-3}$$

Rearranging this equation and substituting values for $[CH_3COOH(aq)]$ and $[CH_3COO^-(aq)]$ gives:

$$[H^+(aq)] = (1.74 \times 10^{-5} \times 0.500)/0.500\ mol\ dm^{-3}$$
$$= 1.74 \times 10^{-5}\ mol\ dm^{-3}$$

This gives a pH for the buffer solution of 4.76.

LEARNING TIP

A potential trap when calculating the pH of a buffer solution is not understanding that when two solutions are mixed there is a dilution. If $50\ cm^3$ of one solution is mixed with $50\ cm^3$ of another solution, then the total volume of the mixture will be $100\ cm^3$. So, the concentration of each solution will be halved on mixing.

HOW DOES THE BUFFER ACTION WORK?

When a small amount of acid is added to the buffer solution, the majority of the H^+ ions added react with the CH_3COO^- ions to form CH_3COOH molecules:

$$CH_3COO^-(aq) + H^+(aq) \rightarrow CH_3COOH(aq)$$

When a little base is added, the majority of the OH^- ions added react with the CH_3COOH molecules:

$$CH_3COOH(aq) + OH^-(aq) \rightarrow CH_3COO^-(aq) + H_2O(l)$$

A new equilibrium mixture will be established in which the concentrations of both CH_3COOH and CH_3COO^- will have changed slightly from their original values. So, there will be a change in pH, but this will be minimal. This is shown by the following argument.

To show that the pH has changed very little, we once again need to make use of the expression for K_a of the acid:

$$K_a = \frac{[CH_3COO^-(aq)][H^+(aq)]}{[CH_3COOH(aq)]}$$

Rearranging this equation gives:

$$[H^+(aq)] = K_a \times \frac{[CH_3COOH(aq)]}{[CH_3COO^-(aq)]}$$

As there are a relatively large number of CH_3COOH molecules in the solution, as the extent of dissociation of the acid is very small, the change in $[CH_3COOH(aq)]$ will be negligible.

Also, the change in $[CH_3COO^-(aq)]$ will be negligible because there are a relatively large number of CH_3COO^- ions present, resulting from the total dissociation of the CH_3COONa.

This means that the ratio $\frac{[CH_3COOH(aq)]}{[CH_3COO^-(aq)]}$ will remain fairly constant.

If this ratio remains fairly constant, then $[H^+(aq)]$ remains fairly constant because, at a given temperature, K_a is also constant.

If $[H^+(aq)]$ remains fairly constant, then the pH remains fairly constant.

To summarise, a buffer solution of a weak acid and its conjugate base maintains a fairly constant pH because the ratio of $[CH_3COOH(aq)]$ to $[CH_3COO^-(aq)]$ remains fairly constant when *small* amounts of either acid or base are added.

The best way to demonstrate the effect of adding a small amount of acid on the pH of a buffer solution is to calculate the pH of the buffer both before and after adding the acid.

We have already calculated the pH of a buffer solution made by mixing equal volumes of 1.00 mol dm^{-3} ethanoic acid and 1.00 mol dm^{-3} sodium ethanoate at 298 K. It is 4.76.

Let us imagine that we have mixed 500 cm^3 of each solution to make 1 dm^3 of buffer solution. The amounts of CH_3COOH and CH_3COO^- in this solution will both be equal to 0.500 mol.

Let us imagine we have 1 dm^3 of this solution and we add 1.00×10^{-2} (0.0100) mol of HCl to it. To make the mathematics easier we will assume that the volume of the solution does not change.

0.0100 mol of HCl will provide 0.0100 mol of H^+ ions. These will react with the CH_3COO^- ions in the buffer in a 1 : 1 molar ratio:

$$CH_3COO^-(aq) + H^+(aq) \rightarrow CH_3COOH(aq)$$

The amount of CH_3COOH present will now have increased from 0.500 to 0.510 mol.

The amount of CH_3COO^- present will have decreased from 0.500 to 0.490 mol.

So, the new concentrations of acid and base present are:

$$[CH_3COOH(aq)] = 0.510\ mol\ dm^{-3}$$

$$\text{and } [CH_3COO^-(aq)] = 0.490\ mol\ dm^{-3}$$

The new hydrogen ion concentration is given by:

$$[H^+(aq)] = 1.74 \times 10^{-5} \times \frac{0.510}{0.490} = 1.81 \times 10^{-5}\ mol\ dm^{-3}$$

The new pH = $-\lg(1.81 \times 10^{-5}) = 4.74$ (to 3 significant figures).

The pH has changed by 0.02 units from 4.76 to 4.74.

To understand how effective the buffer solution is in controlling the pH, let us consider adding 0.0100 mol of H^+ ions to 1 dm^3 of deionised water.

The pH, at 298 K, of deionised water is 7.00 so:

$$[H^+(aq)] = 1.00 \times 10^{7}\ mol\ dm^{-3}$$

Adding 0.0100 mol of H^+ ions gives:

$$[H^+(aq)] = (1.00 \times 10^{-7} + 0.0100) = 1.00001 \times 10^{-2}\ mol\ dm^{-3}$$

The new pH is given by:

$$pH = -\lg(1.00001 \times 10^{-2}) = 2.00$$

The pH has dropped by 5 units as opposed to 0.02 units with the buffer solution. A considerable difference!

HENDERSON–HASSELBALCH EQUATION

The Henderson–Hasselbalch equation can also be used to calculate the pH of a buffer solution.

For a weak acid, HA, and its conjugate base, A^-, the following equation applies:

$$[H^+(aq)] = K_a \times \frac{[HA(aq)]}{[A^-(aq)]}$$

Strictly speaking, the concentration terms in the equation are the concentrations at equilibrium. However, it is reasonable to take the original concentrations, for reasons we have already discussed.

The original concentration of A^- will be the same as that of the salt, provided that a sodium or potassium salt has been used. The equation can now be rewritten as:

$$[H^+(aq)] = K_a \times \frac{[acid]}{[salt]}$$

If we take the logarithm to base 10 of both sides of this equation we get:

$$\lg[H^+(aq)] = \lg K_a + \lg\frac{[acid]}{[salt]}$$

Or:

$$-\lg[H^+(aq)] = -\lg K_a - \lg\frac{[acid]}{[salt]}$$

Or:

$$pH = pK_a - \lg\frac{[acid]}{[salt]}$$

Or:

$$pH = pK_a + \lg\frac{[salt]}{[acid]}$$

This last equation is the most common form of the Henderson–Hasselbalch equation.

LEARNING TIP

Do not worry if you did not fully understand the derivation of the Henderson-Hasselbalch equation. You do not need to learn this equation for the exam, although you may use it if you wish. The equation is also given in another form in which [base] replaces [salt], because A^- is the base in the buffer.

If we use the Henderson–Hasselbalch equation to calculate the pH of our solution containing ethanoic and ethanoate ions, both of concentration 0.500 mol dm^{-3}, we obtain:

$$pH = 4.76 + \lg(0.500/0.500) = 4.76$$

$$[pK_a = -\lg K_a = -\lg(1.74 \times 10^{-5})\ ;\ \lg 1 = 0]$$

This is the same answer as we obtained earlier.

BUFFER MADE FROM A WEAK BASE AND ITS CONJUGATE ACID

The most common example of this type of buffer is ammonia and the ammonium ion. The ammonium ion is usually supplied in the form of ammonium chloride. This mixture will provide a buffer solution with a pH greater than 7.

The most convenient equilibrium to consider is:

$$NH_4^+(aq) \rightleftharpoons NH_3(aq) + H^+(aq)$$

This mixture provides a relatively high concentration of both NH_3 molecules and NH_4^+ ions.

The buffer works in a similar manner to the weak acid–conjugate base system. The addition of acid results in the added H^+ ions reacting with NH_3 molecules:

$$NH_3(aq) + H^+(aq) \rightarrow NH_4^+(aq)$$

whereas the addition of base results in the added OH^- ions reacting with NH_4^+ ions:

$$NH_4^+(aq) + OH^-(aq) \rightarrow NH_3(aq) + H_2O(l)$$

As there is a relatively high concentration of both NH_3 molecules and NH_4^+ ions, the ratio of $[NH_3(aq)]$ to $[NH_4^+(aq)]$ remains relatively constant when we add small amounts of either acid or base. This results in the pH remaining fairly constant because it is given by the equation:

$$pH = pK_a + \lg \frac{[NH_3(aq)]}{[NH_4^+(aq)]}$$

As pK_a is constant at a given temperature, the pH of the solution depends on the ratio of $[NH_3(aq)]$ to $[NH_4^+(aq)]$.

HOW TO MAKE A BUFFER SOLUTION WITH A REQUIRED pH

To make a buffer solution with a pH less than 7, you need to use a mixture of a weak acid and its conjugate base.

In the opposite way, you need to use a mixture of a weak base and its conjugate acid to make a buffer solution with a pH greater than 7.

WORKED EXAMPLE 1

Imagine we have to make a buffer solution of pH 5.00 at a temperature of 298 K.

To make this solution, we need a hydrogen ion concentration, $[H^+(aq)]$, of 1.00×10^{-5} mol dm^{-3}.

The hydrogen ion concentration of a buffer solution of a weak acid and its conjugate base is calculated using the formula:

$$[H^+(aq)] = K_a \times \frac{[\text{acid}]}{[\text{salt}]}$$

If we use ethanoic acid as the weak acid, then:

$$K_a = 1.74 \times 10^{-5} \text{ mol dm}^{-3}$$

If we now substitute our known values into the equation we obtain:

$$1.00 \times 10^{-5} = 1.74 \times 10^{-5} \times \frac{[\text{acid}]}{[\text{salt}]}$$

This gives a value for:

$$\frac{[\text{acid}]}{[\text{salt}]} \text{ of } (1.00 \times 10^{-5} \div 1.74 \times 10^{-5}) = 0.575$$

(to 3 significant figures)

So, if we were supplied with an ethanoic acid solution of concentration 0.575 mol dm^{-3} and a sodium ethanoate solution of 1.00 mol dm^{-3}, we could make a buffer solution of pH 5.00 by mixing equal volumes of the two solutions. This would give us a solution in which the acid concentration was 0.2875 mol dm^{-3} and the salt concentration was 0.500 mol dm^{-3}.

[(0.2875 ÷ 0.500) = 0.575]

WORKED EXAMPLE 2

In what proportions should we mix 0.100 mol dm^{-3} solutions of ammonia and ammonium chloride to obtain a buffer of pH 9.80?

The equation for the dissociation of NH_4^+ in aqueous solution is:

$$NH_4^+(aq) \rightleftharpoons NH_3(aq) + H^+(aq)$$

$$K_a = \frac{[NH_3(aq)][H^+(aq)]}{[NH_4^+(aq)]} = 5.62 \times 10^{-10} \text{ mol dm}^{-3}$$

$$\frac{[NH_3(aq)]}{[NH_4^+(aq)]} = \frac{5.62 \times 10^{-10}}{[H^+(aq)]}$$

$$[H^+(aq)] = 10^{-pH} = 10^{-9.80} = 1.58 \times 10^{-10} \text{ mol dm}^{-3}$$

$$\frac{[NH_3(aq)]}{[NH_4^+(aq)]} = \frac{5.62 \times 10^{-10}}{1.58 \times 10^{-10}} = 3.56$$

Therefore, we have to mix the solutions in a ratio by volume of 3.56 $NH_3(aq)$ to 1 $NH_4Cl(aq)$.

CONTROLLING THE pH OF BLOOD

The human body works within a narrow range of pH values. For example, the pH of arterial blood plasma needs to be in the range of 7.35 to 7.45. The way that the whole body functions would be affected if the pH of this blood plasma were to change significantly, particularly if it were to fall.

The pH of blood in cells is controlled by a mixture of buffers. The most important one is the carbonic acid–hydrogen carbonate buffer mixture.

In this mixture the carbonic acid molecule, H_2CO_3, acts as the weak acid. The hydrogen carbonate ion, HCO_3^-, is the conjugate base of H_2CO_3.

The equilibrium that exists is represented by the equation:

$$H_2CO_3(aq) \rightleftharpoons HCO_3^-(aq) + H^+(aq)$$

Under normal circumstances, the amount of HCO_3^- ion present is approximately 20 times that of H_2CO_3. As normal metabolism produces more acids than bases, this is consistent with the needs of the body.

Any increase in the concentration of hydrogen ions in the blood (for example, by the production of lactic acid in the muscles), results in the equilibrium shown above moving to the *left* as the added H^+ ions react with the HCO_3^- ions.

A variety of respiratory and metabolic factors can also cause the pH of the blood to rise. For example, the overuse of diuretics increases the amount of urine excreted from the body. If the urine contains large amounts of acids, then the pH of the blood will increase. If this happens, the equilibrium shown above will move to the *right* as the H_2CO_3 molecules ionise to increase the H^+ concentration and restore the pH to its normal level.

This is considered to be the most important buffer because it is coupled with the respiratory system of the body. Carbonic acid is not particularly stable, and in aqueous solution it decomposes to form carbon dioxide and water:

$$H_2CO_3(aq) \rightleftharpoons CO_2(aq) + H_2O(aq)$$

It is the respiratory system that is responsible for removing carbon dioxide from the body. Aqueous carbon dioxide exists in equilibrium with gaseous carbon dioxide:

$$CO_2(g) \rightleftharpoons CO_2(aq)$$

Combining these three reactions gives us:

$$CO_2(g) + H_2O(aq) \rightleftharpoons CO_2(aq) + H_2O(aq) \rightleftharpoons H_2CO_3(aq) \rightleftharpoons HCO_3^-(aq) + H^+(aq)$$

When these equilibria shift to the left as a result of an increase in hydrogen ion concentration, the concentration of carbon dioxide in the blood increases. The carbon dioxide leaves the blood in the lungs and is then exhaled, thus maintaining the normal pH of the blood.

BUFFERS IN FOOD

A combination of various factors such as light, oxygen, heat, humidity and/or many kinds of microorganisms (bacteria and fungi) can spoil food. We try to reduce spoilage by keeping certain foods in the dark, in airtight containers and/or in refrigerators.

Spoilage of food by microorganisms depends greatly on the pH value of the food. Most microorganisms thrive when the pH of their surroundings is close to neutral (pH 6.6–7.5). The metabolism of these microorganisms is then greatest and they can multiply quickly. Most bacteria can survive at pH values as low as 4.4 and as high as 9.0. Only specialised bacteria can survive outside this range.

One important factor in the spoilage of food is its **buffer capacity**. Buffer capacity is a measure of the amount of acid or base required to change significantly the pH of the food. The more protein there is in the food, the higher is its buffer capacity. This is because the amino acids present have both acidic and basic properties. This means that it takes longer for the pH of the food to change enough for the bacteria to start multiplying. Bacteria and moulds can also produce waste products that act as poisons and toxins, causing ill-effects.

DID YOU KNOW?

Many processed foods, such as jams, contain buffer systems such as citric acid and sodium citrate. These help to maintain the pH within a range where the growth of microorganisms is very slow or non-existent.

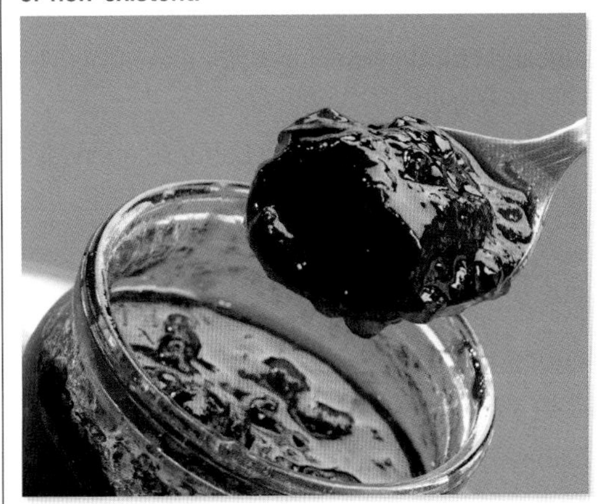

▲ **fig A** Jam is an example of a processed food that can contain citric acid and sodium citrate.

CHECKPOINT

1. (a) Explain what is meant by the term 'buffer solution'.
   (b) Explain how an aqueous solution containing a mixture of methanoic acid, HCOOH, and potassium methanoate, HCOOK, acts as a buffer.
   (c) A buffer solution contains equal concentrations of methanoic acid and potassium methanoate.
   Explain the effect on the pH of this solution of adding some solid potassium methanoate.
   (d) Calculate the pH, at 298 K, of a buffer solution made by mixing equal volumes of 1.00 mol dm^{-3} methanoic acid and 0.500 mol dm^{-3} potassium methanoate.
   [K_a(HCOOH) = 1.79×10^{-4} mol dm^{-3}]

SKILLS CREATIVITY

2. A student prepares two solutions.
   Solution **A** is prepared by mixing 50 cm^3 of 0.100 mol dm^{-3} $CH_3COOH(aq)$ with 25 cm^3 of 0.100 mol dm^{-3} NaOH(aq).
   Solution **B** is prepared by mixing 25 cm^3 of 0.200 mol dm^{-3} $CH_3COOH(aq)$ with 50 cm^3 of 0.100 mol dm^{-3} NaOH(aq).
   Explain why solution **A** is a buffer solution but solution **B** is not.
3. A buffer solution was made by mixing 50 cm^3 of 0.200 mol dm^{-3} aqueous ammonia, $NH_3(aq)$, with 50 cm^3 of aqueous ammonium chloride, $NH_4Cl(aq)$. The pH of the resulting solution was 9.55. Calculate the concentration of the $NH_4Cl(aq)$ used.
   [$K_a(NH_4^+) = 5.62 \times 10^{-10}$ mol dm^{-3}]
4. Calculate the pH of a buffer solution containing 12.20 g of benzoic acid (C_6H_5COOH) and 7.20 g of sodium benzoate (C_6H_5COONa) in 1.00 dm^3 of solution.
   [$pK_a(C_6H_5COOH) = 4.20$]
5. Like water, liquid ammonia undergoes self-dissociation:
   $$2NH_3 \rightleftharpoons NH_4^+ + NH_2^-$$
   (a) Explain why ammonia can be classified as an amphoteric substance.
   (b) For each of the following substances, indicate whether a solution of it in liquid ammonia will be 'acidic', 'basic' or 'neutral'.
   (i) Ammonium chloride, NH_4Cl
   (ii) Sodium amide, $NaNH_2$
   (iii) Potassium hydroxide, KOH

SUBJECT VOCABULARY

buffer solution a solution that *minimises* the change in pH when a *small* amount of either acid or base is added
buffer capacity a measure of the amount of acid or base required to change significantly the pH of food or of a solution of an acid and a base

14B 3 BUFFER SOLUTIONS AND pH CURVES

SPECIFICATION REFERENCE 14.21 CP11

LEARNING OBJECTIVES

- Understand how to use a weak acid–strong base or strong acid–weak base titration curve to:
  (i) demonstrate buffer action
  (ii) determine K_a from the pH at the point where half the acid is neutralised/equivalence point.

BUFFER ACTION DURING A TITRATION

Fig A shows a typical pH curve obtained when a weak acid is titrated against a strong base.

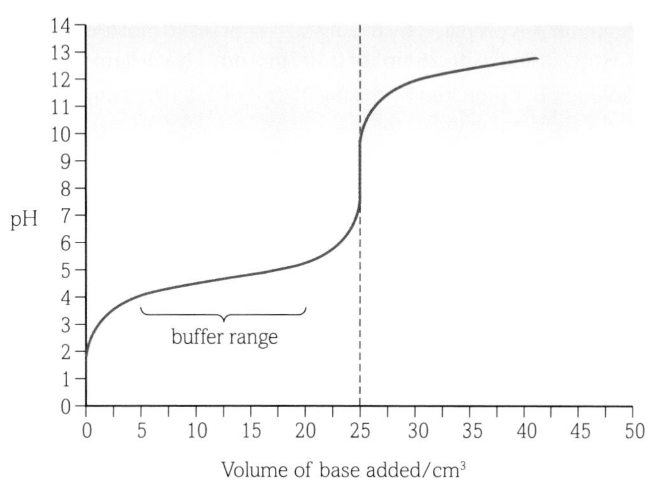

▲ **fig A** Strong base-weak acid pH curve showing buffer range.

EXAM HINT

When a weak acid is titrated against a strong base, there is a fairly rapid change of about 1.5 pH units when the strong base is first added. This is before the buffer range begins. Be sure to show this clearly if you are asked to **sketch** a curve for this titration in an exam question.

In the region marked 'buffer range', the change in pH as the base is added is gradual. Over this range there is a considerable concentration of both acid and conjugate base molecules. This mixture is displaying buffer action.

DETERMINING K_a FROM A pH TITRATION CURVE

This experiment involves performing a titration with an aqueous solution of a weak acid in a conical flask. From a burette, we add a standard solution of a strong base, such as sodium hydroxide. We measure the pH of the solution after each addition.

Then we are able to plot a graph of pH against volume of base added. From this graph we can determine the minimum volume of base required to completely react with all of the acid.

Then we can use the graph to determine the pH at the half-equivalence point. This pH value is equal to the pK_a value of the weak acid. From this it is a simple matter to calculate K_a for the acid.

A typical pH titration curve is shown in **fig B**.

The volume at the equivalence point is 25 cm^3, so the volume at the half-equivalence point is 12.5 cm^3. The pH when 12.5 cm^3 of base is added is 4.80.

The pK_a of the acid = 4.80

K_a for the acid = $10^{-4.80}$ = 1.58×10^{-5} mol dm^{-3}

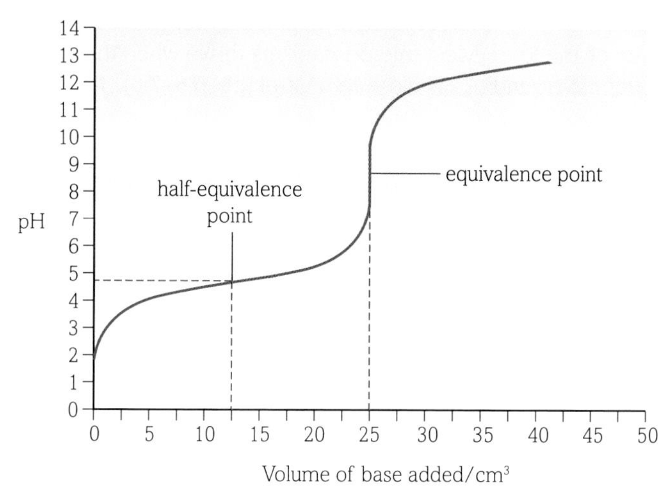

▲ **fig B** Strong base-weak acid pH curve showing equivalence and half-equivalence points.

EXAM HINT

Make sure you use the term half-equivalence point and *not* half-neutralisation point when determining the pK_a of a weak acid.

SUPPORTING THEORY

The theoretical justification for determining K_a for a weak acid by the method shown above is quite straightforward.

The mixture at the half-equivalence point is a buffer solution (as stated above). The pH of a buffer solution is calculated using the following equation:

$$\text{pH} = \text{p}K_a + \lg \frac{[\text{salt}]}{[\text{acid}]}$$

At the half-equivalence point:

[salt] = [acid], so [salt]/[acid] = 1

The logarithm to the base 10 of 1 (lg 1) = 0.

So, the equation becomes:

$$\text{pH} = \text{p}K_a$$

ALTERNATIVE METHOD

This method has been called the 'half-volume method'.

- Using a volumetric pipette, place 25.0 cm^3 of an aqueous solution of the weak acid into a conical flask.
- Add a few drops of phenolphthalein indicator.
- Titrate against a solution of aqueous sodium hydroxide until the end point colour is obtained.
- Note the volume of sodium hydroxide required. This is the minimum volume required to completely react with the acid.
- Use a fresh 25.0 cm^3 sample of the same aqueous solution of the weak acid and the same aqueous solution of sodium hydroxide, but this time do not add the phenolphthalein.
- Add only *half* the volume of sodium hydroxide required to react with the acid.
- Measure the pH of this solution. This pH value is equal to the pK_a value of the acid.

PRACTICAL SKILLS CP11

Note that you also considered this core practical in **Section 14A.4.**

Safety Note: Wear eye protection and avoid skin contact with the solutions. Make sure the top of the burette is below eye level when you fill it.

CHECKPOINT

SKILLS INTERPRETATION

1. A student carried out a titration by adding hydrochloric acid to ammonia solution. A sketch graph of pH against volume of hydrochloric acid added is shown below.

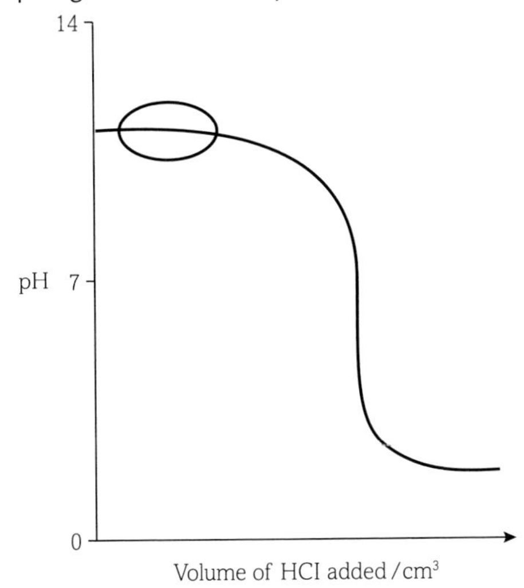

(a) Name the type of solution formed in the region ringed on the sketch graph.

(b) Explain how you deduced your answer to part (a).

SKILLS INTERPRETATION

2. The diagram shows the titration curve for a weak acid.

Use the curve to determine the pK_a value for the acid. Explain how you arrived at your answer.

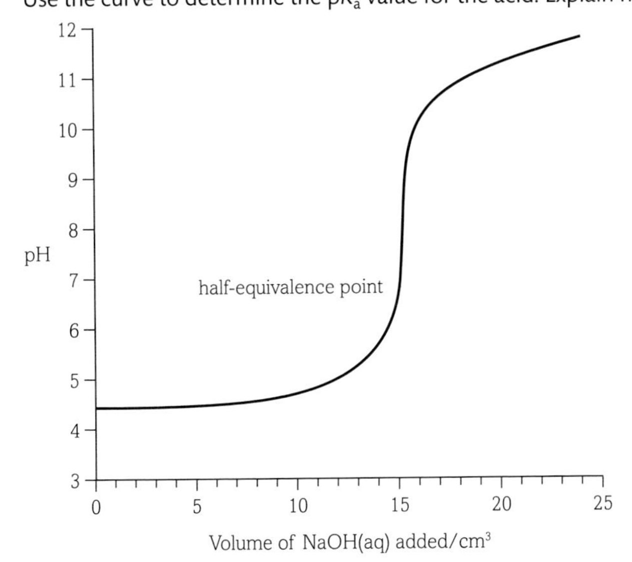

A PROBLEM GROWS

Ocean acidification makes it difficult for mussels to attach to surfaces. This is a worrying prospect in the waters around Puget Sound, USA.

MUSSELS LOSE FOOTING IN MORE ACIDIC OCEAN

fig A Carbon dioxide from greenhouse gas emissions has steadily caused seawater to become more acidic. This has disrupted organisms accustomed to the slightly alkaline waters of the past 20 million years.

PENN COVE, Wash. – Cookie tray in hand and lifejacket around chest, Laura Newcomb looks more like a confused baker than a marine biologist. But the University of Washington marine biologist is dressed for work. Her job: testing how mussels in this idyllic bay are affected by changing ocean conditions, especially warmer and more acidic waters. It's a question that is key to understanding whether climate change threatens mussels around the world, as well as the food chains mussels support and protect in the wild.

'Along the West Coast, mussels are well-known ecosystem engineers,' said Bruce Menge, an Oregon State University researcher who studies how climate impacts coastal ecosystems. 'They provide habitat for dozens of species, they provide food for many predators and occupy a large amount of space, so are truly a dominant species.'

20 million years

Carbon dioxide from greenhouse gas emissions has steadily turned seawater more acidic, disrupting organisms accustomed to the slightly alkaline waters of the past 20 million years. In the case of mussels, an earlier University of Washington lab study found that increased carbon dioxide weakens the sticky fibres, called byssus, that mussels use to survive by clinging to objects like shorelines or ropes.

'If byssal thread weakening does eventually become important,' Menge added, 'the consequences would be major if not catastrophic.'

Newcomb's goal now is to study in the real world what was learned in the lab. 'I use the natural seasonal variation to try to answer the same questions,' Newcomb said.

Newcomb's field office is the rear deck of a boat – right between the toilet and the microwave. The quarters are cramped, but the view is grand: the blue waters of Penn Cove on Washington state's Whidbey Island are set against rolling bluffs and snow-capped mountains.

30 per cent increase in acidity

Placing mussels on her tray, Newcomb samples them for size, thickness and strength. The mussels grow on the long ropes that hang from several dozen rafts in the bay. Newcomb takes samples from two depths: 3 feet and 21 feet. She also samples water temperature and pH levels at those depths.

Prior to the Industrial Revolution and the explosion of anthropogenic CO_2, ocean pH averaged 8.2. Today it's 8.1, a 30 per cent increase in acidity on the logarithmic scale. Computer models peg ocean acidity at 7.8 to 7.7 by the end of the century at the current rate of greenhouse gas emissions.

Washington State is a bit ahead of that curve because ancient carbon stores in the deep ocean are periodically churned up by local currents. The surprising lab discovery was that mussel byssus weakened by 40 per cent when exposed to a pH of 7.5. At Penn Cove, low pH levels are not uncommon – Newcomb has even seen 7.4 in the year that she's been sampling.

'We're worried they're going to see it more frequently,' said Emily Carrington, Newcomb's graduate adviser and leader of the University of Washington team that published the earlier lab results.

From 'The Daily Climate' by Miguel Llanes https://www.dailyclimate.org/-2598789581.html

SCIENCE COMMUNICATION

1 This article is not taken from a scientific journal. Instead it is designed to draw a wider audience's attention to an important environmental problem. Identify writing techniques in the text that the author uses to make the science more accessible.

WRITING SCIENTIFICALLY

Being able to communicate scientific ideas to a wide audience is an important skill, particularly when the science has implications for decision-making in society.

CHEMISTRY IN DETAIL

2 Write an equation for the reaction of carbon dioxide with water to form hydrogen carbonate ions and $H^+(aq)$ ions.

3 (a) Using the relationship $pH = -lg[H^+]$, show that a change in pH from 8.2 to 8.1 is approximately equivalent to a 30% increase in acidity.

(b) Calculate the pH of the ocean if the acidity increases by 100% from a starting pH of 8.2.

4 Many sea organisms form shells made of calcium carbonate. Use the equation below to explain why increasing levels of carbon dioxide in the air are making this process increasingly difficult.

$$CO_2(aq) + CO_3^{2-}(aq) + H_2O(l) \rightleftharpoons 2HCO_3^-(aq)$$

5 Many organisms use intracellular HCO_3^- ions to buffer changes in pH. Use equations to show how HCO_3^- ions can buffer small changes in $H^+(aq)$ and $OH^-(aq)$.

6 The equation for the dissociation of carbonic acid in water can be represented as:

$$H_2CO_3(aq) \rightleftharpoons H^+(aq) + HCO_3^-(aq)$$

(a) Write an expression for K_a for carbonic acid.

(b) Given that the value of this pK_a is 6.3 and that the pH of blood is 7.4, calculate the ratio of hydrogen carbonate ions to non-dissociated carbonic acid molecules in blood plasma.

(c) What assumptions have you made in your answer to question **6b**?

THINKING BIGGER TIPS

Think about how an increase in carbon dioxide would affect the equilibrium of this reaction, and therefore the availability of carbonate ions.

ACTIVITY

The molecule histidine is one of the 20 amino acids that make up all proteins. Histidine residues in haemoglobin molecules are also involved in buffering the blood's pH. Research and create a presentation suitable for 14- to 16-year-old Chemistry students addressing the questions below:

- What does a histidine molecule look like?
- Which part of the molecule acts as the proton donor and which as the proton acceptor?
- Histidine can frequently be found at the catalytic centres of enzymes. Can you suggest why?

DID YOU KNOW?

March 2014 was the first month in which the average atmospheric carbon dioxide levels reached the 400 ppm milestone. But carbon dioxide is not the only cause of climate change. If we factor in oxides of nitrogen and methane and trace amounts of other gases such as SF_6, the 'equivalent CO_2' atmospheric concentration is about 480 ppm. The National Oceanic and Atmospheric Administration (NOAA climate.gov) predicts that if carbon dioxide production continues at current rates then the average ocean pH will be 7.3 by 2100.

14 EXAM PRACTICE

1 In an acid–base titration, a 0.10 mol dm^{-3} solution of a base is added to 25 cm^3 of a 0.10 mol dm^{-3} solution of an acid.

The diagram shows the pH value of the solution plotted against the volume, V, of base added.

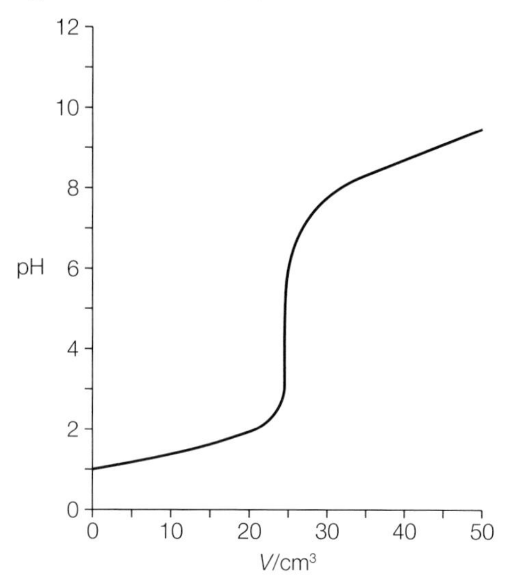

This diagram could represent a titration between:

A $CH_3COOH(aq)$ and $NH_3(aq)$

B $CH_3COOH(aq)$ and $KOH(aq)$

C $HCl(aq)$ and $NH_3(aq)$

D $HCl(aq)$ and $KOH(aq)$ [1]

(Total for Question 1 = 1 mark)

2 Hardness in water can be determined by titrating a sample against a reagent that forms complex ions with dissolved metal ions.

The indicator for this titration requires the pH to be maintained at about 10.

Which of these, in aqueous solution, could be used to do this?

A ammonia and ammonium chloride

B ammonium chloride and hydrochloric acid

C sodium ethanoate and ethanoic acid

D sodium hydroxide and sodium ethanoate. [1]

(Total for Question 2 = 1 mark)

3 A 1.0×10^{-2} mol dm^{-3} aqueous solution of the weak, monoprotic acid HA has a pH of 4.0.

What is the approximate pK_a value for the acid?

A 4.0 **C** 6.0 **C** 7.0 **D** 8.0 [1]

(Total for Question 3 = 1 mark)

4 The following equilibrium exists in a mixture of concentrated nitric acid and concentrated sulfuric acid:

$$HNO_3 + 2H_2SO_4 \rightleftharpoons NO_2^+ + 2HSO_4^- + H_3O^+$$

Which statement is correct?

A HNO_3 and NO_2^+ are a conjugate acid–base pair.

B The nitric acid acts as an oxidising agent.

C The sulfuric acid acts as a dehydrating agent.

D The sulfuric acid acts as a base. [1]

(Total for Question 4 = 1 mark)

5 A sales leaflet claims that 'applications of a solution of ammonium sulfate, which is acidic, improve the growth of acid-loving rhododendron bushes by increasing the availability of nitrogen and also by increasing the pH of the soil'.

What is **wrong** with this statement?

A Aqueous ammonium sulfate is not acidic.

B Ammonium sulfate does not dissolve in water.

C To be a fertiliser, nitrogen is needed in its oxidised form (nitrate) and not in its reduced form (ammonium).

D The pH of the soil will be decreased, not increased. [1]

(Total for Question 5 = 1 mark)

6 A fruit juice contains a monobasic acid HA.

(a) The fruit juice has a hydrogen ion concentration of 2.50×10^{-4} mol dm^{-3}. Calculate the pH of the fruit juice. [2]

(b) A 25.0 cm^3 sample of the fruit juice reacted exactly with 26.70 cm^3 of 0.0100 mol dm^{-3} sodium hydroxide.

(i) Write a chemical equation for the reaction taking place. State symbols are not required. [1]

(ii) Calculate the concentration, in mol dm^{-3}, of HA in the fruit juice. [3]

(iii) Compare and contrast your answer in (b)(ii) with the hydrogen ion concentration. Then, make a deduction about the strength of the acid, HA, in the fruit juice. [2]

(c) (i) Write an equation to represent the dissociation of HA into its ions in aqueous solution. [1]

(ii) Write an expression for the acid dissociation constant, K_a, of HA(aq). [1]

(iii) The value of K_a for HA(aq) is 6.00×10^{-5} mol dm^{-3}. Calculate the concentration of the undissociated acid under these conditions. [2]

(Total for Question 6 = 12 marks)

7 A mixture of ethanoic acid, CH_3COOH, and its sodium salt CH_3COONa, can act as a buffer solution.

(a) State what is meant by the term 'buffer solution'. [2]

(b) Explain how a mixture of ethanoic acid and sodium ethanoate acts as a buffer. [4]

(c) A buffer solution is made by adding a solution of sodium hydroxide, of concentration 1.00 mol dm^{-3}, to a sample of 1.00 mol dm^{-3} ethanoic acid until half of the amount of acid present has reacted. Calculate the pH of this buffer solution.

[K_a of ethanoic acid at the temperature used is 1.70×10^{-5} mol dm^{-3}] [3]

(Total for Question 7 = 9 marks)

8 The values of the ionic product of water, K_w, are 1.00×10^{-14} mol^2 dm^{-6} at 298 K and 5.48×10^{-14} mol^2 dm^{-6} at 323 K.

(a) Calculate the pH of water at each of these two temperatures. [4]

(b) Using your answers to (a), comment on the validity of the following statement:
'Pure water is neutral because it has a pH of 7'. [2]

(c) Show that the data supplied can be used to deduce the sign of ΔH for the dissociation of water into ions. [2]

(Total for Question 8 = 8 marks)

9 In 1923, Johannes Brønsted and Thomas Lowry proposed independently a theory that when an acid reacts with a base the acid forms its conjugate base. The theory is known as the 'Brønsted–Lowry' theory.

(a) State what is meant by the term **conjugate base**. [1]

(b) (i) In each of the two equations, identify the species on the left-hand side of the equation that is behaving as a Brønsted–Lowry acid. [2]

Equation 1: $C_6H_5COO^- + HF \rightleftharpoons C_6H_5COOH + F^-$

Equation 2: $C_6H_5COOH + CN^- \rightleftharpoons C_6H_5COO^- + HCN$

(ii) Explain the relative strengths of the three acids involved in the two equilibria. [Assume that both equilibria lie well to the right-hand side.] [4]

(c) Liquid ammonia, like water, undergoes self-ionisation, according to the following equation:

$NH_3 + NH_3 \rightleftharpoons NH_4^+ + NH_2^-$

For each of the two substances listed, explain whether a solution in liquid ammonia would be 'acidic', 'basic' or 'neutral'.

(i) Ammonium chloride, $NH_4^+Cl^-$

(ii) Sodium amide, $Na^+NH_2^-$ [4]

(Total for Question 9 = 11 marks)

10 Sulfur dioxide reacts with water to produce sulfurous acid, H_2SO_3, which is a weak dibasic acid.

The equation for the first dissociation into ions is:

$H_2SO_3(aq) \rightleftharpoons H^+(aq) + HSO_3^-(aq)$

$K_{a(1)}(298\ K) = 1.20 \times 10^{-2}$ mol dm^{-3}

(a) Calculate the value of $pK_{a(1)}$ for H_2SO_3. [1]

(b) Use $K_{a(1)}$ to calculate the approximate pH of an aqueous solution of 0.500 mol dm^{-3} H_2SO_3 at 298 K. [3]

(c) The measured pH of 0.500 mol dm^{-3} H_2SO_3 is slightly lower than that calculated in part (b). Comment on a possible reason for this difference. [1]

(d) The constant K_w has a value of 1.00 x 10^{-14} mol^2 dm^{-6} at 298 K.

(i) Give the name of the constant K_w. [1]

(ii) Write the expression for K_w. [1]

(e) Potassium hydroxide, KOH, is a strong base in aqueous solution. Calculate the pH of 0.500 mol dm^{-3} KOH. [3]

(Total for Question 10 = 10 marks)

11 The graph shows the change in pH when dilute hydrochloric acid is titrated with sodium carbonate solution.

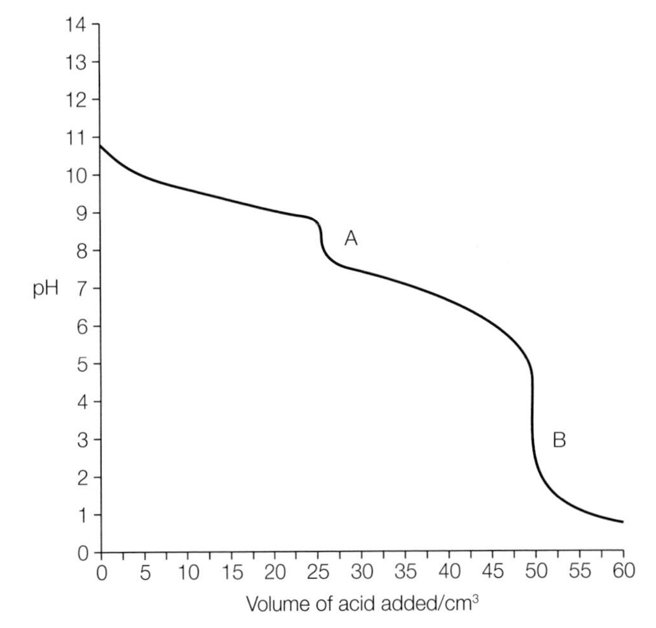

The overall equation for the reaction is:

$Na_2CO_3(aq) + 2HCl(aq) \rightarrow 2NaCl(aq) + CO_2(g) + H_2O(l)$

(a) Suggest suitable indicators to detect the changes at A and at B. [2]

(b) Explain why the change at A would be more difficult to detect than the change at B. [2]

(c) Suggest equations for the reactions taking place at each stage. [2]

(Total for Question 11 = 6 marks)

TOPIC 15 ORGANIC CHEMISTRY: CARBONYLS, CARBOXYLIC ACIDS AND CHIRALITY

A CHIRALITY | B CARBONYL COMPOUNDS | C CARBOXYLIC ACIDS | D CARBOXYLIC ACID DERIVATIVES | E SPECTROSCOPY AND CHROMATOGRAPHY

In **Book 1** you learned about the simplest hydrocarbons (alkanes and alkenes) and how to name them using IUPAC (International Union of Pure and Applied Chemistry) rules. You then considered halogenoalkanes and alcohols, including important practical techniques used to convert one organic compound into another. You learned how the analytical techniques of mass spectrometry and infrared spectroscopy are used to determine the structures of organic compounds.

In this topic, you will extend your understanding of organic chemistry by considering compounds with functional groups such as aldehydes, ketones, carboxylic acids, acyl chlorides and esters. This topic introduces two more modern analytical techniques: nuclear magnetic resonance spectroscopy and chromatography.

Here are some important areas of organic chemistry you will encounter:

- How measurements of optical activity can be used as evidence for organic reaction mechanisms
- How simple chemical tests can be used to identify different functional groups
- The uses of esters in solvents, food flavourings and perfumes.

MATHS SKILLS FOR THIS TOPIC

- Use ratios, fractions and percentages (*i.e. formulae based on experiments*)
- Translate information between graphical, numerical and algebraic forms
- Visualise and represent 2D and 3D forms, including two-dimensional representations of 3D objects
- Understand the symmetry of 2D and 3D shapes

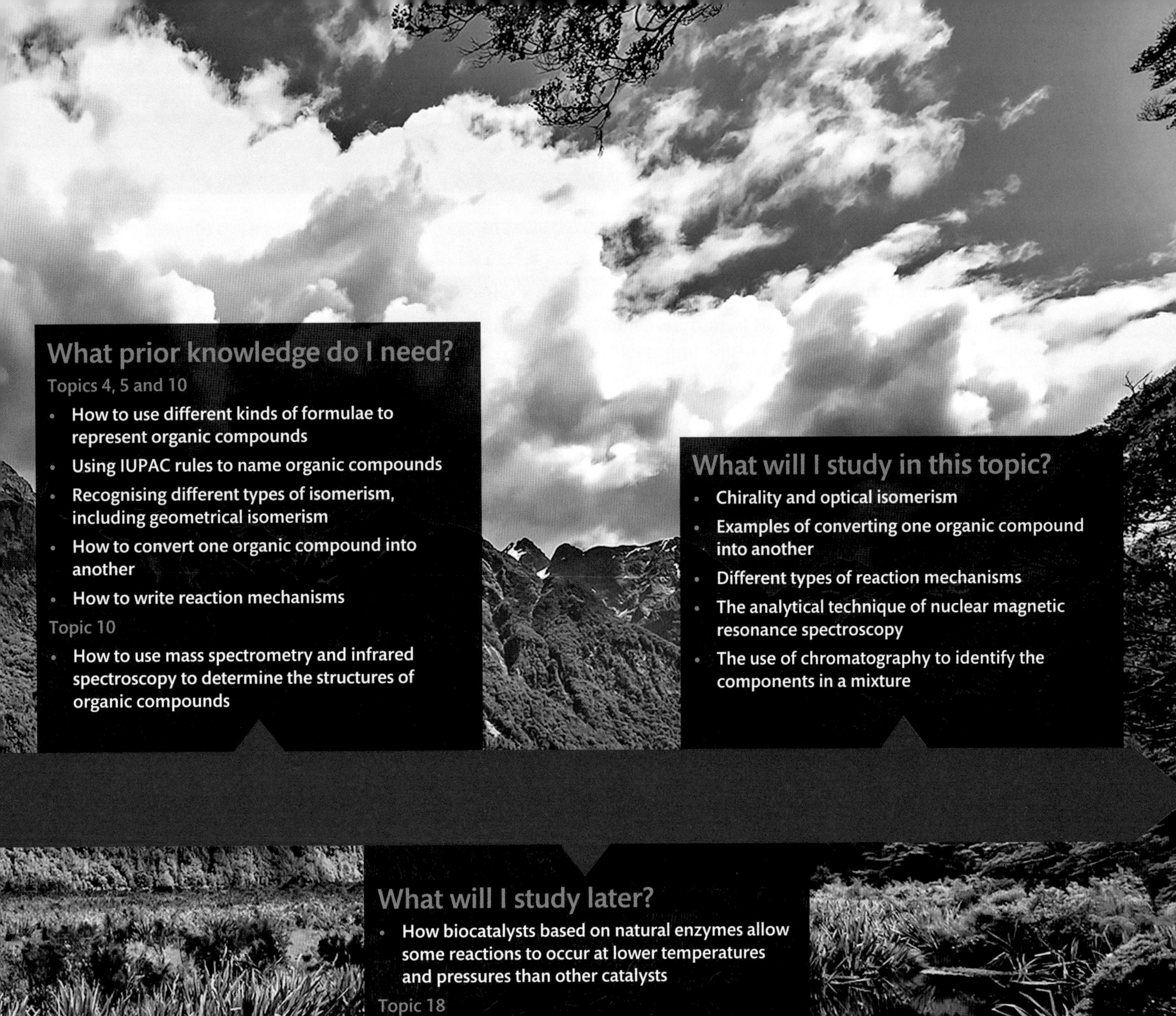

What prior knowledge do I need?

Topics 4, 5 and 10

- How to use different kinds of formulae to represent organic compounds
- Using IUPAC rules to name organic compounds
- Recognising different types of isomerism, including geometrical isomerism
- How to convert one organic compound into another
- How to write reaction mechanisms

Topic 10

- How to use mass spectrometry and infrared spectroscopy to determine the structures of organic compounds

What will I study in this topic?

- Chirality and optical isomerism
- Examples of converting one organic compound into another
- Different types of reaction mechanisms
- The analytical technique of nuclear magnetic resonance spectroscopy
- The use of chromatography to identify the components in a mixture

What will I study later?

- How biocatalysts based on natural enzymes allow some reactions to occur at lower temperatures and pressures than other catalysts

Topic 18

- The chemistry of arenes (aromatic compounds)

Topic 19

- The chemistry of organic nitrogen compounds

Topic 20

- How to convert one compound into another via a series of steps – organic synthesis

Study beyond IAL

- How pharmaceutical companies invest heavily in research and development to design new medicines to treat a wide range of medical problems and illnesses

15A 1 CHIRALITY AND ENANTIOMERS

SPECIFICATION REFERENCE 15.1 15.2

LEARNING OBJECTIVES

- Know that optical isomerism is a result of chirality in molecules with a single chiral centre.
- Understand that optical isomerism results from chiral centre(s) in a molecule with asymmetric carbon atom(s) and that optical isomers (enantiomers) are object and non-superimposable mirror images and be able to draw 3D diagrams of these optical isomers.

DIFFERENT TYPES OF ISOMERISM

In **Book 1**, you learned about structural isomerism and its division into chain isomerism and position isomerism. You then considered a different type of isomerism called 'geometric isomerism', including the terms *E–Z* and *cis–trans* isomerism – this is one type of stereoisomerism.

This chapter looks at a second type of stereoisomerism – optical isomerism. Before we introduce this, it is useful to consider all the types of isomerism that you need to be familiar with, in the form of a family tree.

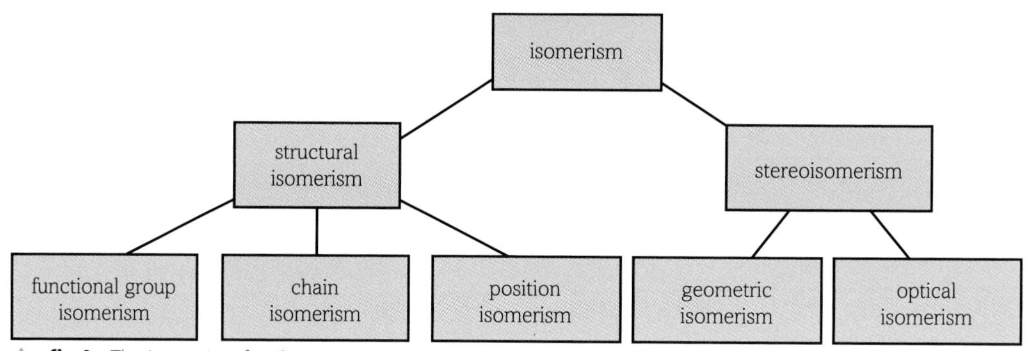

▲ **fig A** The isomerism family tree.

Structural isomers are compounds with the same molecular formula but with different structural formulae.

Functional group isomers differ because they have different functional groups (e.g. aldehydes and ketones, alcohols and ethers).

Chain isomers differ because they have different patterns of branching in their carbon chains.

Positional isomers differ because the same functional group is attached to different carbon atoms in the chain.

Stereoisomers have the same structural formula but differ because their atoms or groups are arranged differently in three dimensions.

Geometric isomers differ because their atoms or groups are attached at different positions on opposite sides of a C=C double bond.

OPTICAL ISOMERISM

How does optical isomerism fit into this family tree? Optical isomers are non-superimposable mirror images of each other but are otherwise the same. This needs careful explanation!

CHIRALITY

This is a term derived from the Greek word for 'hand', and could be translated as 'handedness'. Many objects, including human hands, can be described as **chiral**. Your hands have the same features, and you could describe both hands using the same words. Both consist of a palm with a thumb on one side and four fingers of different sizes. However, if you place one hand on top of the other you can see that they are different – you cannot superimpose one hand on the other.

Now put your hands together – you can see that the thumbs are together, the little fingers are together, and so on – everything corresponds. Next, put your left hand in front of a mirror, then look at the image in the mirror and your right hand at the same time – they should look identical, so your hands can be described as mirror images, or sometimes as an object and its mirror image.

CHIRALITY IN SIMPLE MOLECULES

The key to understanding chirality in molecules is to visualise them in 3D, and ideally use models to help you. Think of a molecule consisting of a single carbon atom joined to four different groups or atoms (represented by W, X, Y and Z). If you represent the molecule in two dimensions, with bond angles of 90°, you cannot easily see why there should be two different arrangements. However, you may remember from **Topic 3 (Book 1: IAS)** on the shapes of molecules that you can attempt to show the 3D nature of molecules using different styles to represent the bonds. Ordinary lines are used to represent bonds in the plane of the paper, tapered or wedge-shaped lines indicate those above the plane of the paper, and dashed lines indicate those below the plane of the paper. Drawing the two structures side-by-side clearly shows how they relate to each other as object and mirror image (**fig B**).

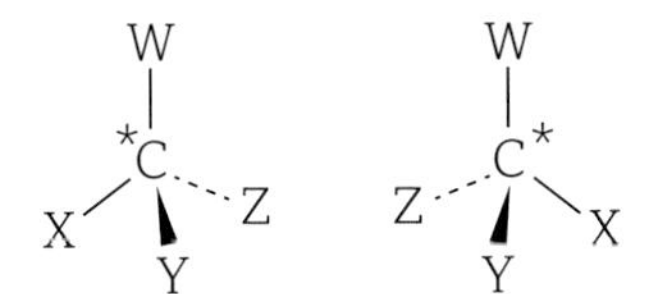

▲ **fig B** This is one way to show optical isomers in 3D.

The asterisk (*) indicates that the carbon atom next to it is a chiral centre, also known as an **asymmetric** carbon atom.

If the attached atoms or groups are W, W, X and Y (so that two are the same) it is possible to show that they relate to each other in the same way (they are mirror images). However, there is no chirality because one can also be superimposed on the other (**fig C**).

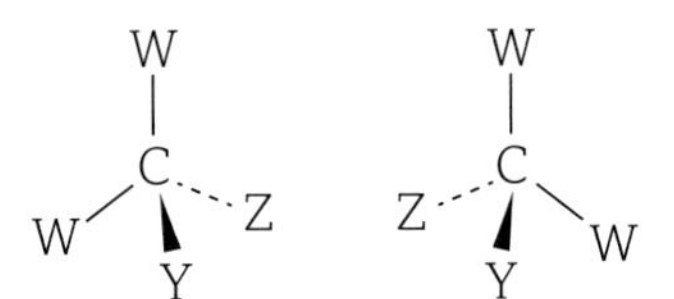

▲ **fig C** These structures are not optical isomers because there are only three different groups joined to the central carbon atom.

HOW TO IDENTIFY CHIRAL CENTRES IN STRUCTURAL FORMULAE

You need to be able to identify chiral centres in molecules such as the examples shown in **table A**. You only need to consider whether there is a carbon atom joined to four different atoms or groups.

STRUCTURE	COMMENTS
CBr_2ClF	The single carbon atom has only three different atoms attached, so there is no chiral centre.
$CH_3CHBrCH_2CH_3$	Carbons 1, 3 and 4 in the chain have two or more hydrogens attached, so are not chiral. Carbon 2 is joined to four different groups (methyl, hydrogen, bromine and ethyl), so it is a chiral centre.
$CH_3CH_2CHBrCH_2CH_3$	Carbons 1, 2, 4 and 5 in the chain have two or more hydrogens attached, so are not chiral. Carbon 3 is joined to only three different groups (two ethyl, one hydrogen and one bromine), so it is not a chiral centre.
$CH_3CH_2CHBrCHBrCH_3$	Carbons 1, 2 and 5 in the chain have two or more hydrogens attached, so are not chiral. Carbon 3 is joined to four different groups (ethyl, hydrogen, bromine and $CHBrCH_3$), so it is a chiral centre. Carbon 4 is joined to four different groups (methyl, hydrogen, bromine and CH_3CH_2CHBr), so it is also a chiral centre.

table A

If a molecule has a chiral centre, then it can exist as optical isomers – each one is known as an **enantiomer**.

LEARNING TIP

It is really worthwhile using molecular models to make some models of simple structures to check the idea of mirror images and superimposability.

CHECKPOINT

1. What is the main similarity and the main difference between optical isomers and geometric isomers?
2. How many chiral centres are there in each of these molecules?
   (a) $CH_3CCl_2CH_3$
   (b) $CH_3CH(OH)COOH$
   (c) $CH_2Cl–CHCl–CHFCH_3$

SUBJECT VOCABULARY

chiral a chiral atom in a molecule is one that allows it to exist as non-superimposable forms. It can also refer to the molecule itself
asymmetric an asymmetric carbon atom in a molecule is one that is joined to four different atoms or groups
enantiomers isomers that are related as object and mirror image

LEARNING OBJECTIVES

- Know that optical activity is the ability of a single optical isomer to rotate the plane of polarisation of plane-polarised monochromatic light in molecules containing a single chiral centre.
- Know what is meant by the term 'racemic mixture'.

PLANE-POLARISED LIGHT

In the previous section we looked at the causes of optical isomerism. Before we consider **optical activity**, we must have some understanding of **plane-polarised light**. One way to consider light is as electromagnetic radiation that travels as a transverse wave. This means that the oscillations exist in planes at right angles to the direction of travel. **Fig A** shows 'normal' **unpolarised light** and plane-polarised light with the oscillations in only one plane.

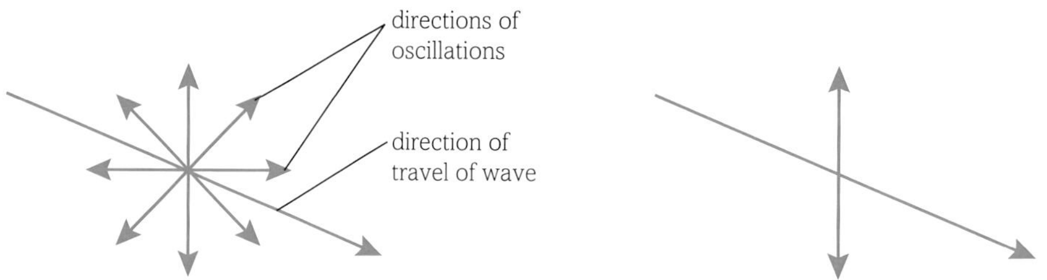

▲ **fig A** In an unpolarised wave, oscillations may occur in any plane, while in a plane-polarised wave they occur in only one plane.

Some natural materials, and synthetic materials such as Polaroid (the material used in some sunglasses), can absorb all of the oscillations except those in a single plane. This means that they convert unpolarised light into plane-polarised light. For convenience, the single plane that remains is often assumed to be the vertical plane, as shown in **fig B**.

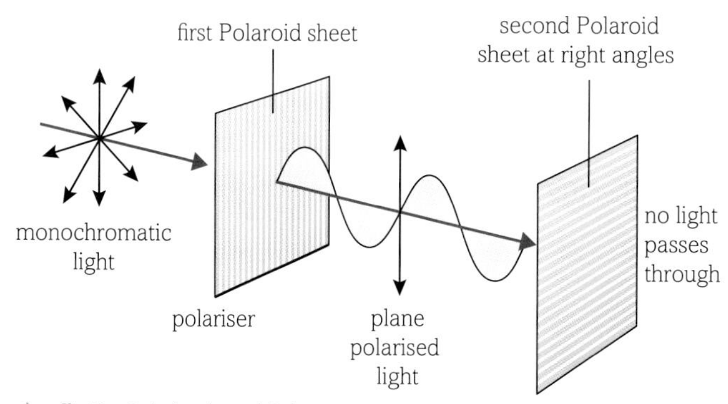

▲ **fig B** Polarisation of light.

Fig B shows that the second sheet of Polaroid has horizontal lines, which means that only horizontally plane-polarised light can pass through. As the first sheet has produced vertically plane-polarised light, then no light can pass through the second sheet.

POLARIMETRY

Polarimetry is the use of a **polarimeter** to measure the amount of optical activity, if any, of a substance. **Fig C** shows how a polarimeter works. A monochromatic light source (light of only one colour or frequency) passes through a polarising filter (such as a piece of Polaroid). This is called the **polariser** because it converts unpolarised light into (vertically) plane-polarised light. The plane-polarised light then passes through a sample tube containing some of the substance in solution. If the substance is optically active (because it contains an enantiomer), then the plane of polarisation will be rotated so that it is no longer vertical. In this example, the rotation is anticlockwise, and the substance can therefore be described as 'laevorotatory'. If the rotation is clockwise, then the substance is 'dextrorotatory'.

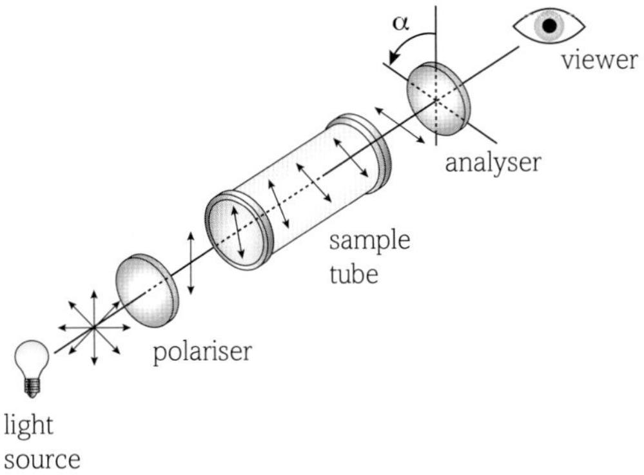

▲ **fig C** You can see how the plane of polarisation of the polarised light is rotated anticlockwise as it passes through the sample tube.

The second polarising filter, known as the **analyser**, is rotated to a position where the maximum light intensity can be seen. The angle of rotation (α) is measured, and is quoted as a positive value if the rotation is clockwise and as a negative value if the rotation is anticlockwise. In **fig C**, the rotation, α, is approximately $-90°$.

PROPERTIES OF ENANTIOMERS

Two enantiomers have identical physical properties, with one exception. The enantiomers rotate the plane of polarisation of plane-polarised light by equal angles, but in opposite directions. So, if the rotation for one enantiomer is $+60°$, then the rotation for the other enantiomer is $-60°$.

Two enantiomers have identical chemical properties, with one exception – the way in which they react with enantiomers of other substances. This chemical property may be different for each enantiomer.

RACEMIC MIXTURES

What happens if a compound with a chiral centre is present as a mixture of both enantiomers? You can imagine that the dextrorotatory and laevoratatory enantiomers have equal but opposite effects on plane-polarised light. This means that the analyser does not need to be rotated to allow the maximum light intensity to be seen. Without knowing anything about the substance, it is not possible to distinguish between a substance that has no optical activity and one that has optically active enantiomers whose effects cancel out. A mixture containing equal amounts of two enantiomers is called a **racemic mixture**.

We will briefly explain here the origin of this scientific term. The Latin word *racemus* means a bunch of grapes. How is this related to optical activity? The French chemist Louis Pasteur achieved a lot in his lifetime (1822–1895), and his name lives on in the word 'pasteurisation', often associated with milk production. He also realised that tartaric acid produced during the fermentation of grapes showed optical activity, but that tartaric acid produced in other ways did not. He eventually realised that a single enantiomer of tartaric acid was produced from fermenting grapes and that a mixture of both enantiomers of tartaric acid was produced in the other ways. Because tartaric acid is present in grapes, the term 'racemic acid' was used, and the 'racemic' part of this term was then used to describe an equimolar mixture of the two enantiomers.

Here are the structures of the two enantiomers of tartaric acid:

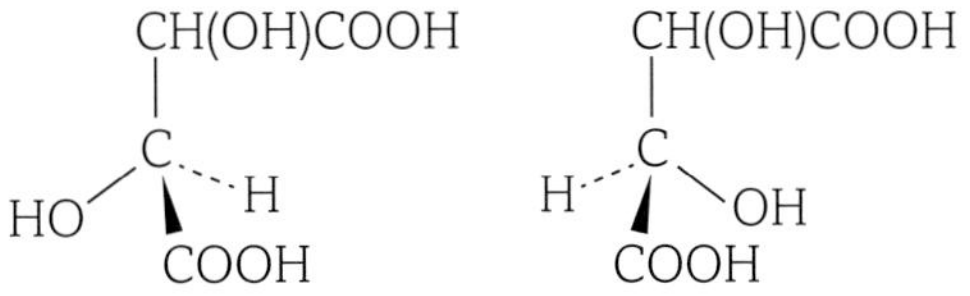

▲ **fig D** The two enantiomers of tartaric acid.

EXAM HINT

You may be asked to **draw** optical isomers in an exam. Make sure that you **identify** the chiral carbon with a * and use the wedge and dotted line notation.

LEARNING TIP

It is worth spending some time making sure you understand the differences between these terms:

unpolarised, polariser, plane-polarised, plane of polarisation.

CHECKPOINT

1. Outline how a polariser works.

SKILLS REASONING

2. A dextrorotatory enantiomer has a rotation of +43°. A mixture of this enantiomer and its laevorotatory enantiomer has a rotation of −10°. What does this information indicate about the composition of this mixture?

SUBJECT VOCABULARY

optical activity a substance shows optical activity if it rotates the plane of polarisation of plane-polarised light
plane-polarised light monochromatic light that has oscillations in only one plane
unpolarised light light that has oscillations in all planes at right angles to the direction of travel
polarimeter the apparatus used to measure the angle of rotation caused by a substance
polariser a material that converts unpolarised light into plane-polarised light
analyser a material that allows plane-polarised light to pass through it
racemic mixture an equimolar mixture of two enantiomers that has no optical activity

carboxylic acid carboxylate ion

▲ **fig A** The structures of a carboxylic acid and its carboxylate ion.

Note that the charge and double bond character are evenly distributed across both oxygen atoms in the carboxylate ion.

PHYSICAL PROPERTIES

Carboxylic acids have distinctive smells and sour tastes. The smells of some of them are considered unpleasant. For example, butanoic acid is responsible for the smell of stale sweat, and hexanoic acid for the characteristic smell of goats. Citric acid gives lemons their very sour taste.

The Latin word for sour is *acidus*, and *acetum* means vinegar, the sour taste of which comes from the acetic acid it contains.

DID YOU KNOW?

Methanoic acid is found in ants and nettle stings. *Formica*, the origin of the name formic acid, is the Latin word for ant.

◀ **fig B** Methanoic acid, HCOOH, makes up more than half of an ant's body mass!

BOILING TEMPERATURES

The presence of three polar bonds, including a polar O—H bond, in the carboxylic acid group means that they have strong intermolecular forces (hydrogen bonding). This means that they have high boiling temperatures compared with other organic compounds with a similar molar mass. As the carbon chain lengthens, the London forces between the non-polar hydrocarbon chains increase. Therefore, boiling temperature increases with increasing molar mass, as shown in **table C**.

NAME	FORMULA	MOLAR MASS/ g mol^{-1}	BOILING TEMPERATURE /°C
methanoic acid	$HCOOH$	46	101
ethanoic acid	CH_3COOH	60	118
propanoic acid	CH_3CH_2COOH	74	141
butanoic acid	$CH_3CH_2CH_2COOH$	88	164

table C

The extent of hydrogen bonding in the shorter-chain carboxylic acids means that they form dimers (double molecules) in the absence of a solvent such as water. **Fig C** shows the hydrogen bonding (dashed lines) between two molecules of ethanoic acid.

▲ **fig C** Hydrogen bonding in ethanoic acid.

EXAM HINT

In some forms of mass spectrometry, the ethanoic acid dimer can be seen at an m/z of 120.

SOLUBILITY IN WATER

The shorter-chain carboxylic acids are soluble in water because they can form hydrogen bonds with water molecules. **Fig D** shows hydrogen bonding between ethanoic acid and water.

▲ **fig D** Hydrogen bonding between ethanoic acid and water.

Solubility decreases with increasing chain length, as the hydrocarbon part of the molecules becomes larger.

LEARNING TIP

Note that the hydrogen bonding between carboxylic acid and water molecules involves:

–C=O ---- H–O–H and –O–H ---- O–H
|
OH

where hydrogen bonds are represented by dashed lines.

CHECKPOINT

1. Write the displayed formula for:
   (a) methanoic acid
   (b) ethanedioic acid.
2. Explain why hexanoic acid is much less soluble than ethanoic acid in water.

15C 2 PREPARATIONS AND REACTIONS OF CARBOXYLIC ACIDS

SPECIFICATION REFERENCE 15.11 15.12(i) 15.12(ii) 15.12(iii) 15.12(iv)

LEARNING OBJECTIVES

- Understand that carboxylic acids can be prepared by the oxidation of alcohols or aldehydes and the hydrolysis of nitriles.
- Understand the reactions of carboxylic acids with:
  (i) lithium tetrahydridoaluminate(III) (lithium aluminium hydride) in dry ether (ethoxyethane)
  (ii) bases to produce salts
  (iii) phosphorus(V) chloride (phosphorus pentachloride)
  (iv) alcohols in the presence of an acid catalyst.

PREPARATION BY OXIDATION

Although carboxylic acids are formed in many chemical reactions, two main methods are used to prepare them in the laboratory.

Oxidation uses either a primary alcohol or an aldehyde as the starting material. The usual oxidising agent is acidified potassium dichromate(VI), and the method is to heat the mixture under reflux. If a primary alcohol is used, it first oxidises to an aldehyde, then to a carboxylic acid, but both oxidations occur inside the apparatus used.

When the oxidation is complete, the reaction mixture is fractionally distilled to obtain a pure sample of the carboxylic acid.

EXAMPLES

Propanoic acid can be prepared from either propan-1-ol or propanal. The equations use [O] to represent the oxygen supplied by acidified potassium dichromate(VI).

From propan-1-ol:

$$CH_3CH_2CH_2OH + 2[O] \rightarrow CH_3CH_2COOH + H_2O$$

From propanal:

$$CH_3CH_2CHO + [O] \rightarrow CH_3CH_2COOH$$

PREPARATION BY HYDROLYSIS

Nitriles are organic compounds containing the CN group. They can be hydrolysed by heating under reflux with either a dilute acid or aqueous alkali. The same apparatus is used for preparation and purification as for oxidation.

In both cases, the C≡N triple bond breaks. The carbon atom remains part of the organic product and the nitrogen atom becomes either ammonia or the ammonium ion.

ACIDIC HYDROLYSIS

Propanoic acid can be prepared from propanenitrile. The equation for the reaction is:

$$CH_3CH_2CN + H^+ + 2H_2O \rightarrow CH_3CH_2COOH + NH_4^+$$

ALKALINE HYDROLYSIS

Butanoic acid can be prepared from butanenitrile. The equation for the reaction is:

$$CH_3CH_2CH_2CN + OH^- + H_2O \rightarrow CH_3CH_2CH_2COO^- + NH_3$$

The product is actually the butanoate ion, but this is easily converted to butanoic acid by adding dilute acid:

$$CH_3CH_2CH_2COO^- + H^+ \rightarrow CH_3CH_2CH_2COOH$$

INTRODUCTION TO REACTIONS

There are four main reactions to consider. These are summarised in **fig A**.

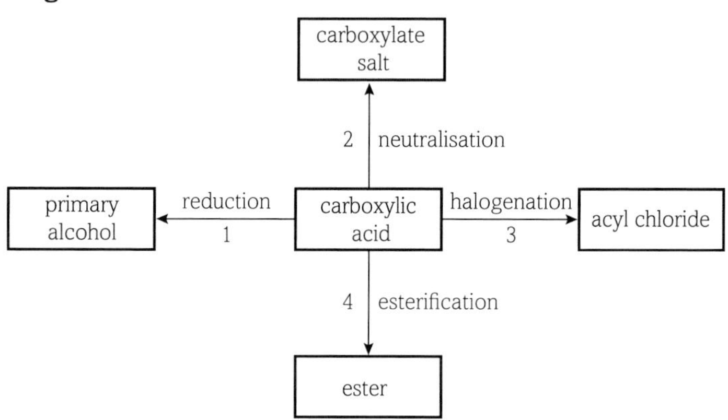

▲ **fig A** These are the four main reactions of carboxylic acids to form other organic compounds.

Some of the products of these reactions, especially acyl chlorides and esters, will be discussed in more detail later on.

1 REDUCTION

You will remember that carboxylic acids are products of oxidation. They can be reduced in the same way as aldehydes and ketones. Carboxylic acids can be reduced to primary alcohols, but not to aldehydes. This is because aldehydes are more easily reduced than carboxylic acids, and so any aldehyde produced during the reduction will be immediately reduced to a primary alcohol. The reducing agent is lithium tetrahydridoaluminate, and is used in a solvent of dry ether. As an example, an equation using butanoic acid is:

$$CH_3CH_2CH_2COOH + 4[H] \rightarrow CH_3CH_2CH_2CH_2OH + H_2O$$

The product is butan-1-ol.

2 NEUTRALISATION

Although carboxylic acids are weak acids, they can be completely neutralised by mixing with aqueous alkali. The products are carboxylate salts, which have a wide range of uses. The commonest, sodium ethanoate, is used in hand warmers and to make the additive that gives the flavour to salt and vinegar potato crisps. An equation for the reaction to make sodium ethanoate occurs between ethanoic acid and sodium hydroxide:

$$CH_3COOH + NaOH \rightarrow CH_3COONa + H_2O$$

Carboxylate salts are ionic, and are sometimes shown with formulae such as $CH_3COO^-Na^+$. Their general formula is RCOONa.

3 HALOGENATION

In this reaction, the OH group is replaced by a halogen atom, usually chlorine, so the functional group becomes COCl. These products are known as acyl chlorides. They are highly reactive compounds with many uses in organic synthesis, as we will see later on. The reagent is phosphorus(V) chloride (old name 'phosphorus pentachloride'). It must be used in anhydrous conditions because both the reagent and the acyl chloride product react with water. The reaction is vigorous (very active), so no heating is required. An equation for the reaction, using propanoic acid as the example, is:

$$CH_3CH_2COOH + PCl_5 \rightarrow CH_3CH_2COCl + POCl_3 + HCl$$

Acyl chlorides are named using the suffix *-oyl chloride*, so this product is propanoyl chloride. Their general formula is RCOCl.

The phosphorus-containing product is phosphorus trichloride oxide (also known as phosphorus oxychloride). It is a liquid that mixes with the acyl chloride, which has therefore to be separated by fractional distillation. The hydrogen chloride gas produced escapes, appearing as misty fumes.

DID YOU KNOW?

Aspirin is a very familiar and inexpensive medication, and it is useful in treating many medical conditions. It is an example of a carboxylic acid.

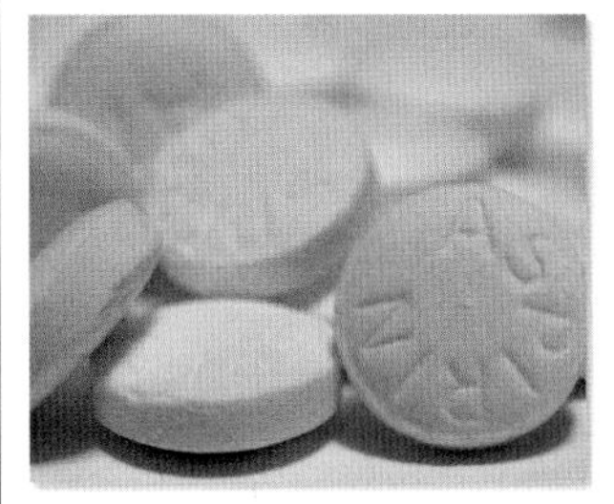

fig B Aspirin tablets

4 ESTERIFICATION

In this reaction, a carboxylic acid is mixed with an alcohol and a small amount of an acid catalyst, often concentrated sulfuric acid. Even with a catalyst, the reactions are slow and reversible.

Esters are used in industry for many purposes, including as solvents and in making polymers (polyesters). They also occur very commonly in nature – they are responsible for the characteristic smells of fruits, and most animal fats and vegetable oils are esters.

An equation for the reaction between methanoic acid and ethanol is:

$$HCOOH + CH_3CH_2OH \rightleftharpoons HCOOCH_2CH_3 + H_2O$$

The names of esters always contain two words – the first is the alkyl group from the alcohol and the second is the carboxylate from the carboxylic acid. So, this product is ethyl methanoate.

LEARNING TIP

With an ester, be careful to make sure that the name and formula match. CH_3COOCH_3 is methyl ethanoate but $HCOOCH_2CH_3$ is ethyl methanoate.

CHECKPOINT

SKILLS DECISION-MAKING

1. Write equations for the preparation of methylpropanoic acid:
   (a) by the oxidation of an alcohol
   (b) by the hydrolysis of a nitrile.
2. Write an equation for the conversion of methylpropanoic acid:
   (a) into an acyl chloride
   (b) into a methyl ester.

15D 1 CARBOXYLIC ACID DERIVATIVES: ACYL CHLORIDES

SPECIFICATION REFERENCE

15.13 PART | 15.14(i) | 15.14(ii) | 15.14(iii) | 15.14(iv)

LEARNING OBJECTIVES

- Understand the nomenclature of acyl chlorides and be able to draw their structural, displayed and skeletal formulae.
- Understand the reactions of acyl chlorides with:
  (i) water
  (ii) alcohols
  (iii) concentrated ammonia
  (iv) amines.

WHAT ARE ACYL CHLORIDES?

These used to be known as acid chlorides, and have the general formula RCOCl. They are derivatives of carboxylic acids, and can be imagined as the result of replacing the OH group in COOH with Cl. Although they contain a carbonyl group (like carboxylic acids, aldehydes and ketones), they are not classed as carbonyl compounds. As with carboxylic acids, having two functional groups sharing the same carbon atom means that the properties are different from both carbonyl compounds and halogenoalkanes.

The commonest example is ethanoyl chloride, CH_3COCl, which is still called by its old name acetyl chloride. You may come across others, such as:

propanoyl chloride CH_3CH_2COCl

and butanoyl chloride $CH_3CH_2CH_2COCl$

DISPLAYED AND SKELETAL FORMULAE

Table A shows the displayed and skeletal formulae of some acyl chlorides.

NAME	DISPLAYED FORMULA	SKELETAL FORMULA
ethanoyl chloride	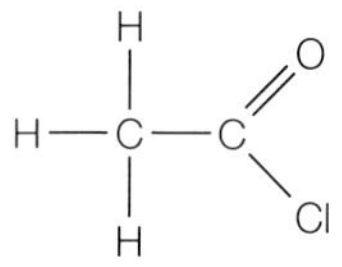	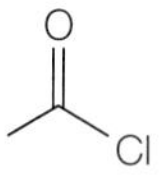
propanoyl chloride	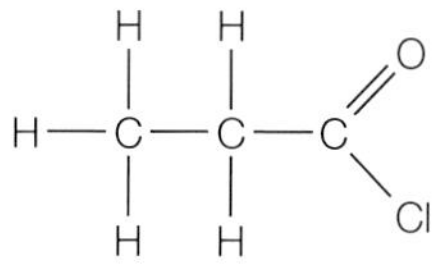	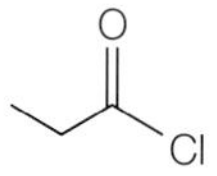
butanoyl chloride	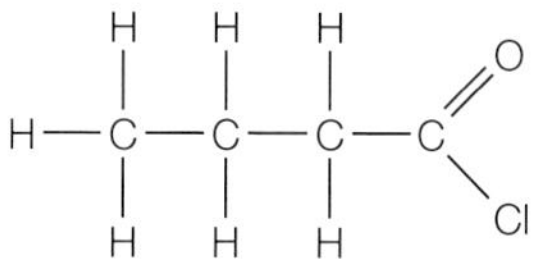	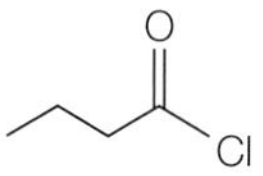

table A

REACTIONS OF ACYL CHLORIDES

The carbon atom in RCOCl is joined to two electronegative atoms, and so is electron-deficient ($\delta+$). You can therefore predict that it will be readily attacked by nucleophiles, such as molecules containing O or N atoms. You need to know about four reactions of acyl chlorides, although no mechanisms are needed for any of them.

REACTION WITH WATER

Acyl chlorides react vigorously (very actively) with cold water, forming a carboxylic acid and releasing hydrogen chloride gas, which appears as misty fumes (**fig A**). The equation for the reaction of ethanoyl chloride with water is as follows:

$$CH_3{-}C(=O){-}Cl + H{-}OH \longrightarrow CH_3{-}C(=O){-}OH + H{-}Cl$$

▲ **fig A** Adding ethanoyl chloride to water produces fumes of hydrogen chloride.

REACTION WITH ALCOHOLS

Acyl chlorides react readily with ethanol to form an ester and hydrogen chloride gas. The equation for the reaction of ethanoyl chloride with ethanol is as follows:

$$CH_3{-}C(=O){-}Cl + CH_3CH_2{-}OH \longrightarrow CH_3{-}C(=O){-}OCH_2CH_3 + H{-}Cl$$

Notice the similarity between these two reactions. In both cases, the H of the second reactant combines with Cl to form HCl, and the other part of the reactant becomes joined to the carbonyl group.

REACTION WITH CONCENTRATED AMMONIA SOLUTION

Similar to alcohols, acyl chlorides react readily with concentrated ammonia solution. You can now predict the reaction that occurs: the H of the second reactant combines with Cl to form HCl, and the other part of the reactant becomes joined to the carbonyl group. The NH_2 group joined to the carbonyl group produces a different functional group – amide (not to be confused with amine). The equation for the reaction of ethanoyl chloride with ammonia is as follows:

$$CH_3{-}C(=O){-}Cl + NH_3 \longrightarrow CH_3{-}C(=O){-}NH_2 + H{-}Cl$$

Unlike with water and alcohols, a further reaction occurs. The reactant is a base and the product is an acidic gas, so these react together to form ammonium chloride:

$NH_3 + HCl \rightarrow NH_4Cl$

You could write a single equation that combines both of these reactions:

$CH_3COCl + 2NH_3 \rightarrow CH_3CONH_2 + NH_4Cl$

You will learn more about amides later in this book.

REACTION WITH AMINES

This reaction is similar to the previous one. A primary amine can be shown as RNH_2, so you can again predict the reaction that occurs: the H of the second reactant combines with Cl to form HCl, and the other part of the reactant (RNH) becomes joined to the carbonyl group. The equation for the reaction of ethanoyl chloride with the simplest primary amine, methylamine, is as follows:

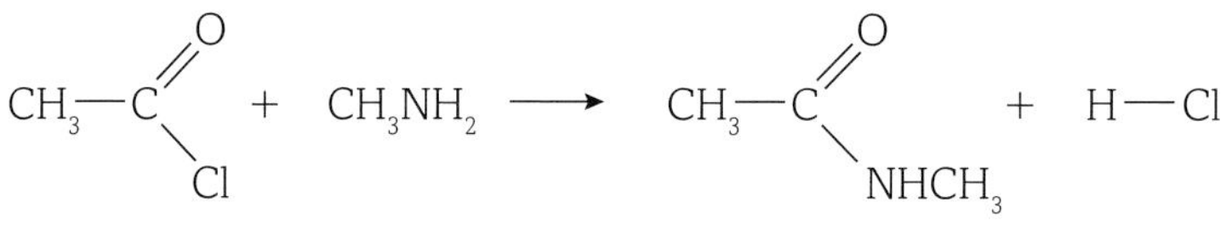

$$CH_3-C(=O)Cl + CH_3NH_2 \longrightarrow CH_3-C(=O)NHCH_3 + H-Cl$$

The product of this reaction is a substituted amide, often called an N-substituted amide. The capital N emphasises that there is an alkyl group attached to N, rather than the usual situation in which it is attached to another C atom. You can see from the product name (N-methylethanamide) that N- is used as a locant, in the same way as in names such as 2-methylbutane – it shows where the methyl group is attached.

Secondary amines (represented as R_2NH) react in the same way, although the product will contain two substituted alkyl groups. A sample equation is:

$CH_3COCl + (CH_3)_2NH \rightarrow CH_3CON(CH_3)_2 + HCl$

The organic product is N,N-dimethylethanamide.

You might be able to predict why this type of reaction does not occur with a tertiary amine (R_3N). With three alkyl groups, there is now no H atom to react with Cl to form hydrogen chloride.

LEARNING TIP

Be sure not to confuse amines (RNH_2) with amides ($RCONH_2$).

CHECKPOINT

SKILLS REASONING

1. Write the names of the organic product formed in the four similar reactions of propanoyl chloride with water, methanol, ammonia and methylamine.
2. Write an equation for butanoyl chloride reacting with:

   (a) propan-1-ol

   (b) ethylamine.

15D 2 CARBOXYLIC ACID DERIVATIVES: ESTERS

SPECIFICATION REFERENCE 15.13 PART 15.15

LEARNING OBJECTIVES

- Understand the nomenclature of esters and be able to draw their structural, displayed and skeletal formulae.
- Understand the hydrolysis reactions of esters, in acidic and alkaline solution.

INTRODUCTION

In previous sections, you have come across esters as the products of reactions of other organic compounds. We will now take a closer look at esters, including their reactions.

NAMING ESTERS

This is straightforward, but you need to remember these points:

- As for carboxylic acids and acyl chlorides (but not amides), an ester name contains two words.
- The first word comes from the alkyl group joined to O.
- The second word comes from the alkyl group joined to C.

Table A shows some examples.

STRUCTURAL FORMULA	NAME
$HCOOCH_3$	methyl methanoate
CH_3COOCH_3	methyl ethanoate
$HCOOCH_2CH_3$	ethyl methanoate
$CH_3COOCH_2CH_3$	ethyl ethanoate
$CH_3COOCH_2CH_2CH_3$	propyl ethanoate

table A

DISPLAYED AND SKELETAL FORMULAE

Table B shows the displayed and skeletal formulae of some esters.

NAME	DISPLAYED FORMULA	SKELETAL FORMULA
methyl methanoate	H—C(=O)—O—CH₃ drawn as: H—C—O—C—H	O
methyl ethanoate	H—C—C—O—C—H	O
ethyl methanoate	H—C—O—C—C—H	O
ethyl ethanoate	H—C—C—O—C—C—H	O
propyl ethanoate	H—C—C—O—C—C—C—H	O

table B

PHYSICAL PROPERTIES

Esters are colourless liquids with relatively low melting and boiling temperatures, and are insoluble in water. All of the hydrogen atoms in their molecules are attached to carbon atoms, so hydrogen bonding is not possible.

The generally pleasant smells of esters are largely responsible for the familiar odours of flowers and fruits. Examples include:

pentyl ethanoate	pears
3-methylbutyl ethanoate	bananas
methyl butanoate	apples

Esters are present in perfumes, food flavourings, solvents, anaesthetics and biofuels.

> **DID YOU KNOW?**
>
> **Benzyl ethanoate is a component of food flavourings that give the taste of apples and pears.**
>
> 
>
> ▲ **fig A** The familiar odours of apples and pears come from the generally pleasant smells of esters.

HYDROLYSIS OF ESTERS

The only reactions of esters that you need to know about are two slightly different types of **hydrolysis**. The term 'hydrolysis' means breaking down (*lysis*) using water (*hydro*). It can be considered as the reverse of esterification. You may remember equations such as this from a previous section:

$$CH_3COOH + CH_3CH_2OH \rightleftharpoons CH_3COOCH_2CH_3 + H_2O$$

This shows the formation of ethyl ethanoate from the corresponding carboxylic acid and alcohol. The reversible arrow indicates that a mixture of ethyl ethanoate and water reacts together to form ethanoic acid and ethanol. This reverse reaction could be described as the hydrolysis of ethyl ethanoate.

HYDROLYSIS IN ACIDIC SOLUTION

We partly covered this type of hydrolysis in the previous paragraph. For a given ester, warming it with water will cause hydrolysis. As with esterification, the reaction is slow. A catalyst such as sulfuric acid will speed up hydrolysis, but will not affect the position of equilibrium, so the reaction will not go to completion:

$$CH_3COOCH_2CH_3 + H_2O \rightleftharpoons CH_3COOH + CH_3CH_2OH$$

HYDROLYSIS IN ALKALINE SOLUTION

The disadvantage of using an alkali such as aqueous sodium hydroxide instead of sulfuric acid is that the reaction produces a carboxylate salt instead of a carboxylic acid. However, the advantage easily outweighs this disadvantage – the reaction goes to completion instead of reaching an equilibrium. The equation for methyl propanoate is:

$$CH_3CH_2COOCH_3 + NaOH \rightarrow CH_3CH_2COO^- + Na^+ + CH_3OH$$

Converting the carboxylate salt into a carboxylic acid only needs a dilute acid to be added:

$$CH_3CH_2COO^- + H^+ \rightarrow CH_3CH_2COOH$$

The final organic products are methanol and propanoic acid.

SAPONIFICATION

The Latin word for soap is *sapo*, so saponification means 'soap-making'. This is a particular example of the alkaline hydrolysis of esters. Soaps and detergents have similar uses, but the main difference is that detergents use organic compounds obtained from crude oil, while soaps use organic compounds obtained from vegetable (and, originally, animal) sources.

Vegetable oils contain large quantities of triglycerides. These are triesters, each of which consists of a large ester molecule that can be hydrolysed to one alcohol and three carboxylic acid molecules. The hydrolysis of a typical triglyceride is shown in **fig B**.

$$\begin{matrix} CH_2-O-C(=O)-C_{17}H_{35} \\ | \\ CH-O-C(=O)-C_{17}H_{35} \\ | \\ CH_2-O-C(=O)-C_{17}H_{35} \end{matrix} + 3NaOH \longrightarrow \begin{matrix} CH_2OH \\ | \\ CHOH \\ | \\ CH_2OH \end{matrix} + 3C_{17}H_{35}C(=O)O^-Na^+$$

a triglyceride: an ester of long-chain carboxylic acids and a triol

propane-1,2,3-triol (glycerol)

sodium octadecanoate (sodium stearate)

▲ **fig B** The hydrolysis of a typical triglyceride is an example of saponification.

In this example, hydrolysis produces an alcohol with three hydroxyl groups (glycerol). Glycerol has many uses, including as a sweetener in foods and toothpastes, and in skin care products. It is also used as a component of the liquid in electronic cigarettes.

The other product of this reaction, commonly known as sodium stearate, is a very common ingredient of most soaps.

LEARNING TIP

Acidic hydrolysis of an ester is a reversible reaction, but alkaline hydrolysis reactions go to completion.

CHECKPOINT

1. What are the names of these esters?
   (a) $CH_3CH_2COOCH_2CH_2CH_3$
   (b) $(CH_3)_2CHCOOCH_3$
2. Write equations for the hydrolysis of propyl butanoate in:
   (a) acidic conditions
   (b) alkaline conditions.

SUBJECT VOCABULARY

hydrolysis the breaking of a compound by water into two compounds

15D 3 CARBOXYLIC ACID DERIVATIVES: POLYESTERS

SPECIFICATION REFERENCE 15.16

LEARNING OBJECTIVES

- Understand how polyesters, such as terylene, are formed by condensation polymerisation reactions.

ADDITION POLYMERISATION

Examples of addition polymerisation were covered in **Topic 5 (Book 1: IAS)**, so may be useful for reminding you of this type of polymerisation. The monomer propene can combine with many thousands of other propene molecules to form a very long chain. This process can be represented as:

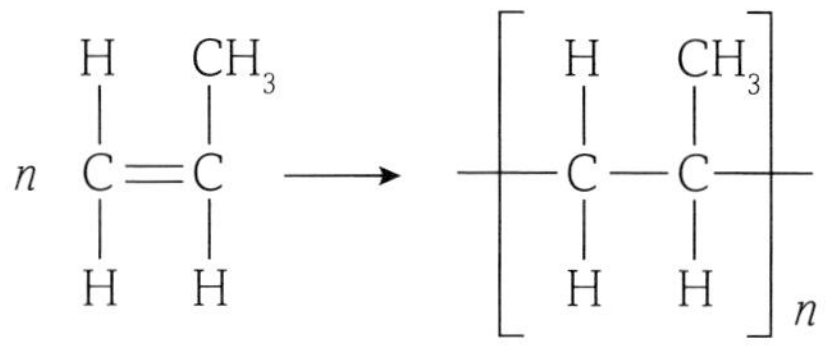

$$n\,\mathrm{CH_2{=}CH(CH_3)} \longrightarrow \left[\!-\mathrm{CH_2{-}CH(CH_3)}-\!\right]_n$$

▲ **fig A** This equation represents the polymerisation of propene.

One thing to notice is that all of the atoms in the monomer molecules end up in the structure of the polymer.

CONDENSATION POLYMERISATION

There are two main differences between addition polymerisation and **condensation polymerisation**. In condensation polymerisation:

- each time two monomer molecules join together, another small molecule is formed
- usually two different monomers react together.

In many examples of condensation polymerisation, the small molecule formed is water, which is the origin of the term 'condensation'. In other examples, the small molecule formed is hydrogen chloride, but the term 'condensation' is still used to describe those examples.

One type of polymer formed in this way is a polyamide, which we will meet in **Topic 19**.

POLYESTERS

The example we will use to introduce condensation polymerisation is the formation of polyesters. You will remember that esters are the products of the reaction between alcohols and carboxylic acids. However, once one molecule of an alcohol has reacted with one molecule of a carboxylic acid, the reaction is complete. The only organic product is a slightly larger ester molecule.

For a reaction to produce a polymer, we need two monomers, each with two reactive groups (one at each end). The alcohol usually involved in this type of reaction, more accurately called a 'diol', has two hydroxyl groups, one at each end:

$$\mathrm{H{-}O{-}CH_2{-}CH_2{-}O{-}H}$$

One of the most common carboxylic acids used (more accurately a dicarboxylic acid) has the structure below. The common name for this compound is terephthalic acid.

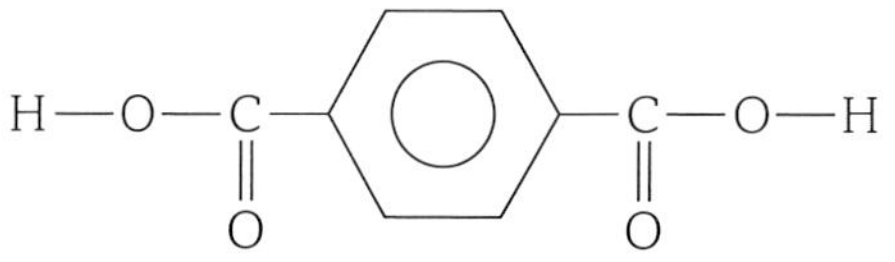

$$\mathrm{H{-}O{-}C({=}O){-}C_6H_4{-}C({=}O){-}O{-}H}$$

Consider how these molecules react together. The OH of the carboxylic acid and the H of the alcohol react together to form the small molecule, water, and the two molecules are linked together by an ester group. This equation shows what happens during the formation of a typical polyester.

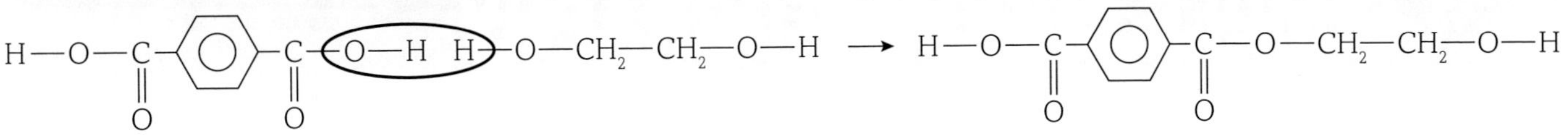

$$H{-}O{-}\underset{\Vert}{\overset{}{C}}(=O){-}C_6H_4{-}C(=O){-}O{-}H \;+\; H{-}O{-}CH_2{-}CH_2{-}O{-}H \longrightarrow H{-}O{-}C(=O){-}C_6H_4{-}C(=O){-}O{-}CH_2{-}CH_2{-}O{-}H$$

This larger molecule that has formed still has reactive groups at both ends. This means it can continue reacting with other molecules in the same way as above, until a very long polymer chain has formed. The structure of the polymer is:

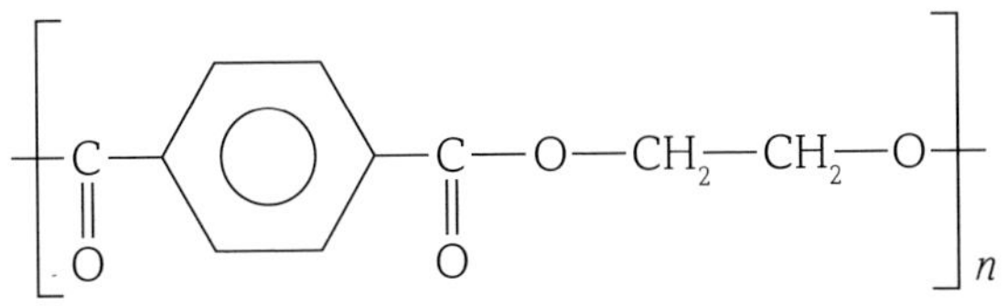

$$\left[{-}C(=O){-}C_6H_4{-}C(=O){-}O{-}CH_2{-}CH_2{-}O{-} \right]_n$$

Because of the original name of the acid (terephthalic acid), the polymer was originally known as Terylene.

Polyesters have a very wide range of uses, including soft drink bottles, food packaging, many types of clothing and duvet fillings.

It is also possible to use a dioyl chloride instead of a dicarboxylic acid, although this is not used as a monomer in industry. The reaction occurs in the same way – the only difference is that the small molecule formed is hydrogen chloride, not water.

EXAM HINT

You may like to think about why dicarboxylic acids rather than dioyl chlorides are used in industry. Hint: look back at **Topic 4 (Book 1: IAS)** and think about hazards and risks.

LEARNING TIP

When showing how a polyester forms, remember that, in the formation of the small molecule, the H comes from the diol and the OH from the dicarboxylic acid (or the Cl from the dioyl chloride).

CHECKPOINT

1. Why is it not possible to make a polymer by reacting together molecules of $HOOCCH_2COOH$ and CH_3OH?
2. Draw the repeat unit of the polymer formed between molecules of HOOCCOOH and $CH_3CH(OH)CH(OH)CH_3$.

SUBJECT VOCABULARY

condensation polymerisation the formation of a polymer, usually by the reaction of two different monomers, when another small molecule is also formed

15E 1 SIMPLE CHROMATOGRAPHY

SPECIFICATION REFERENCE 15.21 15.22

LEARNING OBJECTIVES

- Know that chromatography separates components of a mixture using a mobile phase and a stationary phase.
- Be able to calculate R_f values from one-way chromatograms in paper and thin-layer chromatography (TLC) and understand the reasons for differences in R_f values.

PAPER CHROMATOGRAPHY

You may know about paper chromatography from your early experiences of chemistry. This is a suitable place to remind you of this technique and to explain some of the principles that relate to most types of chromatography. The word 'chromatography' comes from the Greek for colour (*chroma*) and writing (*graphein*) because mixtures of coloured substances were used in early experiments. However, the technique can also be used with colourless substances because these can be detected in other ways without considering their colours. Chromatography is used:

- to separate a mixture into its individual components
- to identify the components of a mixture by considering how far they have travelled up the paper.

If you have synthesised an organic compound, chromatography is a good way to check whether it is pure and, if not, what impurities are present.

APPARATUS

All forms of paper chromatography need the following:

- A container (usually glass) with a lid. The lid is there to prevent evaporation of the solvent. The container could be a beaker or a rectangular tank.
- Paper. In introductory experiments you may have used filter paper or kitchen towel, but specially formulated chromatography paper gives better results. There also needs to be a method of supporting the paper in the container.
- A solvent. In some cases, water works well, but often the solvent is a mixture of organic compounds, and is chosen to fit the characteristics of the components of the mixture.

In paper chromatography, the mixture is spotted onto the paper a short distance from one edge. Often, spots of known substances are spotted separately at the same level as the mixture. A small volume of the chosen solvent is added to the container, and then the paper is inserted and suitably supported. The lid is replaced and the apparatus left until the solvent front has reached almost to the top of the paper. The paper is then removed, the position of the solvent front marked, and the paper left to dry. The resulting dried paper is known as a chromatogram. **Fig A** shows the typical apparatus used to set up a chromatography experiment and the resulting chromatogram.

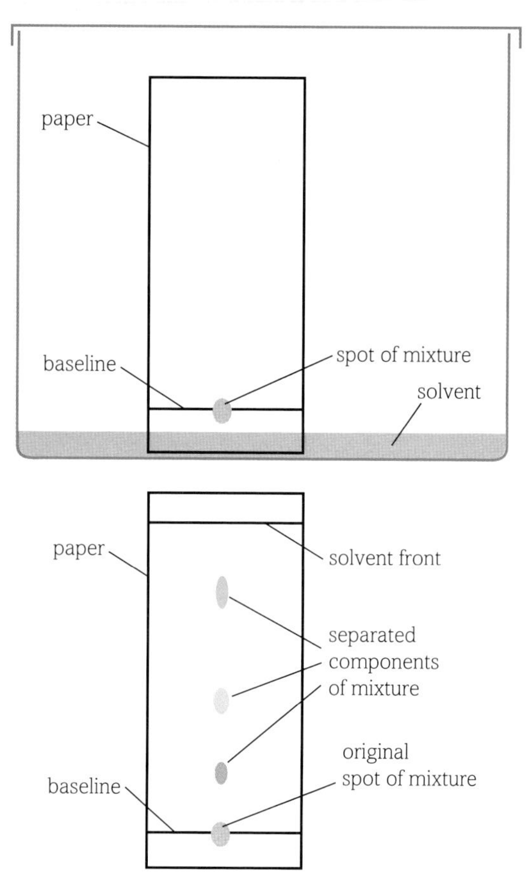

▲ **fig A** The chromatogram here is shown in a rather idealised way – in reality the spots are often elongated and their centres are difficult to identify.

HOW DOES CHROMATOGRAPHY WORK?

All types of chromatography depend on the use of a **stationary phase** and a **mobile phase**. Each component in the mixture is attracted to both phases, but more strongly to one than the other.

- A component that is strongly attracted to the stationary phase but weakly attracted to the mobile phase will not travel very far up the paper.
- A component that is weakly attracted to the stationary phase but strongly attracted to the mobile phase will travel a long way up the paper.

In paper chromatography, the stationary phase is the water trapped in the fibres of the chromatography paper, and the mobile phase is the solvent.

We have used colours to illustrate the different substances in the mixture, but most organic compounds are colourless. Chromatography can still be used, but the components on the chromatogram must be visualised using ultraviolet radiation or by spraying with a chemical reagent that will react with the component to form a coloured product.

THIN-LAYER CHROMATOGRAPHY

The apparatus and method used in thin-layer chromatography (TLC) are very similar to those of paper chromatography. The only difference is that instead of paper, a sheet of glass or plastic coated in a thin layer of a solid such as silica or alumina is used (**fig B**).

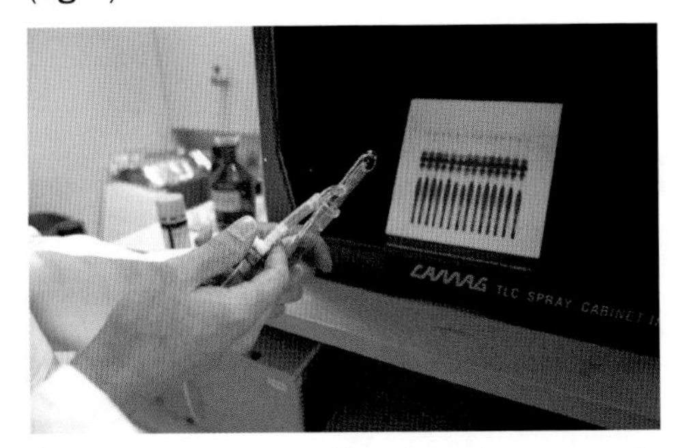

▲ **fig B** This is a TLC plate after a chromatography experiment.

CALCULATING R_f VALUES

The chromatogram is analysed and the distance travelled by the solvent (from the baseline to the solvent front) is measured. For each component in the mixture, the distance it has travelled (starting from the baseline) is measured.

The R_f value is calculated using the expression:

$$R_f = \frac{\text{distance travelled by component}}{\text{distance travelled by solvent}}$$

R_f values have no units because they are a ratio of two distances. The letters in the term refer to 'retardation' (sometimes 'retention') and 'factor'. Each component has a characteristic R_f value. This means that, in theory, different components can be identified by reference to a table of known R_f values. Unfortunately, R_f values depend on the solvent used and other factors.

In practice, paper and thin-layer chromatography are usually carried out with known substances included in the same experiment so that a comparison can be made.

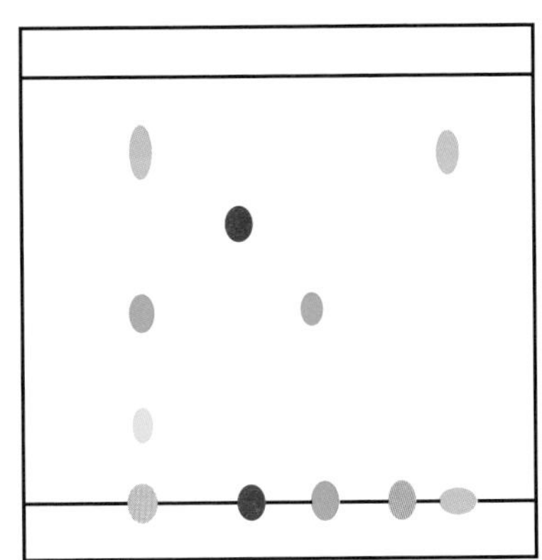

▲ **fig C** Chromatogram.

The chromatogram in **fig C** suggests that the mixture on the left contains three different substances. Two of these three are probably the green and blue substances on the right. The third substance (yellow) cannot be identified by this experiment.

LEARNING TIP

Remember, when calculating R_f values, the distance moved by the solvent is measured from the baseline to the solvent front.

COLUMN CHROMATOGRAPHY

Column chromatography uses the same principles as thin-layer chromatography. The stationary phase is alumina or silica packed into a tube (a burette will be suitable) and soaked in a solvent. The mixture is placed on top of the stationary phase and more solvent (the mobile phase) added on top. When the tap is opened, the solvent drips through the tip and the components of the mixture begin to move down the tube and separate.

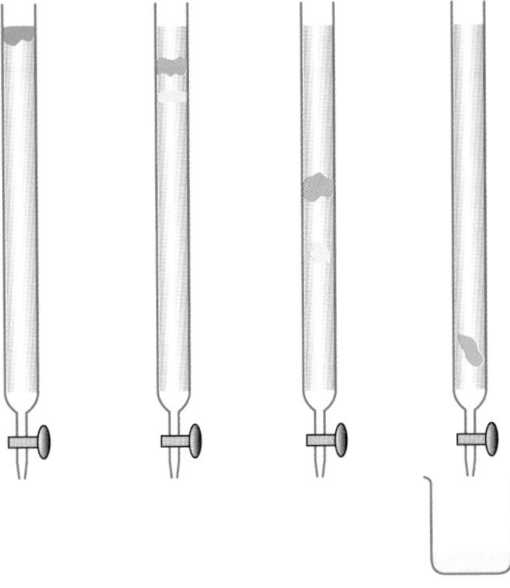

▲ **fig D** Separation of a mixture by column chromatography.

More solvent is added at the top, and eventually one component (yellow in this example) leaves the column and can be collected in a container (**fig D**). If the experiment is continued, the blue component can be collected in the same way.

The advantage of column chromatography is that much larger quantities of material can be separated than with paper chromatography.

CHECKPOINT

1. Suggest why, in paper chromatography, the non-polar substance hexane has a high R_f value.
2. Calculate the R_f value of component X in this diagram.

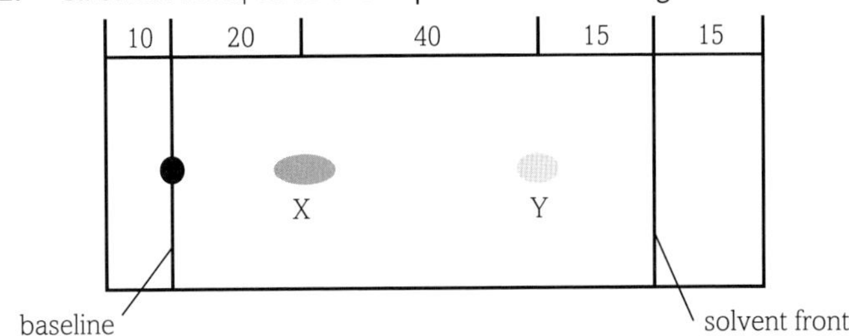

SUBJECT VOCABULARY

stationary phase in chromatography the liquid or solid that does not move
mobile phase in chromatography the liquid that moves through the stationary phase and transports the components

15E 2 DETERMINING STRUCTURES USING MASS SPECTRA

SPECIFICATION REFERENCE 15.17(i) 15.17(ii)

LEARNING OBJECTIVES

- Be able to use data from mass spectra to:
  (i) suggest possible structures of a simple organic compound, given accurate relative molecular masses
  (ii) calculate the accurate relative molecular mass of a compound, given accurate relative molecular masses to four decimal places.

MASS SPECTROMETRY SO FAR

In **Topic 10 (Book 1: IAS)** you learned about two aspects of mass spectrometry. They were:

- determining the relative atomic masses of elements from isotopic abundances
- using fragmentation patterns to determine the structures of organic compounds.

In this section we will learn about another use – determining the molecular formula (although not always the structure or identity) of an organic compound from a precise relative molecular mass obtained from **high resolution mass spectrometry** (HRMS).

HIGH RESOLUTION MASS SPECTROMETRY

Without going into technical detail, there are mass spectrometers that can provide a value for a relative molecular mass to four or more decimal places. Values with this degree of precision sometimes enable a compound to be positively identified from the relative molecular mass alone, without the need for any other information. Consider compounds with $M_r = 58$ (or 58.0). There are several possible structures, including those in **table A**:

FORMULA	STRUCTURE	ACCURATE M_r
C_4H_{10}	$CH_3CH_2CH_2CH_3$ or $(CH_3)_3CH$	58.0780
C_3H_6O	CH_3COCH_3 or CH_3CH_2CHO	58.0417
$C_2H_6N_2$	$H_2NCH{=}CHNH_2$	58.0530

table A

LEARNING TIP

HRMS gives information only about molecular formulae, not about structure. Structures can be deduced from the fragmentation patterns in mass spectrometry.

Once an accurate M_r has been obtained from HRMS, it is possible to decide whether the compound has the molecular formula C_4H_{10} or C_3H_6O or $C_2H_6N_2$. Note that if the accurate M_r is found to be 58.0417, it is still not possible to decide whether the compound is propanone or propanal because they both have the same accurate M_r value.

DID YOU KNOW?

Uses of HRMS include in forensics and environmental analysis.

fig A HRMS equipment looks very similar to other mass spectrometry equipment.

CALCULATING THE ACCURATE RELATIVE MOLECULAR MASS

We can also obtain accurate M_r values by doing some simple calculations. You know that relative atomic masses are relative to the average mass of a carbon atom (^{12}C) taken as exactly 12, which for our purposes means 12.0000. **Table B** shows the accurate atomic masses of some common elements:

ELEMENT	SYMBOL	ACCURATE ATOMIC MASS
hydrogen	H	1.0078
carbon	C	12.0000
nitrogen	N	14.0031
oxygen	O	15.9949

table B

You will need to use information from **table B** in the following examples.

WORKED EXAMPLE 1

A compound is found to have an M_r value of 84.0573. Which of these structures does the compound have?

$CH_2{=}CHCH_2CH_2CH_2CH_3$ or $CH_3CH{=}CHCH_2CHO$

Answer

The molecular formula of $CH_2{=}CHCH_2CH_2CH_2CH_3$ is C_6H_{12} and its relative molecular mass is calculated as follows:

$(6 \times 12.0000) + (12 \times 1.0078) = 84.0936$

The molecular formula of $CH_3CH{=}CHCH_2CHO$ is C_5H_8O and its relative molecular mass is calculated as follows:

$(5 \times 12.0000) + (8 \times 1.0078) + 15.9949 = 84.0573$

This means that the compound is $CH_3CH{=}CHCH_2CHO$.

WORKED EXAMPLE 2

Three compounds have these structures:

A $CH_3CH_2CH_2CH{=}CH_2$
B $CH_3CH(NH_2)CN$
C $CH_2{=}CHCOCH_3$

In a high resolution mass spectrometer, compound Y is found to have an M_r value of 70.0423.

What is the identity of Y?

Answer

The molecular formula of A is C_5H_{10} and its M_r value is:

$(5 \times 12.0000) + (10 \times 1.0078) = 70.0780$

The molecular formula of B is $C_3H_6N_2$ and its M_r value is:

$(3 \times 12.0000) + (6 \times 1.0078) + (2 \times 14.0031) = 70.0530$

The molecular formula of C is C_4H_6O and its M_r value is:

$(4 \times 12.0000) + (6 \times 1.0078) + 15.9949 = 70.0417$

Although none of the calculated M_r values exactly matches the accurate M_r value obtained from the high resolution mass spectrometer, the value is much closer to C than to A or B, so compound Y is C.

WORKED EXAMPLE 3

The high resolution mass spectrum shown in **fig B** has two major peaks.

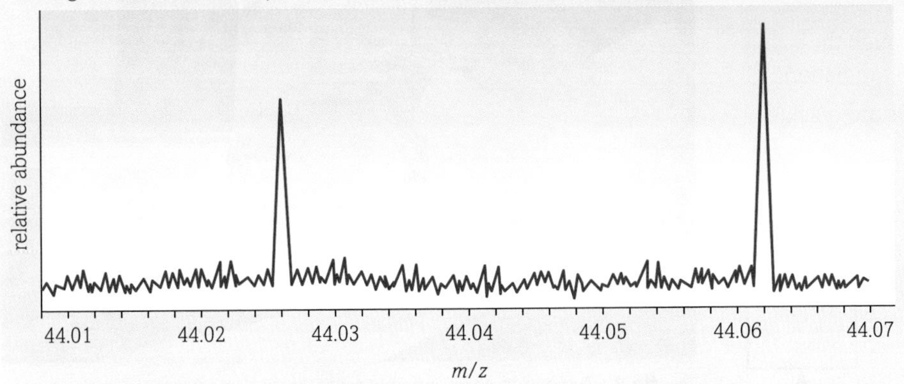

▲ **fig B** A high resolution mass spectrum.

Suggest which of these compounds are responsible for the mass spectrum:

P C_3H_8
Q CH_3CHO
R $O{=}C{=}O$

Answer

M_r of P = $(3 \times 12.0000) + (8 \times 1.0078) = 44.0624$

M_r of Q = $(2 \times 12.0000) + (4 \times 1.0078) + 15.9949 = 44.0261$

M_r of R = $12.0000 + (2 \times 15.9949) = 43.9898$

The two peaks correspond to P and Q, but not to R, so propane and ethanal are present, but not carbon dioxide.

SUBJECT VOCABULARY

high resolution mass spectrometry (HRMS) a type of mass spectrometry that can produce M_r values to several decimal places, usually four or more

CHECKPOINT

SKILLS ANALYSIS

1. Two compounds have the structures $CH_3CH_2CH_2CH_2NH_2$ and $HN{=}CHCOOH$. In a high resolution mass spectrum, one of these compounds has M_r = 73.0812. Explain how you can decide which compound it is. You do not need to show any working.
2. A compound is thought to be either 1,2-diaminoethane or ethanoic acid. Its accurate M_r value is 60.0213. Which compound is it? Show your working.

15E 3 CHROMATOGRAPHY: HPLC AND GC

SPECIFICATION REFERENCE 15.23 PART

LEARNING OBJECTIVES

- Know that high performance liquid chromatography, HPLC, and gas chromatography, GC, are types of column chromatography that separate substances because of different retention times in the column.

HIGH PERFORMANCE LIQUID CHROMATOGRAPHY

High performance liquid chromatography (HPLC) is a refinement of column chromatography, described in **Section 15E.1**. The main differences are:

- The solvent is forced through a metal tube under high pressure, rather than being allowed to pass through by gravity.
- The particle size of the stationary phase is much smaller, which leads to better separation of the components.
- The sample is injected into the column.
- The components are detected after passing through the column, usually by their absorption of ultraviolet radiation.
- The whole process is automated, and the results are quickly available on a computer display.

Fig A shows a typical setup.

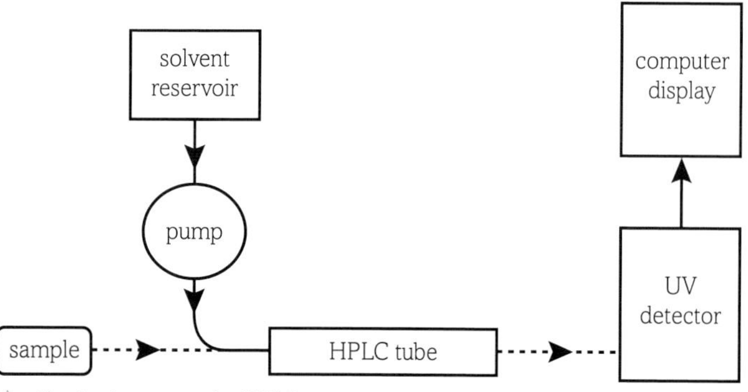

▲ **fig A** Apparatus for HPLC.

As each component reaches the detector, a signal is displayed on the computer screen. Just as retention factor (R_f) values can be obtained from paper and thin-layer chromatography, in HPLC values of **retention time** are obtained. The retention time of a component is the time taken from injection to detection, and these values can be important in identifying the components. However, retention times depend on several variables, including:

- the nature of the solvent
- the pressure used
- the temperature inside the column.

GAS CHROMATOGRAPHY

Gas chromatography (GC) is another refinement of column chromatography. The main differences are:

- The metal tube can be several metres long and is coiled to save space.
- The stationary phase is a solid or liquid coated on the inside of the tube.
- The mobile phase is an inert carrier gas (often nitrogen or helium).
- The sample is injected into the column, as in HPLC.
- The components passing through the column are detected.
- The whole process is automated, and the results are quickly available on a computer display (**fig B**).

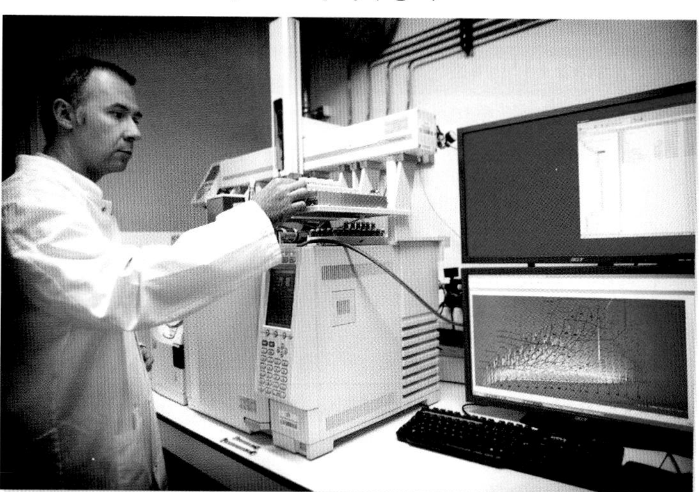

▲ **fig B** The complete GC equipment takes up very little room.

Fig C shows a typical setup.

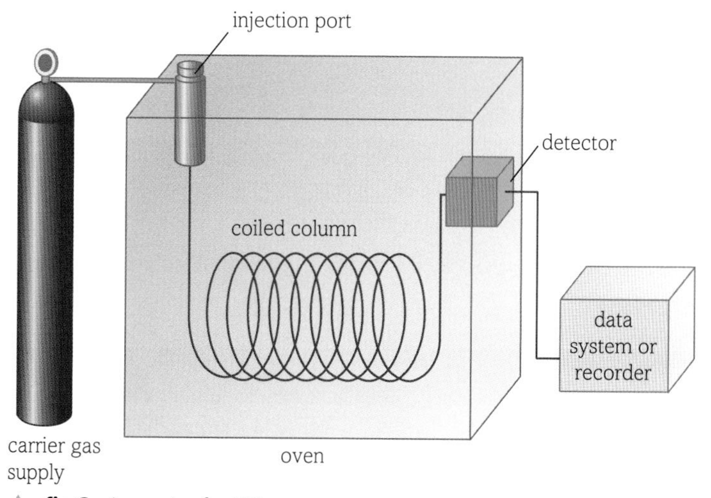

▲ **fig C** Apparatus for GC.

After the sample is injected, the components vaporise and move through the coiled tube with the carrier gas. They move at different speeds, depending on how strongly they are attracted to the stationary phase. Those with weaker attractions move more quickly and have shorter retention times. **Fig D** shows the separation occurring in the coiled tube.

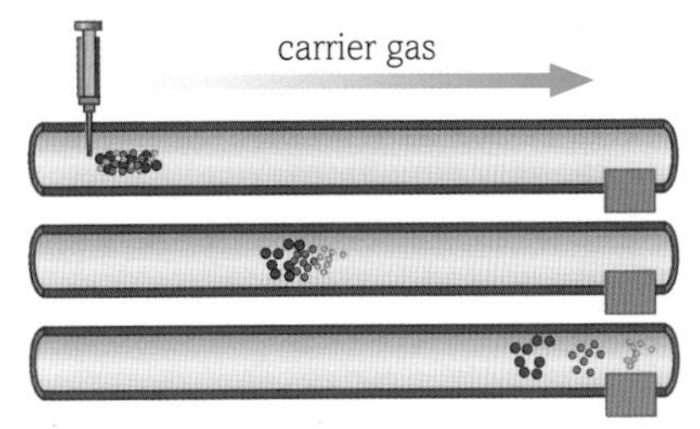

▲ **fig D** A gas chromatograph separating three components.

The relative concentrations of the different components are often displayed on a graph like that in **fig E**.

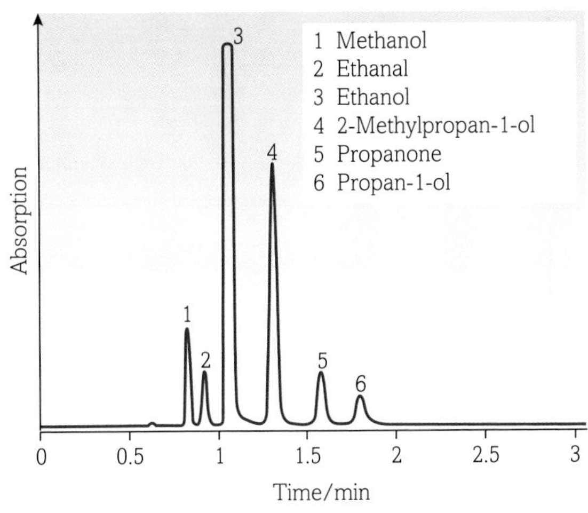

▲ **fig E** The areas under the peaks represent the relative concentrations of the components.

LEARNING TIP

Traditional column chromatography (using a burette, for example) has one advantage over HPLC and GC - reasonable amounts of the components can be collected. In GC the components are often detected in a flame, so they are destroyed during the detection process at the end.

CHECKPOINT

1. How does the movement of the mobile phase in HPLC differ from that in traditional column chromatography?
2. Suggest reasons why retention times for a given substance may be different when obtained from different gas chromatograms.

SKILLS CRITICAL THINKING

SUBJECT VOCABULARY

retention time (of a component) in HPLC and GC, the time taken from injection to detection

15E 4 CHROMATOGRAPHY AND MASS SPECTROMETRY

SPECIFICATION REFERENCE 15.23 PART

LEARNING OBJECTIVES

- Know that high performance liquid chromatography, HPLC, and gas chromatography, GC, are types of column chromatography that separate substances because of different retention times in the column and may be used in conjunction with mass spectrometry, in applications such as forensics or drug testing in sport.

LIMITATIONS OF HPLC AND GC

The two types of chromatography described in **Section 15E.3** are very useful for separating small quantities of components in a mixture. However, they are not very good at positively identifying them. This is partly because it is difficult to control all of the variables (such as solvent, pressure and temperature) and partly because different substances may have the same retention times. Components are often identified by reference to a database of the retention times of known substances. However, if a component has a retention time for which there is no reference, then these techniques provide no useful information.

Here are two areas in which the results of these chromatography methods have to be exactly correct:

- in providing forensic evidence (for use in a court of law)
- in detecting banned drugs in sportsmen and sportswomen and racehorses.

They are used in other areas, including analysis of pollutants in the environment, detecting explosives in airport baggage and in space probes on other planets.

Because HPLC and GC do not give results that are beyond doubt, they are often combined with a different technique that you learned about in **Topics 2** and **10 (Book 1: IAS)**. This is mass spectrometry (MS), a technique that cannot separate a mixture but which can provide information about the structures of compounds. The combinations of these techniques are often abbreviated to HPLC-MS and GC-MS.

▲ **fig A** GC-MS is widely used for forensic purposes.

GC-MS

As the two combined techniques are very similar, we will focus only on GC-MS. **Fig B** shows the stages in a typical setup.

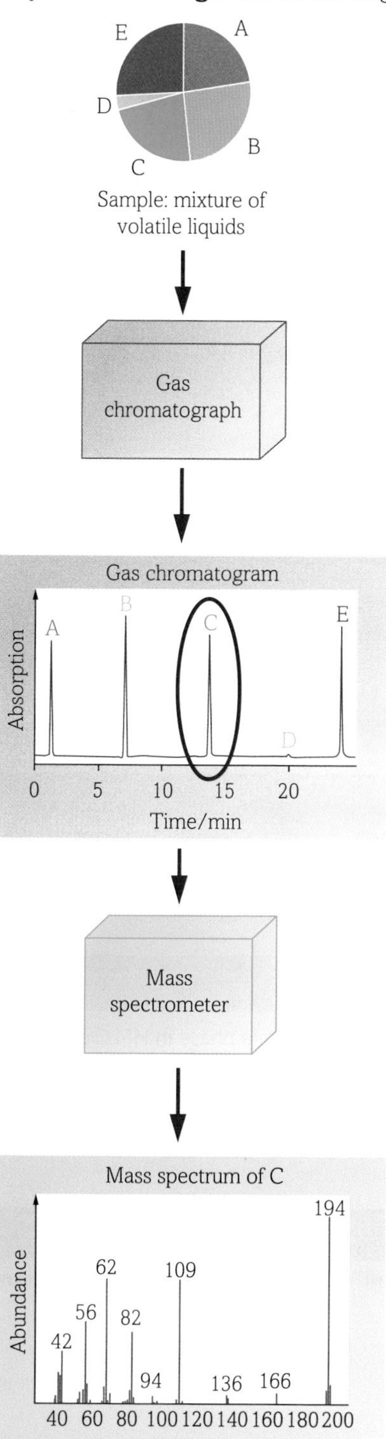

▲ **fig B** Stages in GC-MS.

Imagine a sample containing a mixture of five different substances, A–E.

Stage 1: The mixture is injected into a gas chromatograph.

Stage 2: Each component has a different retention time, so emerges at a different time.

Stage 3: One at a time, each component enters the mass spectrometer.

Stage 4: Component C has its mass spectrum displayed.

Stage 5: The m/z values and relative abundances of components are compared with a database of known substances.

Stage 6: When a match is found, component C has been positively identified.

Very small traces of substances, for example in a human hair or a paint fleck, can be accurately identified by this method.

PROBLEMS WITH DRUG TESTING

Although the use of GC-MS for detecting small amounts of banned drugs is usually accepted as producing reliable results, sometimes a problem arises. One example involves the anabolic steroid called nandrolone. Anabolic steroids have legitimate uses in medicine, especially in the treatment of anaemia, osteoporosis and some forms of cancer. Its effects include muscle growth and increased red blood cell production and bone density.

Mainly because of their effects in promoting muscle growth, sporting authorities have prohibited the use of anabolic steroids in sportsmen and sportswomen, including athletes. When someone takes nandrolone, it is converted in the human body into a similar compound, 19-norandrosterone, which is then excreted in urine. Participants may be routinely tested for the presence of 19-norandrosterone in their urine, and the International Olympic Committee has set a limit of 2 nanograms (ng) per cm^3 of urine, which is only 0.000 000 002 g per cm^3. A participant with a concentration higher than this has 'tested positive'. The skeletal formulae of these two related compounds are:

nandrolone

OH

O

19-norandrosterone

O

HO

▲ **fig C** Skeletal formulae of nandrolone and 19-norandrosterone.

In the 1998 Winter Olympic Games in Nagano, over 600 participants were tested for nandrolone, and none tested positive. In the years that followed, some very high-profile athletes in a range of sports have been banned from competition because of testing positive for nandrolone. However, many scientists believe that the number of positive tests is higher than expected. The suspicion is that nandrolone may be an ingredient in some nutritional supplements that participants are allowed to take. Another theory is that a combination of permitted dietary supplements and exercise may increase the concentration of nandrolone in the human body.

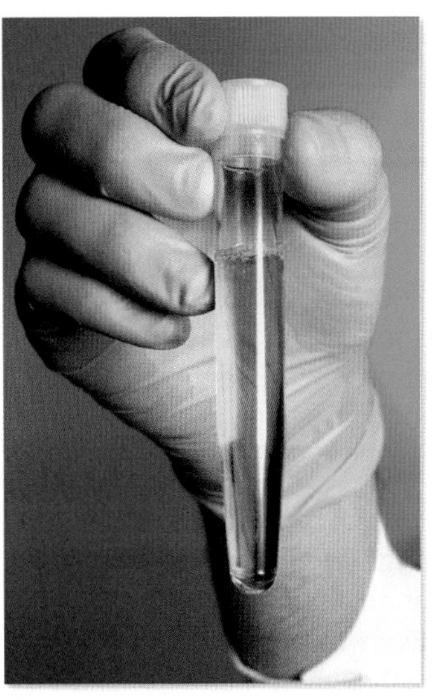

▲ **fig D** A urine sample from an athlete about to be tested for illegal drugs.

LEARNING TIP

In quoting very small concentrations of drugs, remember that a microgram (μg) is one millionth of a gram, 1×10^{-6} g.

CHECKPOINT

1. What is the main function of each technique in HPLC-MS?

SKILLS ETHICS

2. Which of the two compounds nandrolone and 19-norandrosterone would you expect to react with:

(a) bromine water

(b) 2,4-dinitrophenylhydrazine?

15E 5 PRINCIPLES OF NMR SPECTROSCOPY

SPECIFICATION REFERENCE 15.18 PART

LEARNING OBJECTIVES

- Understand the key aspects of NMR spectroscopy.
- Understand how ^{13}C and ^{1}H are detected by NMR.

EXAM HINT: EXTRA MATERIAL

You do not need to revise the principles of NMR spectroscopy, as you will not be asked questions about it in your exam.

WHAT IS NMR?

You learned about two major techniques used in analysing organic compounds in **Topic 10 (Book 1: IAS)**. These are mass spectrometry and infrared (IR) spectroscopy. In **Section 15E.2** you learned more about mass spectrometry (high resolution mass spectrometry).

This section and the following sections in **Topic 15** are about a third major technique, **nuclear magnetic resonance spectroscopy** (NMR). This is important for determining the structures of organic compounds. Every major hospital has a department with an NMR scanner, although they are normally described as MRI (magnetic resonance imaging) scanners.

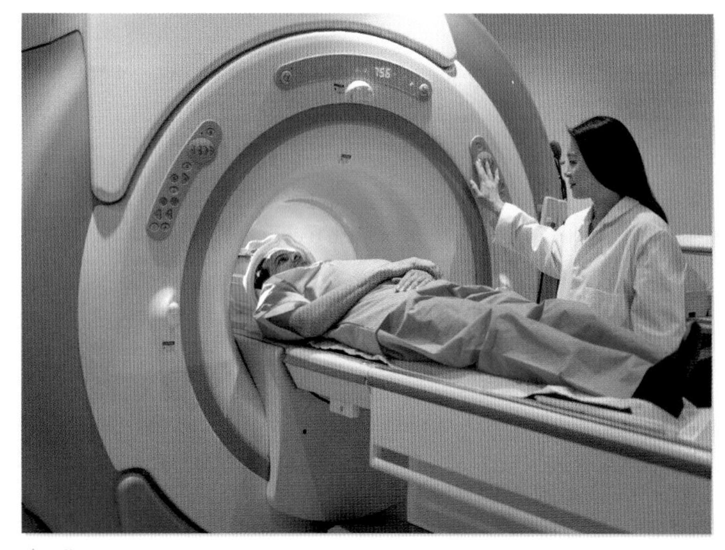

▲ **fig A** A hospital patient about to have an MRI scan. MRI is a development of NMR spectroscopy.

A SIMPLIFIED EXPLANATION

NMR is much more difficult to fully understand than MS or IR spectroscopy. Fortunately, our emphasis will be on interpreting the spectra produced, and not on fully understanding how it works. Even so, it will help if you read a bit of the theory behind it. **Table A** shows some of the key points.

POINT	COMMENT
Nucleons (protons and neutrons) have spin.	You know that electrons in atoms that pair up in orbitals have opposite spins. Well, protons and neutrons in the nucleus also have opposite spins.
Nuclei have either an even or an odd number of nucleons.	In nuclei with an even number of nucleons, these spins cancel out. In nuclei with an odd number of nucleons, these spins cancel out, except for the odd one, which leaves the nucleus with residual spin.
Residual spin causes a tiny magnetic field.	Such nuclei can be imagined as tiny magnets.
These nuclei are affected by an external magnetic field.	Nuclei can be imagined as either lining up or opposing the external magnetic field.
There is a difference in energy between these two different states of a nucleus.	As the spins of nuclei 'flip' between lining up and opposing the external magnetic field, the nuclei can absorb electromagnetic radiation.
The nuclei of two adjacent (neighbouring) atoms in a molecule influence each other, and electrons also have an effect.	This means that electromagnetic radiation is absorbed differently by different atoms in a molecule, and these are detected as different 'signals' at different frequencies.

table A

KEY ASPECTS OF NMR

You need to know about several aspects of NMR.

NMR SPECTRUM

The output of a mass spectrometer or of an infrared spectrometer produces a spectrum (a kind of graph). Similarly, each compound tested using NMR produces a characteristic spectrum. The vertical axis is sometimes labelled 'absorption' (for 'absorption of radio frequency energy') and has no units, although the label is often omitted. The horizontal axis is labelled **chemical shift** (or sometimes with the δ symbol – the same symbol as used for partial charges) and has the units 'ppm' (parts per million). The scale usually starts with a zero value on the right, with the values increasing to the left.

WHICH ATOMS ARE DETECTED BY NMR?

For an atom to show up in an NMR spectrum, it must have an odd number of nucleons (protons and neutrons). That rules out carbon atoms, or at least ^{12}C atoms, as they each have three pairs of protons and three pairs of neutrons. However, in naturally occurring carbon, about 1.1% of the carbon atoms are of the isotope ^{13}C, and they have an odd number of neutrons (seven). So, carbon atoms in molecules can be detected because some of them are ^{13}C atoms. This process is known as carbon-13 or **^{13}C NMR**. Hydrogen atoms are also detected – at least the ^{1}H isotopes that make up 99.985% of all hydrogen atoms. In the context of NMR, ^{1}H hydrogen atoms are often referred to as protons, so an alternative name for the process is **proton NMR**.

We often refer to these two atoms (^{13}C and ^{1}H) as 'producing a signal'. In this book we will consider only these two atoms, although NMR can be successfully used for other atoms, including ^{15}N and ^{31}P.

SUITABLE SOLVENTS

The organic compound being analysed has to be dissolved in a solvent, but most solvent molecules contain carbon and hydrogen atoms. This means that they will produce signals that interfere with those produced by the compound being analysed. The commonest solvent used has the formula $CDCl_3$. The symbol D stands for deuterium, which is the ^{2}H isotope of hydrogen. Deuterium atoms contain a proton and a neutron, and their spins cancel out. This solvent produces no signals that could interfere with the signals from hydrogen atoms. It produces only one signal that can interfere with the signals from carbon atoms, and this can easily be removed from the spectrum.

TMS

Carbon and hydrogen atoms in different chemical environments will produce different signals. Therefore, there needs to be a reference standard that will give a prominent signal. The compound used for this purpose is tetramethylsilane (usually abbreviated to TMS). Its structure is shown below:

CH_3 / H_3C—Si(—CH_3)—CH_3

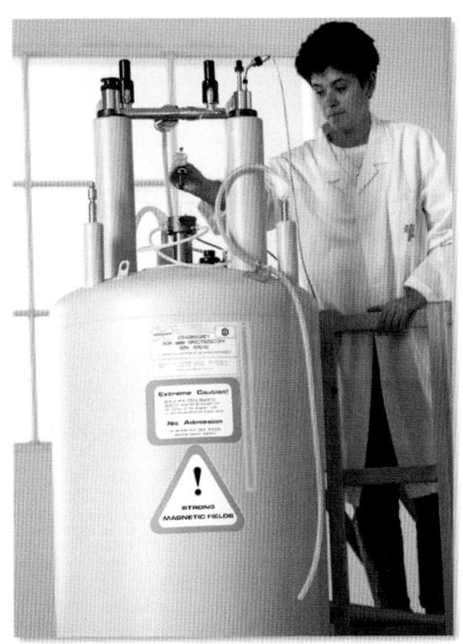

▲ **fig B** The largest part of NMR equipment is often the large vacuum flask containing liquid helium, needed to cool the magnets to below 4 K.

You can see that it resembles a methane molecule, but with silicon instead of carbon and methyl groups instead of hydrogen atoms. SiH_4 is called silane, by comparison with methane, so perhaps you can understand the origin of the name tetramethylsilane. The molecule is suitable because it contains 12 hydrogen atoms, all joined in the same way in a symmetrical arrangement, so it produces a single strong signal that is easy to identify. It is also chemically unreactive, so will not react with most organic compounds.

CHEMICAL SHIFT

Each ^{13}C and ^{1}H atom in a compound produces a signal at a characteristic chemical shift (δ value), depending on the other atoms it is joined to. The reference standard, TMS, is given a chemical shift of zero (δ = 0.0 ppm), and all other atoms have values related to this. Chemical shift can be thought of as the resonant frequency of an atom. In **Sections 15E.6** and **15E.7** you will learn to use a data booklet containing values (or more often, ranges) of chemical shifts to identify specific atoms in molecules. Here are two examples:

carbon in a ketone group	—C—**C**O—C—	δ = 210 ppm
carbon in an alcohol	—**C**—OH	δ = 60 ppm

Note that these values refer only to the carbon atoms shown in bold, and not to any other carbon atoms or to any hydrogen atoms in the molecule.

LEARNING TIP

Do not worry if you think you do not understand the details of NMR theory - by far the most important thing for you to understand is its application.

CHECKPOINT

1. Some organic compounds contain these atoms: ^{16}O, ^{19}F, ^{32}S. Explain which of these are not suitable for use in NMR spectroscopy, and why.
2. Suggest why water (H_2O) is less suitable than TMS as a reference in NMR spectroscopy.

SUBJECT VOCABULARY

nuclear magnetic resonance spectroscopy (NMR) a technique used to find the structures of organic compounds. It depends on the ability of nuclei to resonate in a magnetic field
chemical shift (of a proton or group of protons) a number (in the units ppm) that indicates its behaviour in a magnetic field compared with tetramethylsilane. It can be used to identify the chemical environment of the carbon atoms or of the hydrogen atoms (protons) attached to it
^{13}C NMR the use of NMR spectroscopy to detect ^{13}C nuclei within the molecules of a substance, in order to determine the structure
proton NMR the use of NMR spectroscopy to detect ^{1}H nuclei within the molecules of a substance, in order to determine the structure

15E 6 ^{13}C NMR SPECTROSCOPY

SPECIFICATION REFERENCE

15.18 15.19(i) 15.19(ii)

LEARNING OBJECTIVES

- Understand that ^{13}C NMR spectroscopy provides information about the positions of ^{13}C atoms in a molecule.
- Be able to use data from ^{13}C NMR spectroscopy to:
  (i) predict the different environments for carbon atoms present in a molecule, given values of the chemical shift, δ
  (ii) justify the number of peaks present in a ^{13}C NMR spectrum in terms of the number of carbon atoms in different environments.

WHAT IS SHOWN BY A ^{13}C NMR SPECTRUM?

The actual spectrum of a compound is usually shown as a horizontal line just above the horizontal scale, but with vertical lines showing the signals produced by the ^{13}C carbon atoms in the molecule. These vertical lines are usually referred to as signals or peaks.

- The number of vertical lines tells you the number of different **chemical environments** of carbon atoms in the molecule, but not necessarily the total number of carbon atoms.
- The positions of the vertical lines on the horizontal scale tell you the chemical shifts of each carbon atom and, with reference to a list of chemical shifts, allow you to deduce the chemical environments.

The two isomers of propanol can be used to illustrate these points.

Propan-1-ol has the structure $CH_3—CH_2—CH_2—OH$. Each molecule contains three carbon atoms, all in different chemical environments. You might think that the two carbon atoms in the CH_2 groups are in the same chemical environment, but they are not. The second one is joined to CH_3 and CH_2, but the third one is joined to CH_2 and OH, which means they are in different chemical environments. There should therefore be three peaks in the spectrum.

Propan-2-ol has the structure $CH_3—CH(OH)—CH_3$. Each molecule contains three carbon atoms, but those in the two CH_3 groups are in the same chemical environment. This is because each CH_3 group is joined to CH(OH). This means that there should be two peaks in the spectrum. **Fig A** shows the ^{13}C spectra of these isomers.

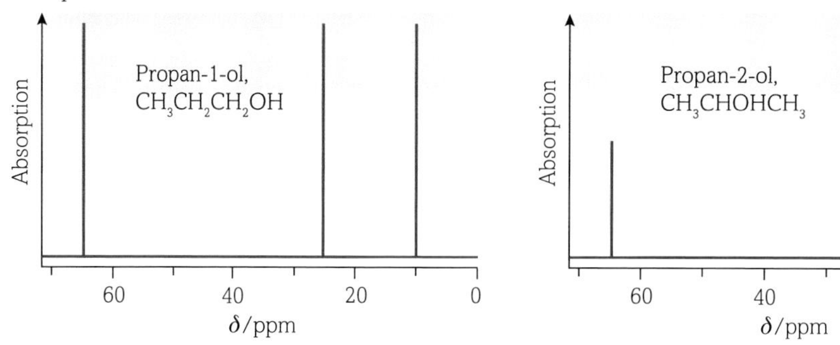

▲ **fig A** ^{13}C NMR spectra of propan-1-ol and propan-2-ol.

The spectra confirm the number of peaks expected, but show two other features.

- The less important feature is that in propan-1-ol the three peaks are of equal height, but in propan-2-ol the two peaks have different heights.
- The more important one is that the peaks have different chemical shifts.

INTERPRETING CHEMICAL SHIFTS

To interpret chemical shifts you need a table of values and corresponding carbon atoms or a chart. The chart in **fig B** will be provided in an examination, and you need to become familiar with using it.

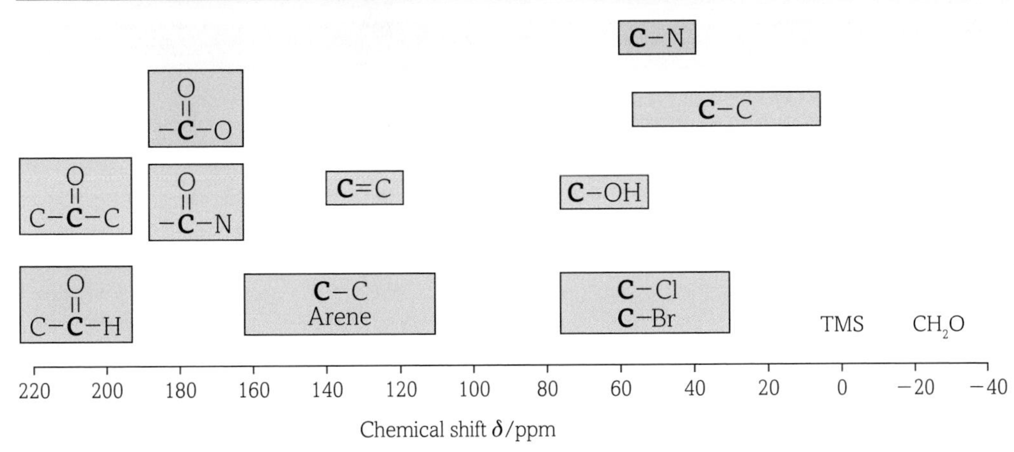

▲ **fig B** ¹³C NMR chemical shift ranges.

We will use this chart to analyse the two spectra for propanol.

Propan-1-ol:

In the spectrum of propan-1-ol the peaks appear at 65, 25 and 10 ppm. The only relevant ranges in the chart are C—C (the range is approximately 60–0 ppm) and C—OH (the range is approximately 75–55 ppm).

This means that the peak at 65 ppm corresponds to C—OH, so this is the carbon in the CH_2OH group. The peaks at 25 and 10 ppm correspond to C—C, so these are the carbons in the CH_3 and CH_2 groups. Notice that it is not possible to decide whether the peak at 25 ppm is due to the CH_3 or the CH_2 group.

Propan-2-ol:

In the spectrum of propan-2-ol the peaks appear at 65 and 27 ppm. The only relevant ranges in the chart are again C—C and C—OH.

This means that the peak at 65 ppm corresponds to C—OH, so this is the carbon in the CHOH group. The peak at 27 ppm corresponds to C—C, so this is due to the carbons in the two CH_3 groups. The peak at 27 ppm is bigger than the one at 65 ppm because it is caused by two carbon atoms, not one.

Fig C shows the interpretation of these two spectra.

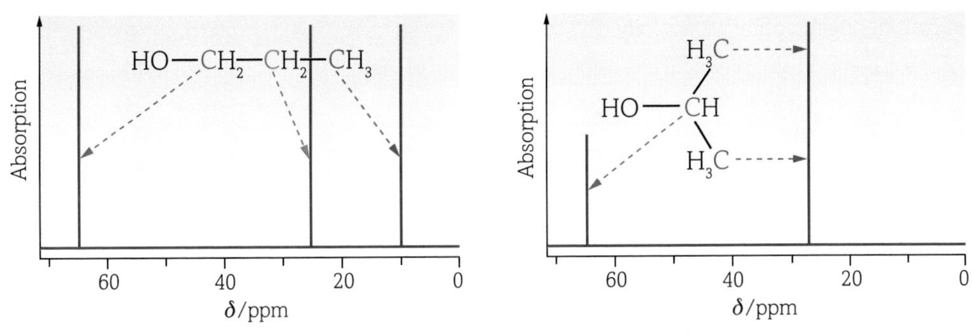

▲ **fig C** Interpreting the ¹³C NMR spectra of propan-1-ol and propan-2-ol.

PRACTICE MAKES PERFECT!

The way to check how well you understand how to use ¹³C NMR spectra is to try lots of examples. In this section we will continue with some examples of varying types.

WORKED EXAMPLE 1

Predicting the number of peaks

You need to consider each carbon atom and whether it is uniquely positioned in the molecule or whether there is one or more carbon atoms identically positioned in the molecule.

For each of these compounds, predict the number of peaks due to the carbon atoms:

A $CH_3-CH_2-CH_2-CH_3$

B $(CH_3)_2-CH-CH_2-CH_3$

C $(CH_3)_4C$

The answers are in **table A**.

COMPOUND	NUMBER OF PEAKS	COMMENTS
A	2	One due to both CH_3 groups (these are both joined to $CH_2CH_2CH_3$, so are identical). One due to both CH_2 groups (these are both joined to CH_3 and CH_2CH_3, so are identical).
B	4	One due to the two left-hand CH_3 groups (these are identically joined to the rest of the molecule) – this should be bigger than the other two peaks because it is due to two carbon atoms. One due to the CH group. One due to the CH_2 group. One due to the right-hand CH_3 group (this is not identical to the other CH_3 groups because it is joined to CH_2 and they are joined to CH).
C	2	One due to the four CH_3 groups, which are all identically joined to the central C atom – this should be much bigger than the other peak as it is due to four carbon atoms. One due to the central C atom.

table A

WORKED EXAMPLE 2

Predicting the appearance of a spectrum

You need to consider all of the carbon atoms, as in Example 1, but this time also use the chart in **fig B** when considering the other atoms joined to them.

For each of these compounds, predict the number of peaks and chemical shift ranges due to the carbon atoms:

A $CH_3-CH_2-CH_2-NH_2$

B $(CH_3)_2CH-COOH$

C $(CH_3)_3N$

The answers are in **table B**.

COMPOUND	NUMBER OF PEAKS	CHEMICAL SHIFT RANGES	COMMENTS
A	3	2 peaks in range 60–0 1 peak in range 63–35	Due to CH_3 and first CH_2 group. Due to CH_2 joined to NH_2.
B	3	1 peak in range 60–0 1 peak in range 60–0 1 peak in range 182–165	Due to the two CH_3 groups. Due to the CH group. Due to the COOH group.
C	1	1 peak in range 63–35	Due to the three CH_3 groups, which are all identical.

table B

WORKED EXAMPLE 3

Interpreting a spectrum

A compound has the molecular formula C_3H_6O and its ^{13}C NMR spectrum is shown in **fig D**.

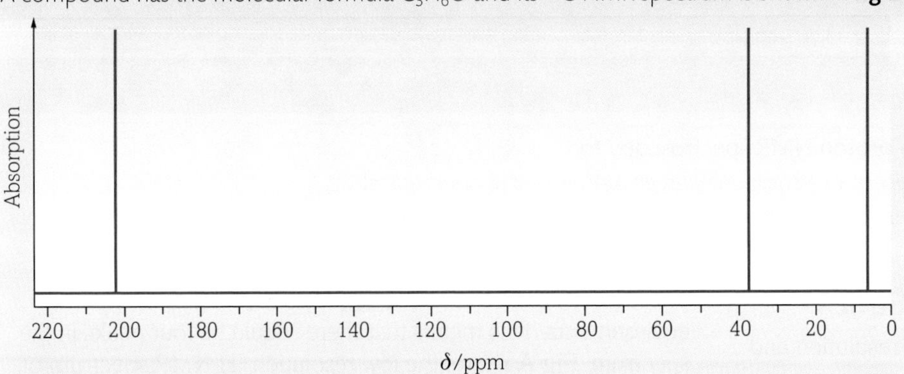

fig D Which compound has this ^{13}C NMR spectrum?

Use the spectrum and the chemical shift chart to explain whether the compound's structure is

$CH_2{=}CH{-}CH_2OH$ or $CH_3{-}CH_2{-}CHO$.

Answer

The peak at δ = 203 ppm is due to a carbon in an aldehyde or ketone.

The peaks at δ = 37 and 6 ppm are due to a carbon joined to another carbon by a single bond.

This interpretation fits $CH_3{-}CH_2{-}CHO$ but not $CH_2{=}CH{-}CH_2OH$.

If the structure were $CH_2{=}CH{-}CH_2OH$, there would have been a peak in the range δ = 140–115 ppm corresponding to C=C, and a peak in the range δ = 75–55 ppm corresponding to C–OH.

LEARNING TIP

When you need to decide how many different chemical environments there are for the carbon atoms in a molecule, it is a good idea to write the structure with the carbon atoms separated, e.g. $CH_3{-}CH(OH){-}CH_3$.

CHECKPOINT

1. Work out the number of different chemical environments for the carbon atoms in the first six straight-chain alkanes.
2. An aromatic compound has the molecular formula $C_8H_8O_2$.

   The compound contains both a phenol functional group and a ketone functional group.

   The ^{13}C NMR spectrum of the compound is shown below.

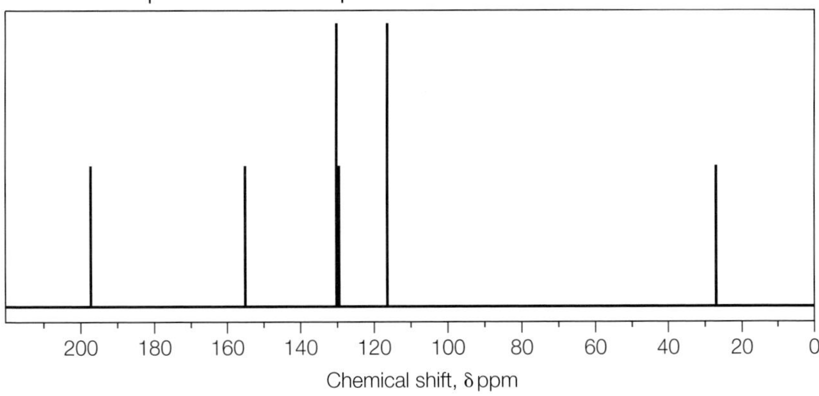

   Use the spectrum to deduce the structure of the compound. Explain your reasoning.

SKILLS ANALYSIS

SUBJECT VOCABULARY

chemical environments the chemical environments of carbon atoms in a molecule are related to whether the carbon atoms are identically, or differently, positioned within a molecule

15E 7 ^{1}H NMR SPECTROSCOPY

SPECIFICATION REFERENCE 15.20(i) 15.20(ii)

LEARNING OBJECTIVES

- Be able to use both low and high resolution proton NMR spectroscopy to:
  (i) predict the different types of protons present in a molecule, given values of the chemical shift, δ
  (ii) relate relative peak areas, or ratio numbers of protons, to the relative numbers of ^{1}H atoms in different environments.

LOW RESOLUTION ^{1}H NMR SPECTROSCOPY

^{1}H NMR spectroscopy comes in two forms – low resolution and high resolution. The low resolution form is no longer used, but if we start with this, it will help you to understand the high resolution form more easily.

As with ^{13}C NMR spectroscopy you may see a peak at $\delta = 0.0$ ppm. This is due to the reference standard tetramethylsilane (TMS) and should be ignored. The chemical shifts are measured relative to the peak for TMS.

WHAT DOES A LOW RESOLUTION ^{1}H NMR SPECTRUM SHOW?

The actual spectrum of a compound is usually shown as a horizontal line just above the horizontal scale, but with **peaks** produced by the ^{1}H hydrogen atoms in the molecule. The horizontal scale often starts from $\delta = 12.0$ ppm on the left to 0.0 ppm on the right. Quite often, only the part of the scale where the peaks appear is shown.

- The number of peaks tells you the number of different chemical environments of the hydrogen atoms in the molecule, but not the total number of hydrogen atoms.
- The positions of the peaks on the horizontal scale tell you the chemical shifts of each hydrogen atom, and, with reference to a list or chart of chemical shifts, these allow you to deduce the types of chemical environment.
- The areas under the peaks represent the relative numbers of hydrogen atoms in each environment.
- Sometimes a number is added at the side of each peak to show the relative area under the peak.

Another way to represent the relative areas is by an **integration trace**. This is a horizontal line that becomes higher as it passes each peak. The increases in height can be measured – they represent the relative areas under the peaks.

The two isomers of propanol can be used to illustrate these points.

Propan-1-ol has the structure CH_3–CH_2–CH_2–OH. There are eight hydrogen atoms, in four different chemical environments. You might think that the two pairs of hydrogen atoms in the CH_2 groups are in the same chemical environment, but they are not. The first CH_2 group is joined to CH_3 and CH_2, but the second one is joined to CH_2 and OH, and these are different chemical environments. This means that there should be four peaks in the spectrum. **Fig A** shows the low resolution ^{1}H NMR spectrum of this isomer with an integration trace.

Propan-2-ol has the structure CH_3–CH(OH)–CH_3. There are eight hydrogen atoms, but those in the two CH_3 groups are in the same chemical environments. This is because each CH_3 group is joined to CH(OH)CH_3. The hydrogen atoms in the CH group and in the OH group are in unique chemical environments. This means that there should be three peaks in the spectrum. **Fig B** shows the low resolution ^{1}H NMR spectrum of this isomer with the ratio of peak areas indicated by circled numbers.

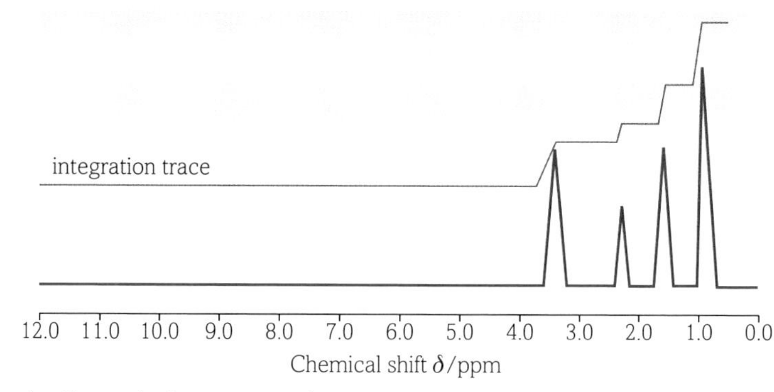

▲ **fig A** The low resolution ^{1}H NMR spectrum of propan-1-ol.

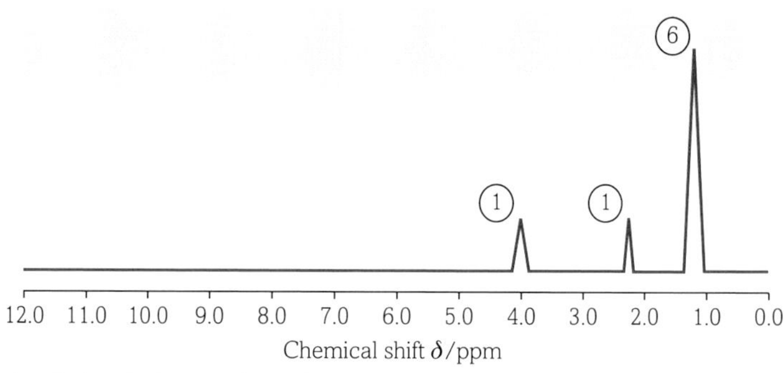

▲ **fig B** The low resolution ^{1}H NMR spectrum of propan-2-ol.

INTERPRETING CHEMICAL SHIFTS

To interpret chemical shifts you need a table of reference values with corresponding hydrogen atoms or a chart. **Fig C** shows a simplified version of the chart that will be provided in an examination, and you need to become familiar with using it.

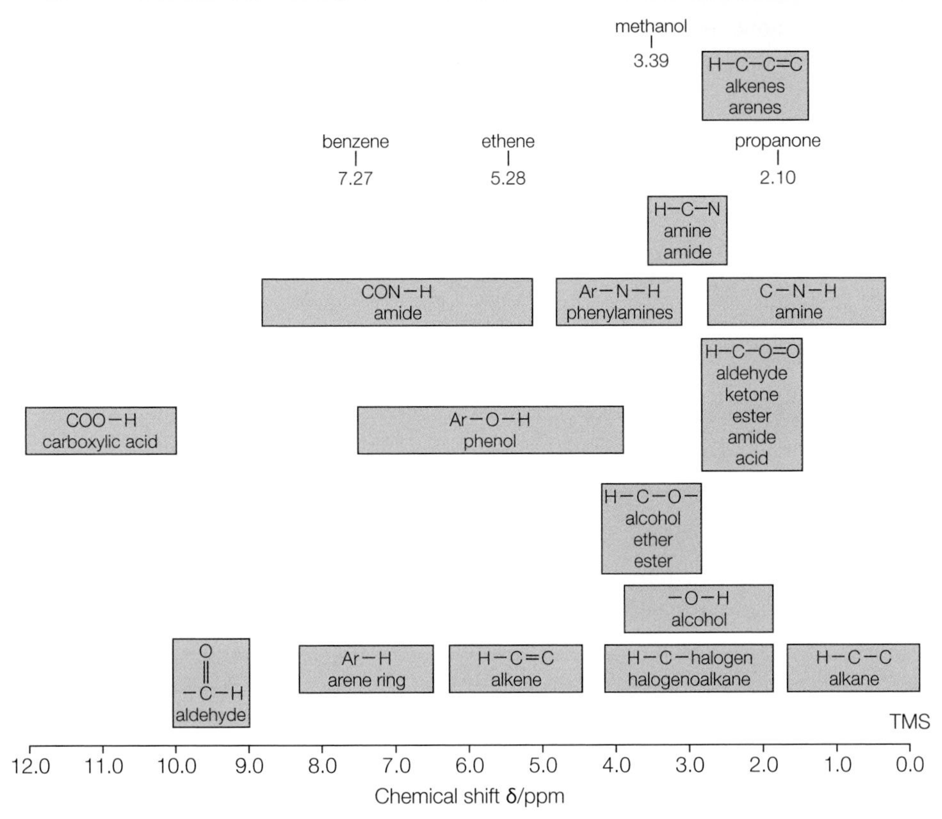

fig C ¹H NMR chemical shift ranges.

For two of the chemical shift ranges in this chart, the compounds included show trends in chemical shift values, as shown here in **table A**.

TYPE OF COMPOUND	TREND
halogenoalkane	$R_2CHF > R_2CHCl > R_2CHBr > R_2CHI$
alkane	$R_3CH > R_2CH_2 > RCH_3$

table A

BUTANAL AND BUTANONE

We will use the chart to interpret the low resolution ¹H NMR spectra of these isomers with the molecular formula C_4H_8O. **Fig D** shows the spectrum for butanal.

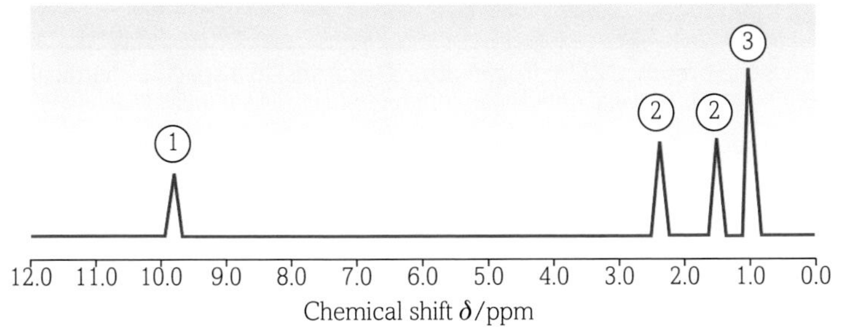

fig D The low resolution ¹H NMR spectrum of butanal.

The peak at $\delta = 9.8$ ppm is within the aldehyde range, so must be due to the H in the CHO group.

The peak at $\delta = 2.4$ ppm is within the range for H—C—C=O, so must be due to the two H atoms in the CH_2 group joined to the CHO group.

The peak at $\delta = 1.6$ ppm is within the range for H—C—C (alkane), so must be due to the two H atoms in the CH_2 group not directly joined to the CHO group. Note that although the spectrum is for an aldehyde, the shift for an alkane is appropriate because this CH_2 group is part of the $CH_3CH_2CH_2$ alkyl group.

The peak at $\delta = 1.0$ ppm is within the range for H—C—C (alkane), so must be due to the three H atoms in the CH_3 group.

EXAM HINT

Remember that the areas under the peaks (integrals) will give you the simplest ratios of the hydrogen environments. Ratios of 3:3:2 could mean hydrogen numbers of 6:6:4 in the actual molecule. You will need to check the molecular ion peak in the mass spectrum to help identify the molecule.

Now for the spectrum of butanone, shown in **Fig E**.

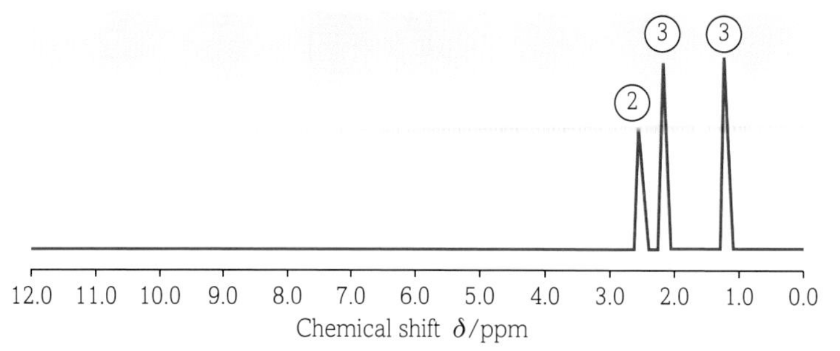

fig E The low resolution ^{1}H NMR spectrum of butanone.

The peak at δ = 2.5 ppm is within the range for H—C—C=O, so must be due to the two H atoms in the CH_2 group joined to the CO group.

The peak at δ = 2.1 ppm is within the range for H—C—C=O, so must be due to the three H atoms in the CH_3 group joined to the CO group.

The peak at δ = 1.1 ppm is within the range for H—C—C (alkane), so must be due to the three H atoms in the CH_3 group not directly joined to the CO group.

LEARNING TIP

Both low resolution and high resolution (see **Topic 15E.8** for more details) ^{1}H NMR spectra indicate three things:

- The number of peaks represents the number of different chemical environments of the hydrogen atoms (protons).
- The areas under the peaks indicate the relative numbers of protons in each chemical environment.
- The chemical shift of a peak can indicate the functional group responsible (e.g. aldehyde or ketone).

CHECKPOINT

SKILLS ANALYSIS

1. Predict the number of peaks in the ^{1}H NMR spectrum of each of the isomers of C_5H_{12}.
2. Why does an integration trace not always indicate the actual numbers of hydrogen atoms in a molecule?
3. How would you easily decide from a ^{1}H NMR spectrum whether a compound is ethylamine ($CH_3CH_2NH_2$) or ethanamide (CH_3CONH_2)?

SUBJECT VOCABULARY

peak a peak in a ^{1}H NMR spectrum shows the presence of hydrogen atoms (protons) in a specific chemical environment

integration trace this shows the relative numbers of equivalent protons (i.e. in the same chemical environment)

15E 8 SPLITTING PATTERNS IN ^{1}H NMR SPECTRA

SPECIFICATION REFERENCE

15.20(iii) 15.20(iv)

LEARNING OBJECTIVES

- Be able to use both low and high resolution proton NMR spectroscopy to:
  (i) deduce the splitting patterns of adjacent, non-equivalent protons using the (n+1) rule and hence suggest the possible structures for a molecule
  (ii) predict the chemical shifts and splitting patterns of the 1H atoms in a given molecule.

HIGH RESOLUTION ^{1}H NMR SPECTROSCOPY

In **Section 15E.7** we introduced NMR by starting with the low resolution form, although only the high resolution form is actually used now. Both forms have two features in common:

- Peaks at different chemical shift values help to identify the different chemical environments of the hydrogen atoms. If the environments are identical, then the protons are often described as being equivalent (i.e. they are **equivalent protons**).
- The relative peak areas (i.e. the ratio of peak areas) help to decide the numbers of hydrogen atoms responsible.

A high resolution ^{1}H NMR spectrum has one extra feature not seen in a low resolution spectrum – the peaks may have a **splitting pattern**. This means that the peak is split into a group of smaller peaks (sub-peaks) grouped very closely together. When considering these split peaks:

- you can take the chemical shift to be at the centre of the group of sub-peaks
- the relative peak areas are shown in the same way as in a low resolution spectrum – by an integration trace or by showing numbers next to the peaks
- remember that the numbers indicate the ratios of the numbers of protons – they may happen to be the actual numbers of protons, but this is not always the case.

WHAT CAUSES THE SPLITTING OF PEAKS?

Although the most important thing is for you to be able to interpret the splitting patterns, it will help if you understand how the splitting is caused. The fundamental point to be aware of is that hydrogen atoms (we will call them protons from now on) joined to a carbon atom have an influence on the protons on adjacent or neighbouring carbon atoms. Consider a molecule of butanone, in which the protons are labelled with letters:

$$\overset{a}{CH_3}-\underset{\underset{O}{\parallel}}{C}-\overset{b}{CH_2}-\overset{c}{CH_3}$$

fig A The three different chemical environments for the protons in butanone.

The protons labelled a are not influenced by any other protons because there are none on the adjacent carbon atom. The protons labelled b are influenced by the protons labelled c and vice versa. These influences are what cause splitting of the peaks (sometimes referred to as spin–spin coupling), usually considered in terms of the n+1 rule. Very simply, if a carbon atom has n protons, then the peaks on adjacent carbon atoms are split into n+1 sub-peaks. The split peaks are sometimes referred to as **multiplets**, but more often by a term indicating the number of sub-peaks.

If the peak is not split, it is called a singlet.

If the peak is split into two sub-peaks, it is called a doublet.

If the peak is split into three sub-peaks, it is called a triplet.

If the peak is split into four sub-peaks, it is called a quartet.

The sub-peaks in a split peak have distinctive shapes, owing to splitting into different sub-peak areas, as shown in **table A**.

N	N+1	MULTIPLET	RATIO OF PEAK AREAS
0	1	singlet	1
1	2	doublet	1 : 1
2	3	triplet	1 : 2 : 1
3	4	quartet	1 : 3 : 3 : 1

table A

LEARNING TIP

The splitting pattern of a group of protons indicates nothing about the group of protons being split, but relates only to the number of protons on an adjacent atom.

UNDERSTANDING THE SPECTRUM OF BUTANONE

We can apply the information provided above to explain the spectrum of butanone in **table B**.

PROTONS	MULTIPLET	EXPLANATION
a	singlet	There are no protons on the adjacent carbon atom (C=O), so there is no splitting.
b	quartet	There are three protons on the adjacent CH_3 group, so applying the n+1 rule, the peak for the two protons is split into a quartet.
c	triplet	There are two protons on the adjacent CH_2 group, so applying the n+1 rule, the peak for the three protons is split into a triplet.

table B

TWO MORE POINTS TO REMEMBER

- We have said that protons cause splitting of peaks on adjacent atoms. While this is generally true, it does not apply in cases where the molecule is completely symmetrical. For example, there is splitting of peaks for 1,1-dichloroethane, $CHCl_2$—CH_3. The single proton splits the peak for the CH_3 group into a doublet, and the three protons in the CH_3 group split the CH peak into a quartet, following the n+1 rule. However, in 1,2-dichloroethane, CH_2Cl—CH_2Cl, neither of the CH_2 groups affects the protons of the other CH_2 group because all of the protons are equivalent (the CH_2 groups are in the same chemical environment).
- Now consider 1-chloropropane, CH_3—CH_2—CH_2—Cl. This time, the two CH_2 groups are in different chemical environments (so the protons are not equivalent). One is joined to CH_3 and CH_2, and the other is joined to CH_2 and Cl, so there will be splitting of peaks by the CH_2 groups. However, this example has been included to make a different point. The first CH_2 group would be influenced by the CH_3 group on its left and by the CH_2 group on its right. Using the n+1 rule, you would make the correct prediction that the CH_3 and the CH_2 group acting together would cause splitting into six sub-peaks – this splitting pattern would be described as a sextet. In other compounds, a quintet would be formed by four protons on adjacent carbon atoms – these could be two CH_2 groups or one CH_3 group and one proton.

PRACTICE MAKES PERFECT, AGAIN!

The best way to check how well you understand how to use ^{1}H NMR spectra is to try lots of examples. We will use three examples, all of which require you to use the chart of chemical shifts in **Section 15E.7**.

WORKED EXAMPLE 1

Predicting the appearance of a spectrum (chemical shifts and splitting patterns)

What would you predict about the appearance of the ^{1}H NMR spectrum of methyl propanoate ($CH_3CH_2COOCH_3$)?

Answer

The CH_3 in the ethyl group is part of an alkyl group and is joined only to a hydrocarbon (CH_2) group, so it should have a peak in the range δ = 1.8–0.1 ppm, which is split into a triplet because of the two protons in the adjacent CH_2 group.

The CH_2 in the ethyl group is part of an alkyl group and is joined to a hydrocarbon (CH_3) group but also to the C of the COO (ester) group, so it should have a peak in the range δ = 2.9–1.8 ppm, which is split into a quartet because of the three protons in the adjacent CH_3 group.

The CH_3 in the methyl group is joined only to an oxygen atom in the COO group, so it should have a peak in the range δ = 4.2–2.8 ppm, which is a singlet. This peak is not split because there are no protons on the adjacent atom.

Fig B shows the actual spectrum of methyl propanoate, so you can check whether the predictions are correct.

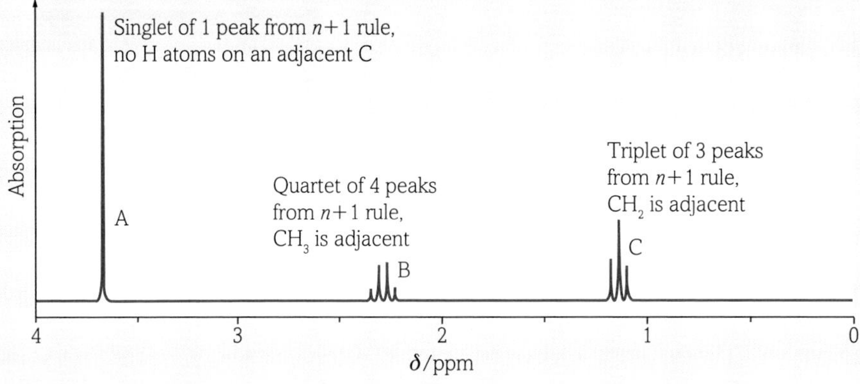

▲ **fig B** The ^{1}H NMR spectrum of methyl propanoate.

WORKED EXAMPLE 2

Predicting the structure of a compound from its spectrum

Fig C shows the ^1H NMR spectrum of a compound with the molecular formula $C_4H_8O_2$.

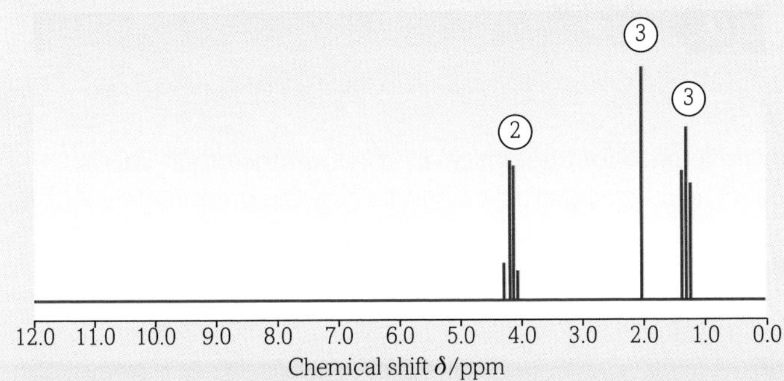

▲ **fig C** The ^1H NMR spectrum of a compound with molecular formula $C_4H_8O_2$.

What is its structure?

Remember that the spectrum shows four features:

- the number of different hydrogen environments
- relative peak areas
- chemical shifts
- splitting patterns

all of which need to be considered together.

Answer

There are three peaks, so there are three different hydrogen environments. The peak on the left is a quartet at $\delta = 4.1$ ppm representing two protons. This suggests a CH_2 group next to a CH_3 group and an O in an alcohol, ether or ester, so $O{-}CH_2{-}CH_3$.

The peak in the middle is a singlet at $\delta = 2.0$ ppm representing three protons. This suggests a CH_3 group joined to C=O in an aldehyde, ketone, ester, amide or acid, so $CH_3{-}CO{-}$.

The peak on the right is a triplet at $\delta = 1.3$ ppm representing three protons. This suggests a CH_3 group next to a CH_2 group, so $-CH_2{-}CH_3$.

The final step is to put all these interpretations together. This is often easy to do but difficult to explain. The only structure that fits the interpretations is $CH_3{-}CO{-}O{-}CH_2{-}CH_3$, which is the ester ethyl ethanoate.

WORKED EXAMPLE 3

Another prediction of the structure of a compound from its spectrum

Fig D shows the ^1H NMR spectrum of a compound with the molecular formula $C_4H_8O_2$ (the same molecular formula as in Example 2).

Another prediction of the structure of a compound from its spectrum

Fig D shows the ^1H NMR spectrum of a compound with the molecular formula $C_4H_8O_2$ (the same molecular formula as in Example 2).

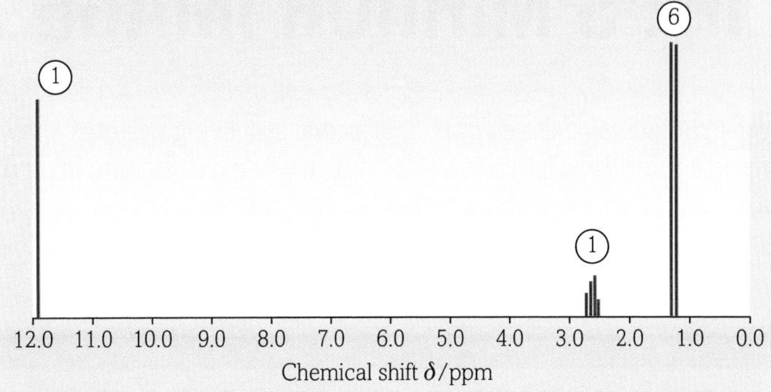

▲ **fig D** The ^1H NMR spectrum of a compound with the formula $C_4H_8O_2$.

What is its structure?

Answer

The peak on the left is a singlet at $\delta = 11.9$ ppm representing one proton. This suggests a proton in a COOH group, so COOH.

The peak in the middle is a quartet (or multiplet) at $\delta = 2.6$ ppm representing one proton. This suggests a CH group joined to a CH_3 group, so $CH_3{-}CH{-}$.

The peak on the right is a doublet at $\delta = 1.2$ ppm representing six protons. This suggests two CH_3 groups next to a CH group, so $(CH_3)_2{-}CH{-}$.

The final step is to put all these interpretations together. Again, this is easy to do but difficult to explain. The only structure that fits the interpretations is $(CH_3)_2{-}CH{-}COOH$ - this is the acid methylpropanoic acid.

CHECKPOINT

SKILLS REASONING

1. A compound has the molecular formula C_4H_8O. Its ^1H NMR spectrum contains only three singlets, with chemical shift values of 4.0, 3.4 and 2.1 ppm. The integration trace shows the ratios 2 : 3 : 3. What is its structure?
2. A compound has the molecular formula $C_5H_{10}O_2$. Its ^1H NMR spectrum contains a singlet with a chemical shift of 2.3 ppm, a quartet with a chemical shift of 5.0 ppm and a doublet with a chemical shift of 1.2 ppm. The integration trace shows the ratios 3 : 1 : 6. What is its structure?

SUBJECT VOCABULARY

equivalent protons hydrogen atoms in the same chemical environment

splitting pattern the appearance of a peak as a small number of small sub-peaks very close to each other

multiplets the different splitting patterns observed (singlets, doublets, triplets or quartets) in a high resolution ^1H NMR spectrum

15 THINKING BIGGER

LIFE'S MIRROR IMAGES

SKILLS INITIATIVE, SELF-DIRECTION

Some molecules exist as optical isomers, and living systems show a clear preference for one optical isomer over the other. This is important for the pharmaceutical industry when designing drug molecules. The following article suggests how the predominance of a particular isomer may have occurred early in the history of planet Earth.

ENRICHING THE ORIGIN OF LIFE THEORY

An enantioenrichment of the amino acid valine, which could shed light on the origin of chirality on Earth, has been achieved by scientists in Spain.

Chirality is important as amino acids and sugars are present as one single enantiomer in all organisms, and enzymes and receptors are chiral too. As organisms do not have the same response to different enantiomers and many active components of medicines are chiral, understanding how pure enantiomers form can help us understand how we evolved and help develop the medicines of the future.

Cristobal Viedma, from the Complutense University in Madrid, and colleagues, have been able to amplify the purity of one enantiomer over another in valine, one of life's building blocks. They chose valine because it was found in meteorite samples in a non-racemic form, and the enantiomeric excesses matched the values obtained under simulated interstellar conditions.

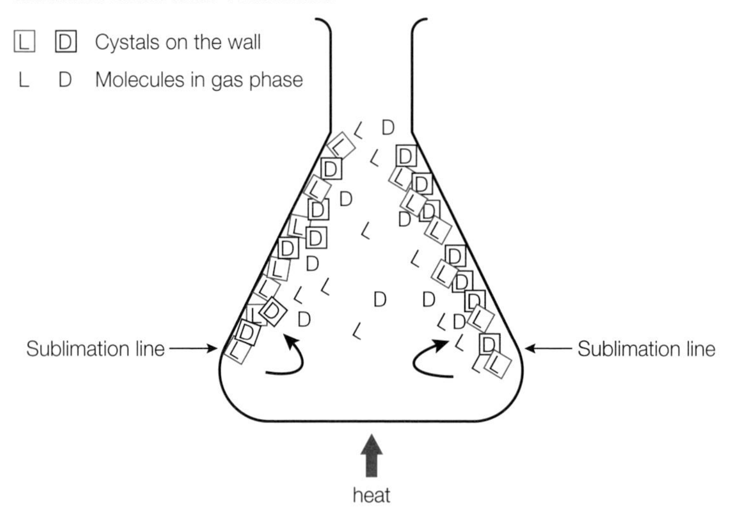

▲ **fig A** Scientists observe the enantiomeric enrichment of the amino acid valine in a sublimation experiment. (Note: the letters L and D refer to the two different optical isomers of the valine molecule. The L and D notation for optical isomers is beyond the aims of this book, but you are encouraged to look it up.)

Viedma's team sublimated a racemic mixture of valine (**fig A**). When the valine condensed, it formed a conglomerate. A conglomerate is a mixture of crystals formed by pure enantiomers – the pure crystals can be extracted from the mixture. The group found that when they continued to heat the flask, this amplified the initial enantiomeric excess. "One can imagine similar processes occurring near volcanoes, where temperature gradients are enormous, and such scenarios would have been plausible on primitive Earth," comments Viedma.

Up until now, enantioenrichment has been studied with solid–liquid systems, even though gas–solid transformations may have been plausible on both our primeval planet and in star-forming regions. "It is true that gas–solid transformations may not be as general as solid–liquid equilibria, but you never know what is going to turn out to be useful," says Viedma. "This is an interesting project in terms of the intriguing conundrum of the origin of biological chirality as well as extensions to the manufacture of chiral substances."

"I have no doubt that it will change my understanding of how crystals grow in a profound way," says Michael McBride, who studies the chemical reactivity and physical properties of organic solids at Yale University, USA. "Understanding this behaviour might expand the range of what we can control in chemistry and materials science."

From an article in *Chemistry World*. 'Enriching the origin of life theory' by Joanne Thomson, 26 November 2010. https://www.chemistryworld.com/news/enriching-the-origin-of-life-theory/3001444.article

SCIENCE COMMUNICATION

1 Communicating ideas about three-dimensional (3D) molecules using books and leaflets can be difficult. Modern 3D molecular modelling software lets us display molecules in different ways, and allows us to rotate molecules virtually. Download and install a 3D software package (such as Jmol). Then prepare a presentation (e.g. using PowerPoint) in which you display and rotate a valine molecule.

INTERPRETATION NOTE

Why do you think it is difficult to write about and explain 3D molecules?

CHEMISTRY IN DETAIL

2 The amino acid valine mentioned in the article has the IUPAC name 2-amino-3-methylbutanoic acid.

(a) Draw both optical isomers of valine, clearly indicating the chiral carbon.

(b) Draw the zwitterion formed by valine in aqueous solution.

(c) The isoelectric point (pI) of valine at 25 °C is 6.02. Explain what this means.

3 An equimolar mixture of both optical isomers of valine is optically inactive. Explain why.

4 Suggest why pharmaceutical molecules with one or more chiral carbons may have different biological activities.

THINKING BIGGER TIP

Look up the structure of the amino acid called glycine, for example in a dictionary or online. Can you explain why glycine does not form optical isomers?

ACTIVITY

Valine is an example of an essential amino acid. Use the Internet to help you classify each of the 20 'biogenic' amino acids as either essential or non-essential. Then choose **two** of the essential amino acids and find out a possible dietary source for each of them.

15 EXAM PRACTICE

1 Which amino acid contains two chiral carbon atoms?

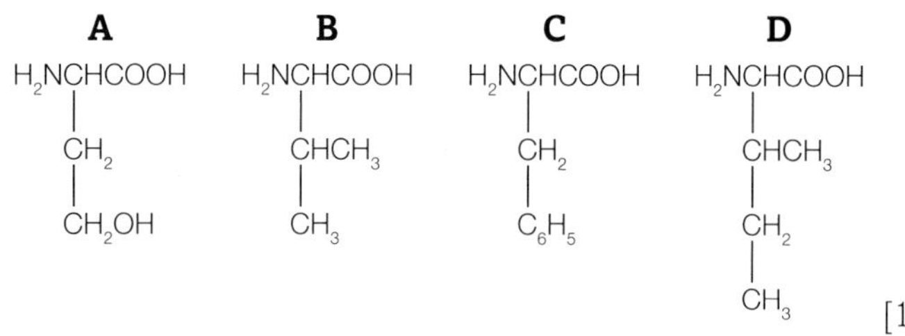

[1]

(Total for Question 1 = 1 mark)

2 The same carboxylic acid is obtained either by the hydrolysis of a nitrile ***P*** or by the oxidation of an alcohol ***Q***.

Which pair could be ***P*** and ***Q***?

	P	*Q*
A	CH_3CH_2CN	CH_3CH_2OH
B	$(CH_3)_2CHCN$	$(CH_3)_3COH$
C	$C_6H_5CH(CH_3)CN$	$C_6H_5CH_2CH(OH)CH_3$
D	$C_6H_5CH_2CN$	$C_6H_5CH_2CH_2OH$

[1]

(Total for Question 2 = 1 mark)

3 Which pair of reactants produces an ester with the formula $C_3H_7COOC_2H_5$?

A C_2H_5Cl and C_3H_7COOH

B C_2H_5OH and C_3H_7COOH

C C_3H_7OH and C_2H_5COCl

D C_3H_7OH and C_2H_5COOH [1]

(Total for Question 3 = 1 mark)

4 One of the constituents of beeswax has the formula $CH_3(CH_2)_{24}CO_2(CH_2)_{29}CH_3$.

What are the products of its alkaline hydrolysis?

A $CH_3(CH_2)_{24}COOH$ and $CH_3(CH_2)_{29}O^-$

B $CH_3(CH_2)_{24}COOH$ and $CH_3(CH_2)_{29}OH$

C $CH_3(CH_2)_{24}COO^-$ and $CH_3(CH_2)_{29}OH$

D $CH_3(CH_2)_{24}COO^-$ and $CH_3(CH_2)_{29}O^-$ [1]

(Total for Question 4 = 1 mark)

5 An organic compound has the following properties:

- it gives a positive tri-iodomethane (iodoform) test
- it gives a yellow precipitate with 2,4-dinitrophenylhydrazine
- it does not react with either Tollens' reagent or Fehling's solution

Which compound gives these results?

A CH_3CHO

B CH_3CH_2OH

C $CH_3CH_2COCH_3$

D $CH_3CH_2CH_2CHO$ [1]

(Total for Question 5 = 1 mark)

6 What are the principal inorganic and organic products when propanal reacts with Tollens' reagent?

A Ag and CH_3CH_2COOH

B Ag and $CH_3CH_2COO^-$

C Ag_2O and CH_3CH_2COOH

D Ag_2O and $CH_3CH_2COO^-$ [1]

(Total for Question 6 = 1 mark)

7 The structures of three organic compounds, **A**, **B** and **C**, are shown.

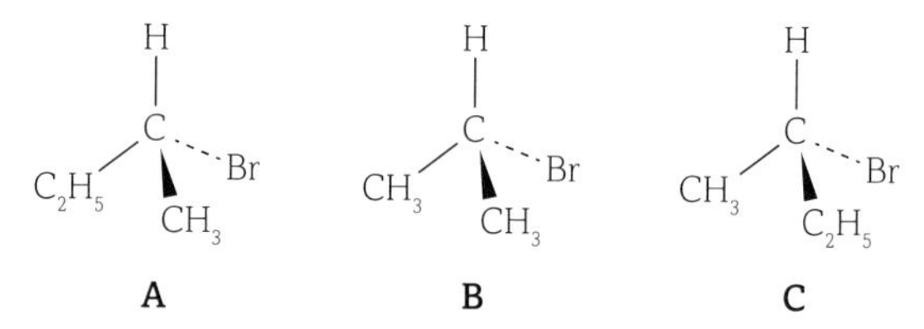

(a) Explain which two of these compounds are enantiomers. [2]

(b) Explain why a mixture containing equal numbers of these enantiomer molecules cannot be distinguished from the other compound of the three in a polarimeter. [2]

(c) Compound **A** reacts with aqueous sodium hydroxide. Explain how measurements of the optical activity of compound **A** and the organic product of the reaction can be used to confirm whether the reaction occurs by an S_N1 mechanism or an S_N2 mechanism. [6]

(Total for Question 7 = 10 marks)

8 The structures of two carbonyl compounds are:

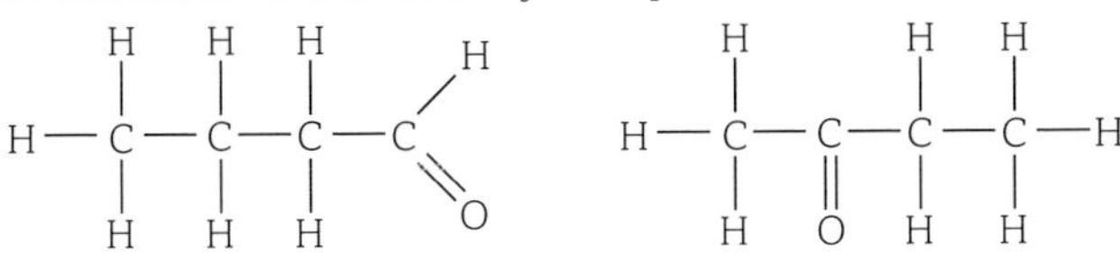

(a) State why both compounds are soluble in water. [1]

(b) (i) Name the mechanism of the reaction between each of the compounds and hydrogen cyanide. [1]

(ii) Both compounds react with 2,4-dinitrophenylhydrazine.

Describe how the organic product of each of these reactions could be used to identify the carbonyl compounds. [2]

(c) Give the structure of the organic product of the reaction of each compound with lithium tetrahydridoaluminate, followed by water. [2]

(d) Only one of the two compounds reacts on warming with Tollens' reagent.

Write an equation for the reaction that occurs, using [O] to represent Tollens' reagent, and state the observations that will be made when a positive result is obtained. [2]

(Total for Question 8 = 8 marks)

9 (a) The structures of three alcohols are:

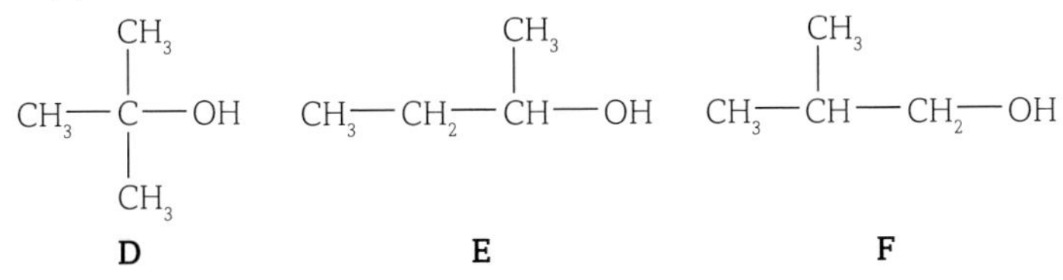

Explain which of these alcohols reacts on warming with acidified potassium dichromate(VI) to form a carboxylic acid. [3]

(b) Draw the displayed formula of the nitrile that can be hydrolysed to the carboxylic acid with the formula $(CH_3)_2CHCH_2COOH$. [1]

(c) The diagram summarises a sequence of two reactions.

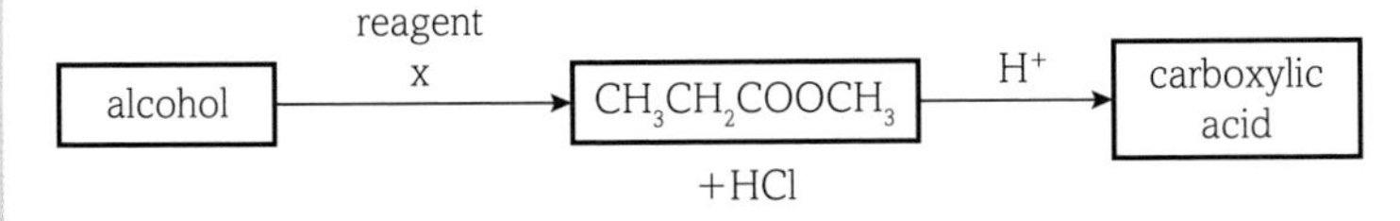

(i) Deduce the name of the alcohol. [1]

(ii) Explain how the homologous series to which reagent X belongs can be deduced from the diagram. [2]

(iii) Deduce the structural formula of the carboxylic acid. [1]

(iv) The yield of carboxylic acid in the second step can be improved by using a different reagent before the acid in a two-step process.

Write an equation for each of these steps. [2]

(Total for Question 9 = 10 marks)

10 The structures of two compounds are:

$CH_3C(CH_3)(OH)CH_3$ $\qquad$ $CH_2(OH)CH(CH_3)CH_3$

(a) Explain why infrared spectroscopy is not used to distinguish these two compounds. [2]

(b) Explain how the number of peaks in the ^{13}C NMR spectra of these compounds can be used to identify which spectrum is obtained from each compound. [4]

(c) One of these compounds has a ^1H NMR spectrum that shows only two singlets.

Explain which compound can be identified from this information. [3]

(d) A simplified mass spectrum of each of the compounds is shown below.

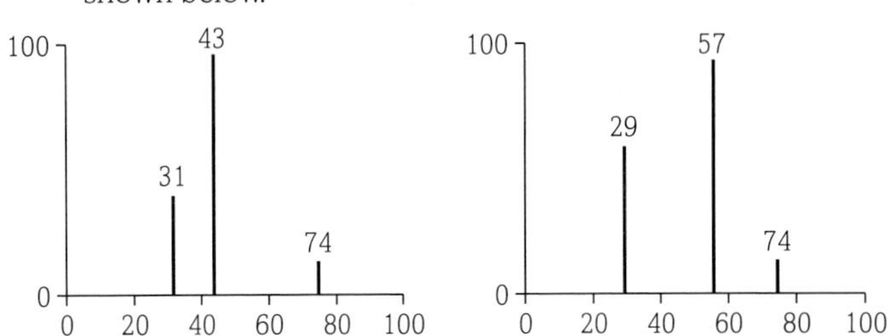

Explain how this information can be used to identify each compound. [6]

(Total for Question 10 = 15 marks)

TOPIC 16 REDOX EQUILIBRIA

A STANDARD ELECTRODE POTENTIAL | B REDOX IN ACTION

In **Topic 8 (Book 1: IAS)** you learned about one of the main types of reactions: redox reactions. Now that we have studied equilibrium and thermodynamics we can investigate these reactions in more depth. Redox reactions keep us alive. They capture the energy of the Sun by photosynthesis, and we can then use that energy. However, redox reactions can also kill, because, for example, they can result in explosions. Redox reactions are used throughout the chemical industry, for example, to extract metals from their ores. However, redox reactions are also responsible for the corrosion of metals. What redox reactions can achieve, they can also destroy.

In **Topic 12** we saw that some chemical reactions take place spontaneously. In this topic we will look at the branch of chemistry known as electrochemistry, which deals with the use of spontaneous reactions to produce electricity.

In **Topic 14**, where you developed your knowledge of acid-base equilibria, we saw that a great deal of chemistry can be discussed in terms of the transfer of protons. In this topic we will see that another large area of chemistry can be understood in terms of the transfer of another fundamental particle, the electron. Proton transfer is the basis of acid-base reactions; electron transfer is the basis of redox reactions.

MATHS SKILLS FOR THIS TOPIC

- Recognise and make use of appropriate units in calculation
- Recognise and use expressions in decimal and ordinary form
- Use an appropriate number of significant figures
- Change the subject of an equation
- Solve algebraic equations

Problems associated with hydrogen–oxygen fuel cells

Hydrogen explodes easily when ignited, so great care has to be taken when transporting it. We will consider the following options in turn.

- compressing the gas
- adsorbing it onto the surface of a suitable solid material
- absorbing it into a suitable material.

Compressing the gas has disadvantages in terms of safety. Transporting hydrogen under pressure is particularly hazardous.

Many metals adsorb hydrogen, but so far no firm conclusion has been reached as to the best one to use. One of the most promising areas of research is the use of carbon nanotubes to adsorb hydrogen.

Many metal hydrides absorb hydrogen, which can be released under the correct conditions. The problem is that high temperatures are required to release the hydrogen.

Another problem is the supply of hydrogen. At the moment, most of the hydrogen is produced from methane, which is a finite or limited source. It can be produced using renewable sources, but it is an expensive process. Integrated wind-to-hydrogen (power-to-gas) plants, using electrolysis of water, are exploring technologies to deliver costs low enough, and quantities great enough, to compete with traditional energy sources.

Fig B shows a prototype car that uses fuel cell technology.

DID YOU KNOW?

Fuel cells have been used in various kinds of vehicles, including forklifts, especially in indoor applications where their clean emissions are important to air quality, and in space applications. The first commercially produced hydrogen fuel cell automobile, the Toyota Mirai was introduced in 2015, after which Hyundai and Honda entered the market. Fuel cells are also being developed and tested in trucks, buses, boats, motorcycles and bicycles, among other kinds of vehicles.

▲ **fig B** A car that uses a hydrogen–oxygen fuel cell.

CHECKPOINT

1. Nickel–cadmium (NiCd) cells are used in many electronic devices because they are very reliable, economical and easy to use.

The standard electrode potentials for the redox systems in NiCd cells are:

$Cd(OH)_2(s) + 2e^- \rightleftharpoons Cd(s) + 2OH^-(aq) \quad E^\ominus = -0.88\,V$

$NiO(OH)(s) + H_2O(l) + e^- \rightleftharpoons Ni(OH)_2(s) + OH^-(aq) \quad E^\ominus = +0.52\,V$

(a) What is the standard cell potential, $E^\ominus_{cell}$, of a NiCd cell?

(b) Construct the overall reaction that takes place when a NiCd cell is producing an electric current.

(c) Use oxidation numbers to show which element is oxidised and which is reduced during the discharge of a NiCd cell.

2. Fuel cells are being developed as an alternative to the use of petrol and diesel in cars. The most common fuel used in these cells is hydrogen, but cells using other fuels are being developed.

In a methanol fuel cell, the fuel is supplied at the negative electrode, with oxygen at the positive electrode.

The ionic half-equation for the reaction of oxygen at the positive electrode is:

$\frac{1}{2}O_2(g) + 2H^+(aq) + 2e^- \rightarrow H_2O(l)$

(a) Write a chemical equation for the complete combustion of methanol in oxygen.

(b) Use the equation in (a), and the ionic half-equation for the reaction of oxygen at the positive electrode, to deduce an ionic half-equation for the reaction that takes place at the negative electrode.

SKILLS ETHICS, INNOVATION

(c) Give two advantages of using fuel cells compared with the combustion of petrol or diesel.

(d) Give an advantage of using methanol, rather than hydrogen, in a fuel cell for use in cars.

3. (a) What is the essential difference between a fuel cell and a conventional electrochemical cell?

SKILLS ETHICS, INNOVATION

(b) State two ways that hydrogen might be stored as a fuel for use in cars.

(c) Suggest why some scientists consider that using hydrogen as a fuel for cars uses up more energy than using petrol or diesel as the fuel.

16B 2 REDOX TITRATIONS

LEARNING OBJECTIVES

- Be able to carry out both structured and unstructured titration calculations involving redox reactions, including iron(II) ions and potassium manganate(VII) and sodium thiosulfate and iodine.
- Understand the methods used in redox titrations.

REDOX TITRATIONS WITH POTASSIUM MANGANATE(VII), $KMnO_4$

Potassium manganate(VII) is a powerful oxidising agent. It is used for the quantitative estimation of many reducing agents, especially compounds of iron(II) and for ethanedioic acid and its salts.

In acidic solution, the half-equation for the reduction of the manganate(VII) ion is:

$$MnO_4^-(aq) + 8H^+(aq) + 5e^- \rightarrow Mn^{2+}(aq) + 4H_2O(l)$$

In alkaline solution, potassium manganate(VII) yields manganese(IV) oxide as a brown precipitate, and this interferes with the end point colour. For this reason, therefore, potassium manganate(VII) is almost always used in solutions that are sufficiently acidic to prevent the formation of manganese(IV) oxide.

As the titration proceeds, manganese(II) ions, Mn^{2+}, accumulate but, at the dilution used, they give a colourless solution. As soon as the potassium manganate(VII) is in excess, the solution becomes pink. It therefore acts as its own indicator – the end point is the first permanent pink colour.

TITRATION OF POTASSIUM MANGANATE(VII) WITH IRON(II) IONS

In this reaction, the iron(II) ions are oxidised and the manganate(VII) ions are reduced.

The ionic half-equations involved are:

$$MnO_4^-(aq) + 8H^+(aq) + 5e^- \rightarrow Mn^{2+}(aq) + 4H_2O(l)$$

$$Fe^{2+}(aq) \rightarrow Fe^{3+}(aq) + e^-$$

The overall equation for the reaction is:

$$MnO_4^-(aq) + 8H^+(aq) + 5Fe^{2+}(aq) \rightarrow Mn^{2+}(aq) + 4H_2O(l) + 5Fe^{3+}(aq)$$

An aqueous solution containing MnO_4^- ions is purple. In reasonably concentrated solutions containing Fe^{2+}, Fe^{3+} and Mn^{2+}, the ions have the following colours:

$Fe^{2+}(aq)$ – pale green

$Fe^{3+}(aq)$ – yellow

$Mn^{2+}(aq)$ – pale pink

However, the solutions used in titrations are usually so dilute that they appear colourless.

The method of performing the titration is:

- Pipette an accurately measured volume, usually 25.0 cm^3, of the iron(II) solution into a conical flask and then add a small volume of dilute sulfuric acid.
- Slowly add potassium manganate(VII) solution of accurately known concentration from a burette and swirl the mixture.
- The potassium manganate(VII) solution will turn colourless until all of the iron(II) ions have been oxidised.
- The addition of one more drop of potassium manganate(VII) solution will turn the mixture pale pink (**fig A**).

▲ **fig A** End point colour in a potassium manganate(VII) titration.

EXAM HINT

Could acidified potassium dichromate be used as an oxidising agent for quantitative redox titrations? In theory, the answer is yes. However, in practice, the answer is no. This is because the colour change from orange to green has no clear end point.

WORKED EXAMPLE

25.0 cm^3 of iron(II) sulfate solution reacts with 22.40 cm^3 of 0.0200 mol dm^{-3} potassium manganate(VII) solution. Calculate the concentration of the iron(II) sulfate.

$5Fe^{2+}(aq) \equiv MnO_4^-(aq)$

25.0 cm^3 of x mol dm^{-3} $Fe^{2+} \equiv$ 22.40 cm^3 of 0.0200 mol dm^{-3} MnO_4^-

$$\frac{25.0x}{22.40 \times 0.0200} = \frac{5}{1}$$

$$x = \frac{5 \times 22.40 \times 0.0200}{25.0} = 0.0896$$

$$[FeSO_4(aq)] = 0.0896 \text{ mol dm}^{-3}$$

TITRATION OF POTASSIUM MANGANATE(VII) WITH ETHANEDIOIC ACID

In this reaction, the ethanedioic acid is oxidised to carbon dioxide. The ionic half-equation for this oxidation is:

$$H_2C_2O_4(aq) \rightarrow 2H^+(aq) + 2CO_2(g) + 2e^-$$

The overall equation for the reaction is:

$$2MnO_4^-(aq) + 6H^+(aq) + 5H_2C_2O_4(aq) \rightarrow 2Mn^{2+}(aq) + 8H_2O(l) + 10CO_2(g)$$

Aqueous ethanedioic acid is pipetted into the conical flask (Care: ethanedioic acid is extremely poisonous, so a pipette filler *must* be used) and is then acidified with dilute sulfuric acid. Aqueous potassium manganate(VII) is added from the burette. However, the reaction is very slow, so the ethanedioic acid is heated to around 60 °C before starting the titration. Even then, the reaction is slow to begin with and the pink colour does not immediately disappear when the first sample of potassium manganate(VII) is added. This may lead you to erroneously think that the titration is complete. After the initial sample is added, the Mn^{2+} ions produced act as a catalyst and the reaction speeds up, allowing you to titrate normally and obtain an accurate end point.

Potassium manganate(VII) can also be used to estimate the amount of ethanedioate ion present in a solution. The equation for the reaction is:

$$2MnO_4^-(aq) + 16H^+(aq) + 5C_2O_4^{2-}(aq) \rightarrow 2Mn^{2+}(aq) + 8H_2O(l) + 10CO_2(g)$$

DID YOU KNOW?

A reaction in which a product acts as a catalyst is said to be autocatalysed. Autocatalysis is discussed in Topic 11.

LEARNING TIP

The dilute sulfuric acid is boiled to remove dissolved oxygen that might otherwise oxidise some of the iron(II) ions to iron(III).

EXAM HINT

In your exam you may be asked to **calculate** the measurement uncertainty for a titration. You can find out more about this in **Topic 8 (Book 1: IAS)**.

PRACTICAL SKILLS CP13A

A redox titration: the determination of the number of moles of water of crystallisation per mole of hydrated iron(II) sulfate crystals

Note that this suggested practical uses techniques and skills similar to those needed in the **Core Practical 13a: Redox titrations with iron (II) ions and potassium manganate (IV)** in your Lab Book.

Procedure

1. Bring to the boil 100 cm^3 of dilute sulfuric acid and then allow it to cool to room temperature.
2. Weigh accurately a weighing bottle containing approximately 7 g of hydrated iron(II) sulfate crystals.
3. Dissolve the crystals in the cool acid and reweigh the bottle. Transfer the solution to a 250 cm^3 volumetric flask. Add several washings from the beaker and then make up to the mark. Shake well.
4. Titrate a 25.0 cm^3 sample of the iron(II) solution, as before, with the standard potassium manganate(VII) solution.
5. Perform repeat titrations until you obtain concordant results.

Calculations

1. Calculate the mean titre from your concordant results.
2. Calculate the concentration of the iron(II) sulfate solution.
3. Calculate the mass of iron(II) sulfate in the 250 cm^3 of solution.
4. Calculate the mass of water in the mass of crystals taken and hence calculate the number of moles of water per mole of salt.

Sample results

Concentration of potassium manganate(VII) solution = 0.0198 mol dm^{-3}

Mass of hydrated iron(II) sulfate = 6.84 g Mean titre = 24.95 cm^3

[Molar masses: $FeSO_4$ = 151.9 g mol^{-1}; H_2O = 18.0 g mol^{-1}]

$$5FeSO_4 \equiv KMnO_4$$

25.0 cm^3 of x mol dm^{-3} $FeSO_4 \equiv$ 24.95 cm^3 of 0.0198 mol dm^{-3} $KMnO_4$

$$\frac{25.0x}{24.95 \times 0.0198} = \frac{5}{1}$$

$[FeSO_4(aq)]$ = 0.0988 mol dm^{-3}

Mass of $FeSO_4$ in 250 cm^3

$$x = \frac{5 \times 24.95 \times 0.0198}{25.0} = 0.0988$$

Mass of water = (6.84 − 3.75) = 3.09 g

If the formula of the hydrated salt is $FeSO_4.yH_2O$, then

$$\text{Mass of } FeSO_4 \text{ in } 250\ cm^3 = 0.0988 \times 151.9 \times \frac{250}{1000} = 3.75\ g$$

$$\frac{18.0y}{151.9} = \frac{3.09}{3.75}$$

$$y = \frac{3.09 \times 151.9}{18.0 \times 3.75}$$

$$= 6.95$$

∴ The formula of hydrated iron(II) sulfate is $FeSO_4.7H_2O$

Safety Note: Wear eye protection and avoid skin contact with the solutions. Make sure the top of the burette is below eye level when you fill it.

REDOX TITRATIONS WITH IODINE AND SODIUM THIOSULFATE

Thiosulfate ions reduce iodine to iodide ions. The ionic half-equations involved are:

$$2S_2O_3^{2-}(aq) \rightarrow S_4O_6^{2-}(aq) + 2e^-$$

$$I_2(aq) + 2e^- \rightarrow 2I^-(aq)$$

The overall equation for the reaction is:

$$2S_2O_3^{2-}(aq) + I_2(aq) \rightarrow S_4O_6^{2-}(aq) + 2I^-(aq)$$

The reaction can be used for the direct estimation of iodine, or for the estimation of a substance that can take part in a reaction that produces iodine.

The indicator used in titrating iodine with sodium thiosulfate is starch solution. With free iodine it produces a deep blue-black colour. The blue-black colour disappears as soon as sufficient sodium thiosulfate has been added to react with all of the iodine.

The procedure is as follows.

- Add sodium thiosulfate solution from the burette to the iodine solution until the original brown colour of the iodine changes to pale yellow.
- Add a few drops of starch solution to produce a blue colouration.
- Add the sodium thiosulfate solution drop by drop until the blue-black solution turns colourless.

If the starch is added too early, it adsorbs some of the iodine and reduces the accuracy of the titration.

If starch were not added, it would be very difficult to detect the end point. The colour of the iodine solution becomes very faint towards the end of the reaction, and this makes it difficult to accurately assess when the end point has been reached.

PRACTICAL SKILLS CP13B

A redox titration: determination of the percentage of copper(II) ions in a sample of hydrated copper(II) sulfate

Note that this suggested practical uses techniques and skills similar to those needed in the **Core Practical 13b: Redox titrations with sodium thiosulfate and iodine** in your Lab Book.

The copper(II) sulfate solution must be free from anything but a trace of mineral acid, otherwise the end point is not accurate. This solution produces iodine when potassium iodide is added. The ionic equation for this reaction is:

$$2Cu^{2+}(aq) + 4I^-(aq) \rightarrow 2CuI(s) + I_2(aq)$$

The iodine produced is then titrated with standard sodium thiosulfate solution.

Procedure

1. Weigh accurately a weighing bottle containing approximately 6 g of hydrated copper(II) sulfate crystals.
2. Transfer the crystals to a 250 cm^3 volumetric flask and reweigh the bottle.
3. Add sodium carbonate solution until a *slight* permanent bluish precipitate of copper(II) carbonate is formed.
4. Acidify the mixture with a little dilute ethanoic acid until a clear blue solution is formed. Make up the solution to 250 cm^3 with deionised (or distilled) water and shake well.
5. Pipette 25.0 cm^3 of the solution into a conical flask and add about 1.5 g of solid potassium iodide.
6. Titrate the iodine produced with sodium thiosulfate solution, adding starch solution just as the colour of the mixture changes to pale yellow. The end point is achieved when you can see a white suspension of insoluble copper(I) iodide, CuI(s), in a colourless solution.
7. Perform repeat titrations until you obtain concordant results.

Calculation

Calculate the percentage of copper(II) ions in the sample of hydrated copper(II) sulfate crystals.

Sample results

Concentration of sodium thiosulfate solution = 0.0995 mol dm^{-3}

Mass of hydrated copper(II) sulfate crystals = 5.85 g

Mean titre = 23.45 cm^3

[Molar mass of Cu^{2+} = 63.5 g mol^{-1}]

$$2Cu^{2+} \equiv I_2 \equiv 2Na_2S_2O_3$$

$$\therefore \quad Cu^{2+} \equiv Na_2S_2O_3$$

$$25.0\ cm^3 \text{ of } x \text{ mol dm}^{-3}\ Cu^{2+} \equiv 23.45\ cm^3 \text{ of } 0.0995 \text{ mol dm}^{-3}\ Na_2S_2O_3$$

$$\therefore \quad 25.0x = 23.45 \times 0.0995$$

$$\therefore \quad x = 0.0933$$

$$\text{Solution} = 0.0933 \text{ mol}^{-1} \text{ dm}^{-3} \times 63.5 \text{ g mol}^{-1} \times \frac{250}{1000} = 1.48 \text{ g}$$

$$\text{Percentage of } Cu^{2+} \text{ in the sample} = \frac{1.48}{5.85} \times 100 = 25.3\%$$

Safety Note: Wear eye protection and avoid skin contact with the solids and liquids, especially the hot acid. Boil the acid gently for a short time in a larger beaker to avoid it boiling over and to cool it faster. Make sure the top of the burette is below eye level when you fill it.

CHECKPOINT

1. An iron nail of mass 1.50 g was reacted with dilute sulfuric acid. The resulting solution was transferred to a volumetric flask and made up to 250 cm^3 with deionised water.

   A 25.0 cm^3 sample of this solution reacted with exactly 24.40 cm^3 of 0.0218 mol dm^{-3} potassium manganate(VII) solution.

   The ionic equation for the reaction between the iron nail and the acid is:

   $Fe(s) + 2H^+(aq) \rightarrow Fe^{2+}(aq) + H_2(g)$

   Calculate the percentage by mass of iron in the nail.

   [Molar mass of iron = 55.8 g mol^{-1}]

2. A 50 cm^3 volume of a solution of hydrogen peroxide was diluted with water to 1.00 dm^3. A 25.0 cm^3 sample of this solution was acidified and then titrated against 0.0200 mol dm^{-3} potassium manganate(VII). The mean titre value was 23.90 cm^3. Calculate the concentration, in mol dm^{-3}, of the original solution of hydrogen peroxide.

   The ionic half-equation for the oxidation of hydrogen peroxide is:

   $H_2O_2(aq) \rightarrow 2H^+(aq) + O_2(g) + 2e^-$

3. Calculate the volume of a 0.0200 mol dm^{-3} solution of potassium manganate(VII) required to completely oxidise 1.00 g of iron(II) ethanedioate (FeC_2O_4).

   [Molar mass of FeC_2O_4 = 143.8 g mol^{-1}]

SKILLS PROBLEM-SOLVING CREATIVITY

4. An aqueous solution contains a mixture of iron(II) ions and iron(III) ions. A 25.0 cm^3 volume of the solution required 21.60 cm^3 of 0.0210 mol dm^{-3} potassium manganate(VII) for oxidation.

   A separate 25.0 cm^3 sample of the solution was reacted with an excess of zinc amalgam in order to reduce the iron(III) ions to iron(II). This sample, after filtration to remove the excess zinc, required 44.40 cm^3 of 0.0210 mol dm^{-3} potassium manganate(VII) solution for complete oxidation.

   Calculate the concentrations, in mol dm^{-3}, of both the iron(II) and iron(III) ions in the aqueous solution.

5. When bleaching powder reacts with a dilute acid, chlorine is produced. This chlorine is available for bleaching and is known as 'available' chlorine.

   A 2.50 g sample of bleaching powder is added to an excess of potassium iodide solution and the mixture is then acidified. The solution is placed in a volumetric flask and made up to 250 cm^3 with deionised water. A 25.0 cm^3 volume of this solution required 23.20 cm^3 of 0.105 mol dm^{-3} sodium thiosulfate solution to oxidise the iodine present.

   Calculate the percentage of 'available' chlorine in the 2.50 g sample of bleaching powder.

   $Cl_2(aq) + 2I^-(aq) \rightarrow 2Cl^-(aq) + I_2(aq)$

   [Molar mass of Cl_2 = 71.0 g mol^{-1}]

6. A 25.0 cm^3 sample of of 0.0210 mol dm^{-3} potassium peroxydisulfate, $K_2S_2O_8$, was treated with excess potassium iodide. The iodine produced reacted with 21.00 cm^3 of 0.0500 mol dm^{-3} sodium thiosulfate solution.

   Suggest a likely ionic equation for the reaction between $S_2O_8^{2-}$ ions and I^- ions.

ELECTROLYTE EVOLUTION

One of the reasons that your laptops and smartphones are significantly lighter and slimmer than those of ten years ago is not so much because of advances in computing power but because of advances in rechargeable battery technology.

BATTERY POWER

Primary cells

These are the disposable cells – they are discharged once and discarded – that for over a century powered small and portable equipment. Nowadays, as we become more waste conscious, we find that many of these cells are giving way to rechargeable (secondary) cells and batteries.

The most common primary cells are based on the zinc–manganese dioxide couple: either zinc–carbon cells or alkaline manganese cells. For a short period, some manufacturers offered mercury cells, which were replaced by zinc–air batteries, and some companies now produce 3V lithium–MnO_2 cells.

Zinc–carbon

This cell became commercially available in the late 1800s and was a dry cell version of the original wet Leclanché cell. (The latter was made up of a conducting solution (electrolyte) of ammonium chloride with a negative terminal of zinc and a positive terminal of manganese dioxide.) Replacing the liquid electrolyte with a gel and then sealing the whole cell made it suitable for domestic applications where portability was a key feature. These cells were developed and marketed by Ever Ready for use in radios and torches and are still popular today, though Ever Ready has now changed to Energizer and many other makes have become available.

The dry cell zinc–carbon chemistry performs well in applications where there is intermittent use, such as flashlights, but performance is not so good in devices that put a heavy drain upon the cell. In such applications, polarisation and loss of output results. However, the cell chemistry recovers when left idle for a while – the well-known case of the dead battery that comes back to life. Polarisation is caused by the products of the electrode reactions building up on the electrode surface and preventing new reactants arriving. On standing, diffusion occurs and new reactants reach the electrode as the products disperse.

Anode: Zn metal

Cathode: MnO_2 powder with graphite powder for electrical conduction

Electrolyte: $NH_4Cl(aq)$ and/or $ZnCl_2(aq)$

In earlier cells, the zinc doubled as the outer can, which tended to leak as the cell ran down and the zinc passed into solution.

$$Zn(s) + 4NH_3(aq) + 2MnO_2(s) + 2H_3O^+(aq) \rightarrow [Zn(NH_3)_4]^{2+}(aq) + 2Mn(OH)_3(s)$$

▲ **fig A** Alkaline-manganese batteries.

Alkaline–manganese

These batteries developed from the zinc–carbon cell and became available for domestic use in the 1960s and quickly gained in popularity because they were less prone to polarisation, had greater capacity and were less likely to leak.

Here the anode is powdered zinc, which provides a greater reactive surface area and thus more power. The electrolyte is an alkali, in contrast to the previous cell, which was acidic.

Anode: Zn metal powder

Cathode: MnO_2 powder with graphite powder for electrical conduction

Electrolyte: KOH(aq)

$Zn + 2MnO_2 \rightarrow ZnO + Mn_2O_3 \quad E = 1.5\,V$

Development of these batteries continues, and Panasonic introduced a vacuum-forming process to compact manganese dioxide and graphite powders, which the company claims gives improved capacity.

From 'After Battery Power' by Tony Hargreaves. https://eic.rsc.org/section/feature/battery-power/2020096.article

SCIENCE COMMUNICATION

1 In the final sentence of the extract, the author tells us that Panasonic claim their vacuum forming process for cathodes of alkaline–manganese batteries gives improved capacity. What stages would the technology need to go through before Panasonic's claim was proven?

INTERPRETATION NOTE

Think about the role of the scientific community in validating new knowledge and ensuring integrity.

CHEMISTRY IN DETAIL

2 Write half-equations for the reactions taking place at the anode and the cathode of the zinc–carbon cell.

3 The electrolyte for this cell is $NH_4Cl(aq)$. Why would you expect this electrolyte to be acidic? You may use equations to help explain your answer.

4 Give details of the bonding in the $[Zn(NH_3)_4]^{2+}$ complex and explain why this species is colourless in solution.

5 In the alkaline–manganese cell, the standard electrode potential for Zn is −0.76 V. Calculate the standard electrode potential for the reduction of manganese(IV) oxide to manganese(II) hydroxide given that the E_{cell} = 1.50 V.

WRITING SCIENTIFICALLY

Equations, diagrams and calculations can all be useful tools to help justify an answer, particularly when a question uses the command words 'explain' or 'justify'.

THINKING BIGGER TIP

Think about the electronic configuration of the Zn^{2+} ion.

ACTIVITY

One of the major technological developments over the last few years has been the lithium-ion battery. This has enabled mobile devices to be rechargeable and light-weight. Produce a presentation on the technology behind lithium-ion batteries. Your presentation should be 5–8 slides in length and should include:

- the reactions involved in generating the voltage
- the advantages of lithium-ion batteries over earlier secondary cells
- any disadvantages associated with lithium-ion batteries.

DID YOU KNOW?

In 2013, around five billion lithium-ion batteries were sold to supply electronic devices such as phones, laptops, tablets, cameras, power tools and even electric cars.

16 EXAM PRACTICE

1 $Mg^{2+}(aq) + 2e^- \rightleftharpoons Mg(s)$ $E^\ominus = -2.37$ V

$Cu^{2+}(aq) + 2e^- \rightleftharpoons Cu(s)$ $E^\ominus = +0.34$ V

What is the standard cell potential, $E^\ominus_{cell}$, of the following cell?

$Mg(s) \mid Mg^{2+}(aq) \parallel Cu^{2+}(aq) \mid Cu(s)$

A 2.71 V **B** 2.37 V **C** 2.03 V **D** 0.34 V [1]

(Total for Question 1 = 1 mark)

2 $I_2(aq) + 2e^- \rightleftharpoons 2I^-(aq)$ $E^\ominus = +0.54$ V

$2H^+(aq) + O_2(g) + 2e^- \rightleftharpoons H^2O^2(aq)$ $E^\ominus = +0.68$ V

$H_2O_2(aq) + 2H^+(aq) + 2e^- \rightleftharpoons 2H_2O(l)$ $E^\ominus = +1.77$ V

What is observed when a few drops of acidified hydrogen peroxide solution are added to an excess of aqueous potassium iodide?

A The solution turns brown and effervescence occurs.

B The solution turns brown with no effervescence.

C The solution does not change colour and effervescence occurs.

D The solution does not change colour and no effervescence occurs. [1]

(Total for Question 2 = 1 mark)

3 From the following two standard electrode (redox) potentials:

$MnO_2(s) + 4H^+(aq) + 2e^- \rightleftharpoons Mn^{2+}(aq) + 2H_2O(l)$ $E^\ominus = +1.23$ V

$Cl_2(g) + 2e^- \rightleftharpoons 2Cl^-(aq)$ $E^\ominus = +1.36$ V

it can be predicted that chloride ions will **not** be oxidised by solid manganese(IV) oxide.

In fact, chlorine can be prepared by heating manganese(IV) oxide with concentrated hydrochloric acid.

Why does this prediction fail?

A The reaction does not produce manganese(II) ions.

B The standard electrode (redox) potentials apply only if one mole of manganese(IV) oxide is used.

C The manganese(IV) oxide needs to be in alkaline solution.

D Standard electrode (redox) potentials apply only under specified conditions of concentration and temperature. [1]

(Total for Question 3 = 1 mark)

4 The standard electrode (redox) potential of four half-cells are listed below:

$Cu^{2+}(aq) + 2e^- \rightleftharpoons Cu(s)$ $E^\ominus = +0.34$ V

$AgCl(s) + e^- \rightleftharpoons Ag(s) + Cl^-(aq)$ $E^\ominus = +0.22$ V

$H^+(aq) + e^- \rightleftharpoons \frac{1}{2}H_2(g)$ $E^\ominus = +0.00$ V

$Zn^{2+}(aq) + 2e^- \rightleftharpoons Zn(s)$ $E^\ominus = -0.76$ V

Which standard cell potential, $E^\ominus_{cell}$, could be obtained by combining two of these standard half-cells?

A 0.42 V **B** 0.54 V **C** 0.56 V **D** 0.98 V [1]

(Total for Question 4 = 1 mark)

5 The standard electrode potentials of $Ag^+(aq)|Ag(s)$ and $Zn^{2+}(aq)|Zn(s)$ are +0.80 V and −0.76 V, respectively.

Which conclusion can be made from these data?

A Silver displaces zinc from a solution containing zinc ions.

B Silver is an oxidising agent.

C Zinc has a greater tendency than silver to form positive ions.

D Zinc ions can act as a reducing agent. [1]

(Total for Question 5 = 1 mark)

6 The table shows the formulae of some vanadium ions that can exist in aqueous solution, together with the colour of each ion.

	$VO_3^-(aq)$	$VO^{2+}(aq)$	$V^{3+}(aq)$	$V^{2+}(aq)$
oxidation number of vanadium				
colour of ion	yellow	blue	green	violet

(a) Complete the table by inserting the oxidation numbers. [1]

(b) Iron powder is added to an acidified solution of ammonium vanadate(V), NH_4VO_3, and the mixture is shaken until no further change takes place.

Use the following data to explain what you would expect to observe in the sequence of reactions that take place. Your answer should consider **all** of the electrode reactions listed. [5]

	Electrode reaction	$E^\ominus$/V
1	$V^{2+}(aq) + 2e^- \rightleftharpoons V(s)$	−1.18
2	$V^{3+}(aq) + e^- \rightleftharpoons V^{2+}(aq)$	−0.26
3	$VO^{2+}(aq) + 2H^+(aq) + e^- \rightleftharpoons V^{3+}(aq)$	+0.34
4	$VO_3^-(aq) + 4H^+(aq) + e^- \rightleftharpoons VO^{2+}(aq) + 2H_2O(l)$	+1.00
5	$Fe^{2+}(aq) + 2e^- \rightleftharpoons Fe(s)$	−0.44

(Total for Question 6 = 6 marks)

7 (a) Draw a suitably labelled diagram to show how you would measure the standard electrode potential of the $Cu^{2+}(aq) \mid Cu(s)$ half-cell. [6]

(b) The table contains some standard electrode (redox) potentials involving copper and its ions.

	Electrode reaction	$E^\ominus$/V
1	$Cu^{2+}(aq) + e^- \rightleftharpoons Cu^+(aq)$	+0.15
2	$Cu^{2+}(aq) + 2e^- \rightleftharpoons Cu(s)$	+0.34
3	$Cu^+(aq) + e^- \rightleftharpoons Cu(s)$	+0.52

(i) Use the data in the table to explain why the following reaction is likely to occur. [2]

$2Cu^+(aq) \rightarrow Cu^{2+}(aq) + Cu(s)$

(ii) Use oxidation numbers to explain why this reaction is classified as a disproportionation reaction. [3]

(Total for Question 7 = 11 marks)

8 Chlorine gas can be prepared in the fume cupboard of a laboratory by reacting hydrochloric acid with potassium manganate(VII). The two standard electrode (redox) potentials that relate to this reaction are:

$\frac{1}{2}Cl_2(g) + e^- \rightleftharpoons Cl^-(aq)$ $E^\ominus = +1.36\,V$

$MnO_4^-(aq) + 8H^+(aq) + 5e^- \rightleftharpoons Mn^{2+}(aq) + 4H_2O(l)$ $E^\ominus = +1.52\,V$

(a) State what is meant by the term **standard electrode potential**. [2]

(b) Calculate the standard emf, $E^\ominus_{cell}$, for an electrochemical cell constructed from these two redox systems. [1]

(c) (i) Construct an ionic equation for the reaction that takes place between hydrochloric acid and potassium manganate(VII). Include state symbols. [2]

(ii) Use oxidation numbers to explain which species is acting as the reducing agent in this reaction. [2]

(d) When very dilute solutions of potassium manganate(VII) and hydrochloric acid are added together, there is no visible change. Explain, using the changes in electrode potentials that occur, why no reaction takes place when the solutions are diluted. [3]

(Total for Question 8 = 10 marks)

9 Vanadium(V), in the form of vanadate(V) ions, can be reduced in acidified solution by adding sulfur dioxide.

A student carries out an experiment to determine the new oxidation state of the vanadium after the reaction has taken place. He uses this method:

- He dissolves 2.24 g of ammonium vanadate(V), NH_4VO_3, (molar mass = 116.9 g mol^{-1}) in water and makes up the solution to 250 cm^3.
- In a fume cupboard, he bubbles sulfur dioxide through the solution until there is no further colour change.
- He then boils the solution for five minutes and allows the solution to cool.
- He labels this solution '**Solution Y**'.
- He then titrates 25.0 cm^3 of solution Y against 0.0200 mol dm^{-3} $KMnO_4$ solution. He finds that 38.40 cm^3 of $KMnO_4$ are required to oxidise the vanadium.

In this titration:

- The manganate(VII) ions are reduced from the +7 to the +2 oxidation state.
- The vanadium in solution Y is oxidised to the +5 oxidation state.

(a) Give a reason why the student carried out the experiment in a fume cupboard. [1]

(b) Use the data to show that the oxidation state of the vanadium in solution Y is +3. [5]

(c) The student repeats the experiment with the same mass of ammonium vanadate(V), but this time he boils the solution for only one minute. He finds that the volume of potassium manganate(VII) solution required in the titration has increased. Explain this observation. [2]

(Total for Question 9 = 8 marks)

10 The diagram shows a hydrogen–oxygen fuel cell.

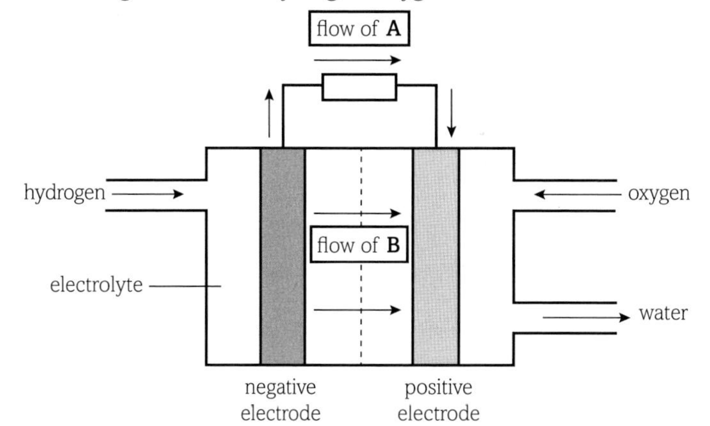

(a) Identify the particles represented by **A** and **B**. [2]

(b) The two electrodes are coated with a metal catalyst. Name a suitable metal to use as the catalyst. [1]

(c) State two advantages of fuel cells over the use of fossil fuels in cars. [2]

(d) Apart from cost, state two disadvantages of using fuel cells rather than fossil fuels in cars. [2]

(e) The zinc–silver oxide cell is used in button cells for watch batteries. It is based on the following half-cell reactions:

$Zn^{2+}(aq) + 2e^- \rightleftharpoons Zn(s)$ $E^\ominus = -0.76V$

$Ag_2O(s) + H_2O(l) + 2e^- \rightleftharpoons 2Ag(s) + 2OH^-(aq)$ $E^\ominus = +0.34V$

(i) Write the cell diagram for this cell. [2]

(ii) Write the equation for the overall reaction taking place in the cell when it is in use. Include state symbols. [2]

(iii) Calculate the standard emf, $E^\ominus_{cell}$, of the cell. [1]

(iv) Use your answer to part (iii) to calculate the equilibrium constant, *K*, for the reaction at 298 K. [R = 8.31 J mol^{-1} K^{-1}] [2]

(Total for Question 10 = 14 marks)

TOPIC 17 TRANSITION METALS AND THEIR CHEMISTRY

A PRINCIPLES OF TRANSITION METAL CHEMISTRY I
B TRANSITION METAL REACTIONS I
C TRANSITION METALS AS CATALYSTS

In **Book 1** you learned about the Periodic Table, and in particular the elements of Groups 1, 2 and 7. These three groups show typical properties of s-block and p-block elements and clear trends in physical and chemical properties. For example, in Group 1, the first ionisation energy decreases down the group, and in Group 7 the reactivity decreases down the group. You learned to understand simple chemical reactions of these elements and their compounds and how to test for ions such as chloride and iodide.

In this topic, you will learn about transition metals, in particular those in the first row between the s-block element calcium and the p-block element gallium. Transition metals have some distinctive properties and uses not generally found in s-block and p-block elements, for example:

- the elements have important uses as industrial catalysts
- they help to remove pollutants from car exhausts
- they are important structural metals, such as iron and copper
- they are used in jewellery, especially platinum, gold and silver
- their compounds have bright colours, such as copper(II) sulfate
- they form ions and compounds with different oxidation numbers, such as iron(II) and iron(III) compounds
- they form complexes, some of which are vital to life, such as haemoglobin, and *cis*-platin, which is used in cancer treatment.

Many of these properties will be explained, especially those that involve complexes. The term 'complex' will be carefully explained. Remember that 'complex' does not mean 'complicated'!

MATHS SKILLS FOR THIS TOPIC

- Appreciate angles and shapes in regular 2D and 3D structures

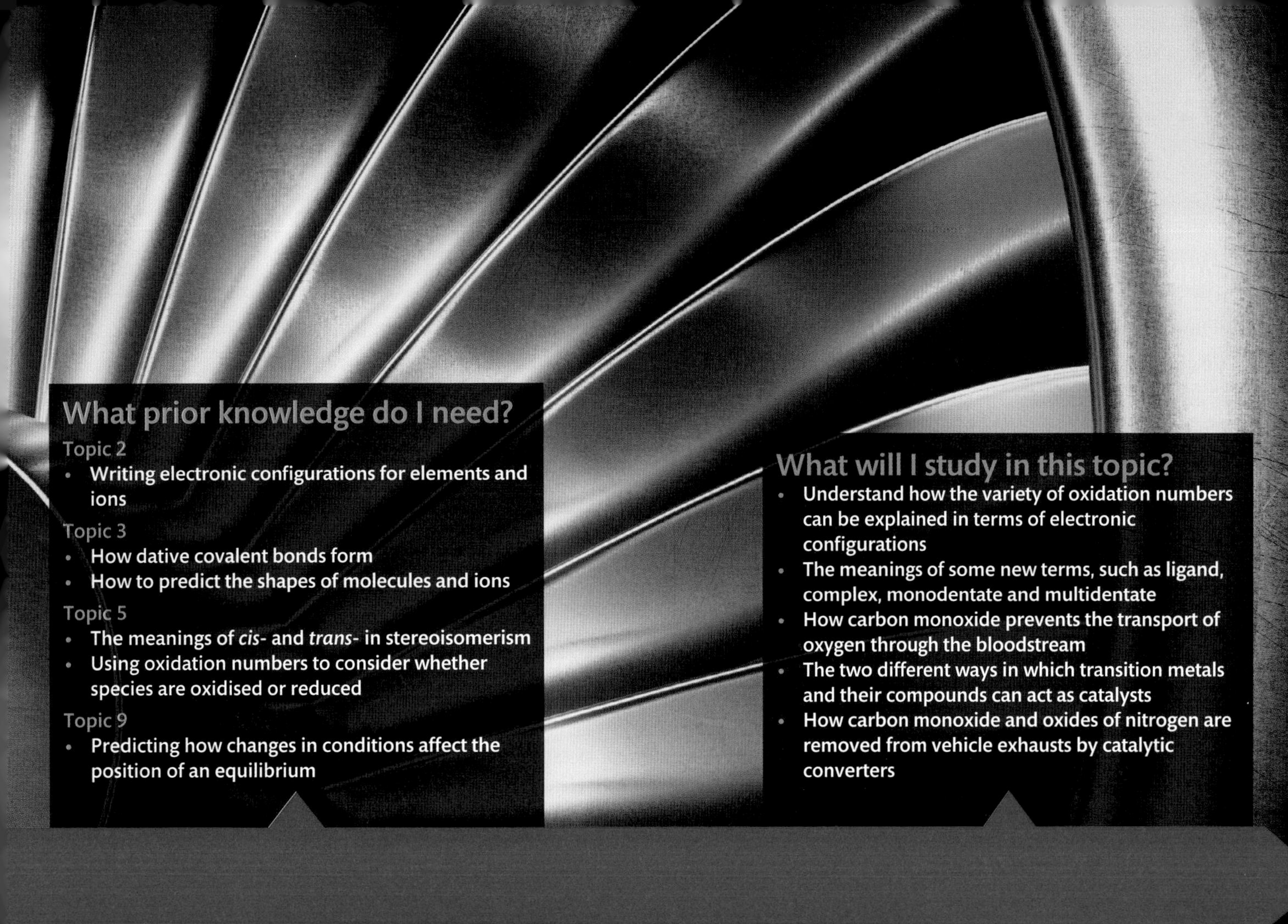

What prior knowledge do I need?

Topic 2

- Writing electronic configurations for elements and ions

Topic 3

- How dative covalent bonds form
- How to predict the shapes of molecules and ions

Topic 5

- The meanings of *cis*- and *trans*- in stereoisomerism
- Using oxidation numbers to consider whether species are oxidised or reduced

Topic 9

- Predicting how changes in conditions affect the position of an equilibrium

What will I study in this topic?

- Understand how the variety of oxidation numbers can be explained in terms of electronic configurations
- The meanings of some new terms, such as ligand, complex, monodentate and multidentate
- How carbon monoxide prevents the transport of oxygen through the bloodstream
- The two different ways in which transition metals and their compounds can act as catalysts
- How carbon monoxide and oxides of nitrogen are removed from vehicle exhausts by catalytic converters

What will I study later?

Study beyond IAL

- The f-block elements, also known as the inner transition elements or rare earth elements
- Uses of f-block elements in modern technology, such as the use of neodymium in headphone speakers, hard drives, electric and hybrid cars and wind turbines

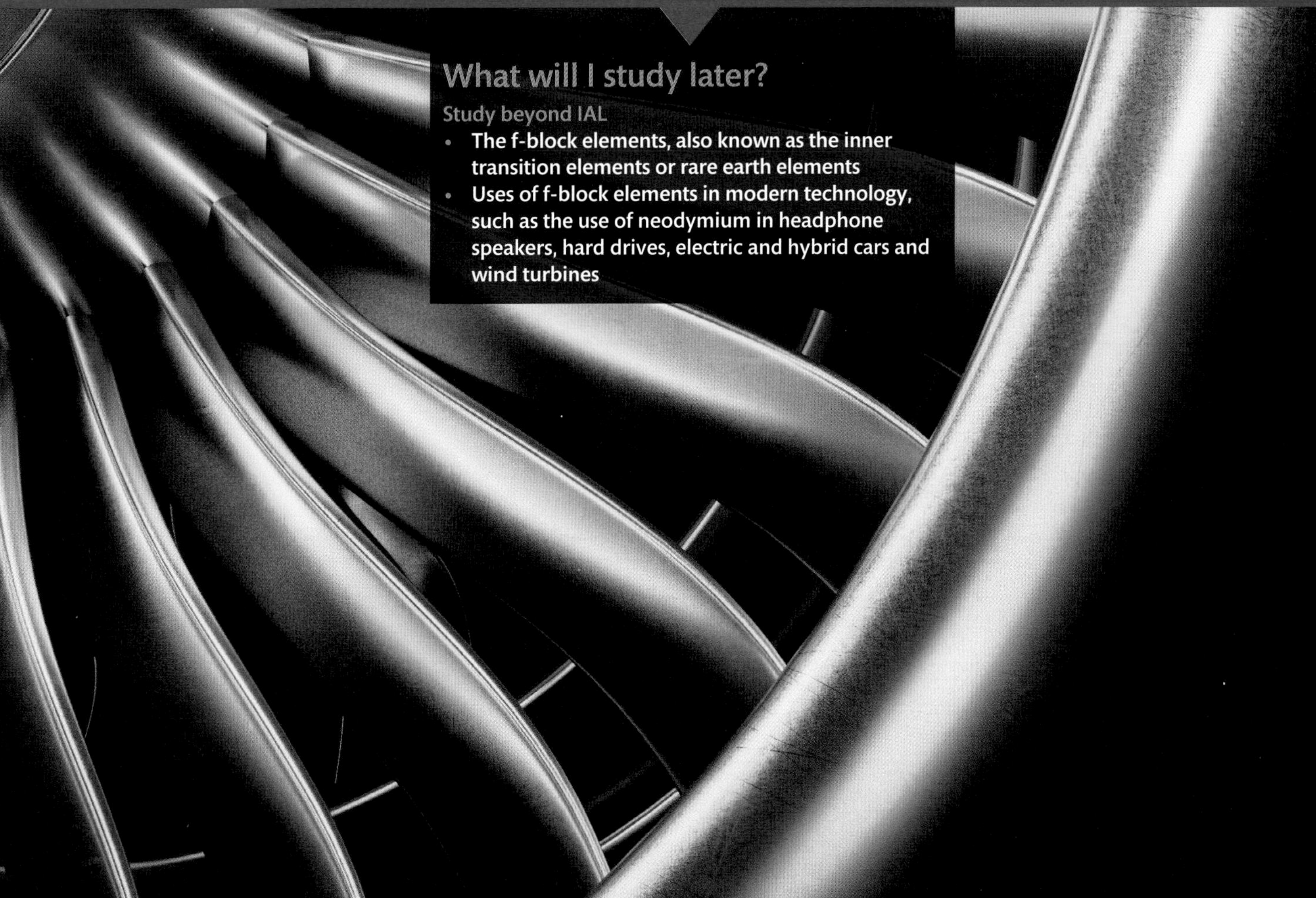

17A 1 TRANSITION METAL ELECTRONIC CONFIGURATIONS

SPECIFICATION REFERENCE 17.1 17.2 17.3

LEARNING OBJECTIVES

- Know that transition metals are d-block elements that form one or more stable ions with incompletely filled d-orbitals.
- Be able to deduce the electronic configurations of atoms and ions of the d-block elements of Period 4 (Sc–Zn) given their atomic number and charge (if any).
- Understand why transition metals show variable oxidation numbers.

WHICH ELEMENTS ARE THE TRANSITION METALS?

WHERE ARE THEY FOUND IN THE PERIODIC TABLE?

You will be familiar with the Periodic Table and the numbers of the different groups. Group 1 is on the far left. On the far right are the noble gases, which are in a group numbered 0 (or sometimes 8). When you studied periodicity using the elements in Periods 2 and 3, you ignored the big gap between Groups 2 and 3. However, when you look at Period 4 you can see elements between Groups 2 and 4. Some of these elements are **transition metals**.

You will also be familiar with the idea of classifying elements as belonging to one of the blocks (s, p, d or f) of the Periodic Table. The transition metals are in the d-block of the Periodic Table (**fig A**). In this topic we will concentrate on the 10 elements in Period 4.

																	0(8)
1	2		H 1									3	4	5	6	7	He 2
Li 3	Be 4											B 5	C 6	N 7	O 8	F 9	Ne 10
Na 11	Mg 12											Al 13	Si 14	P 15	S 16	Cl 17	Ar 18
K 19	Ca 20	Sc 21	Ti 22	V 23	Cr 24	Mn 25	Fe 26	Co 27	Ni 28	Cu 29	Zn 30	Ga 31	Ge 32	As 33	Se 34	Br 35	Kr 36
Rb 37	Sr 38	Y 39	Zr 40	Nb 41	Mo 42	Tc 43	Ru 44	Rh 45	Pd 46	Ag 47	Cd 48	In 49	Sn 50	Sb 51	Te 52	I 53	Xe 54
Cs 55	Ba 56	Lanthanide series	Hf 72	Ta 73	W 74	Re 75	Os 76	Ir 77	Pt 78	Au 79	Hg 80	Tl 81	Pb 82	Bi 83	Po 84	At 85	Rn 86
Fr 87	Ra 88	Actinide series															

the d-block elements

▲ **fig A** The position of the transition metals in the periodic table.

WHAT IS A TRANSITION METAL?

It is easier to list the characteristics of transition metals than to explain fully what they are. To describe them as being in the d-block indicates only their location and not their characteristics. Most definitions of a transition metal refer to the electronic configurations of their atoms or ions (see the Subject Vocabulary for a simple definition).

A fuller definition is quoted in the Edexcel specification:

'Transition metals are d-block elements that form one or more stable ions with incompletely filled d-orbitals.'

CHARACTERISTICS OF TRANSITION METALS

Transition metals:

- are hard solids
- have high melting and boiling temperatures
- can act as catalysts
- form coloured ions and compounds
- form ions with different oxidation numbers
- form ions with incompletely filled d-orbitals.

Of the ten d-block elements in Period 4, the first and the last (scandium and zinc) do not have some of these characteristics. So, although it is correct to describe them as d-block elements, they are not considered to be transition metals. In particular, each forms only one ion and their compounds are not coloured.

▲ **fig B** Transition metals include precious metals such as silver and gold, along with familiar metals such as copper, iron and zinc. Each transition metal has more than one oxidation state, and their salts form coloured solutions in water.

ELECTRONIC CONFIGURATIONS

THE ELEMENTS

You learned in **Topic 2 (Book 1: IAS)** how to write the electronic configurations of the elements from hydrogen to krypton, which includes the ten d-block elements in Period 4. For reference, we show these in full in **table A** using the spdf notation, with electrons-in-box diagrams for the electrons in the 4s and 3d orbitals. Copper and chromium do not follow the expected pattern. Where irregularities in the spdf notation exist, they are shown in red. In the electrons-in-box diagrams, paired electrons are shown with a green background and unpaired electrons with a blue background.

It is sometimes acceptable to use an abbreviated shorter form of electronic configuration that represents the complete shells with reference to the previous noble gas. For example, the electronic configuration of titanium is shown in full as $1s^22s^22p^63s^23p^63d^24s^2$, but can be abbreviated as $[Ar]3d^24s^2$.

THE IONS

When these elements form ions, they lose one or more electrons. Scandium loses both of its 4s electrons and its only 3d electron, forming the Sc^{3+} ion. Zinc loses both of its 4s electrons and none of its 3d electrons, forming the Zn^{2+} ion. Because these two elements form only one ion each, and these ions have no incompletely filled d-orbitals, they are not classified as transition metals.

LEARNING TIP

Remember that in **Topic 2 (Book 1: IAS)** you learned that the 3d orbital starts to fill before the 4s and that the 4s electrons have higher energy than the electrons in the 3d orbitals. This is why the 4s electrons are lost first when the atoms form ions.

ELEMENT	*Z*	ELECTRONIC CONFIGURATION	ELECTRONS-IN-BOXES DIAGRAM FOR 3d AND 4s ELECTRONS					
scandium	21	$1s^2\,2s^2\,2p^6\,3s^2\,3p^6\,3d^1\,4s^2$	↑					↑↓
titanium	22	$1s^2\,2s^2\,2p^6\,3s^2\,3p^6\,3d^2\,4s^2$	↑	↑				↑↓
vanadium	23	$1s^2\,2s^2\,2p^6\,3s^2\,3p^6\,3d^3\,4s^2$	↑	↑	↑			↑↓
chromium	24	$1s^2\,2s^2\,2p^6\,3s^2\,3p^6\,3d^5\,4s^2$	↑	↑	↑	↑	↑	↑
maganese	25	$1s^2\,2s^2\,2p^6\,3s^2\,3p^6\,3d^5\,4s^2$	↑	↑	↑	↑	↑	↑↓
iron	26	$1s^2\,2s^2\,2p^6\,3s^2\,3p^6\,3d^7\,4s^2$	↑↓	↑	↑	↑	↑	↑↓
cobalt	27	$1s^2\,2s^2\,2p^6\,3s^2\,3p^6\,3d^8\,4s^2$	↑↓	↑↓	↑	↑	↑	↑↓
nickel	28	$1s^2\,2s^2\,2p^6\,3s^2\,3p^6\,3d^8\,4s^2$	↑↓	↑↓	↑↓	↑	↑	↑↓
copper	29	$1s^2\,2s^2\,2p^6\,3s^2\,3p^6\,3d^{10}\,4s^1$	↑↓	↑↓	↑↓	↑↓	↑↓	↑
zinc	30	$1s^2\,2s^2\,2p^6\,3s^2\,3p^6\,3d^{10}\,4s^2$	↑↓	↑↓	↑↓	↑↓	↑↓	↑↓

table A Electronic configurations of the elements scandium to zinc.

When the other eight elements form ions, they lose their 4s electrons before their 3d electrons. Each element can lose a variable number of electrons, and so forms ions with different oxidation numbers. The higher oxidation numbers are not found in simple ions. For example, manganese forms an ion with oxidation number +2 (the Mn^{2+} ion), but the manganese ion with oxidation number +7 is MnO_4^-, not Mn^{7+}. Transition metal ions involving higher oxidation numbers usually contain an electronegative element, often oxygen. **Table B** shows the most common oxidation numbers and examples of compounds formed.

ELEMENT	COMMON OXIDATION NUMBERS							COMPOUNDS		
Ti			+3	+4				Ti_2O_3	$TiCl_4$	
V		+2	+3	+4	+5			VCl_3	V_2O_5	
Cr			+3			+6		$CrCl_3$	K_2CrO_4	
Mn		+2		+4			+7	$MnCl_2$	MnO_2	$KMnO_4$
Fe		+2	+3					$FeCl_2$	Fe_2O_3	
Co		+2	+3					$CoSO_4$	$CoCl_3$	
Ni		+2						$NiSO_4$		
Cu	+1	+2						Cu_2O	$CuSO_4$	

table B The most common oxidation numbers and compounds for Period 4 transition elements.

You can see that the highest common oxidation number increases from Ti (+4) to Mn (+7) as all the 4s and 3d electrons become involved in bonding. From Fe to Cu, the increasing nuclear charge means that the electrons are attracted more strongly and are less likely to be involved in bonding. Therefore, ions with higher oxidation numbers are less common.

CHECKPOINT

1. Write out the full and abbreviated electronic configurations for the element chromium.
2. Describe, with reference to its electrons, how an atom of manganese becomes an ion with an oxidation number of +4.

SUBJECT VOCABULARY

transition metal an element that forms one or more stable ions with incompletely filled d-orbitals

17A 2 LIGANDS AND COMPLEXES

SPECIFICATION REFERENCE 17.4 17.5 17.6 17.10

LEARNING OBJECTIVES

- Know what is meant by the term 'ligand'.
- Understand that dative (coordinate) covalent bonding is involved in the formation of complex ions.
- Know that a complex ion is a central metal ion surrounded by ligands.
- Understand the meaning of the term 'coordination number'.

SYMBOLS AND EQUATIONS WITHOUT LIGANDS

An example of the symbol of a metal ion is Na^+, the sodium ion. You will also have used the symbol $Na^+(aq)$, which shows that the sodium ion is dissolved in water. The process of dissolving solid sodium chloride in water may have been explained in terms of the slightly negative ($\delta-$) oxygen atoms in the water molecules being attracted to the positively charged sodium ions and keeping them in solution (**fig A**). You may have a mental picture of Na^+ being surrounded by an unspecified number of water molecules, with these water molecules leaving and being replaced by others.

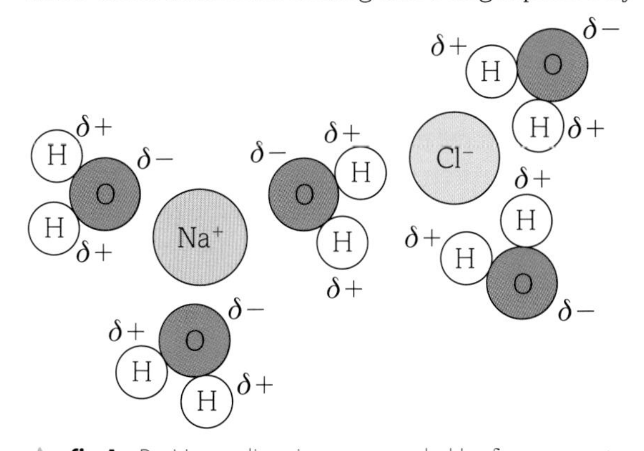

fig A Positive sodium ions surrounded by $\delta-$ oxygen atoms in water molecules, and negative chloride ions surrounded by $\delta+$ hydrogen atoms in water molecules.

When you write equations for the reactions of transition metal ions you can sometimes use the same method. For example, the iron(II) ion can be shown as Fe^{2+}, or as $Fe^{2+}(aq)$ to show that it is dissolved in water.

The equation for a displacement reaction involving a transition metal ion is:

$$Mg(s) + Fe^{2+}(aq) \rightarrow Mg^{2+}(aq) + Fe(s)$$

This equation clearly shows that the reaction is an example of redox, as magnesium transfers electrons to iron(II) ions.

However, it is often better to consider the reactions of transition metal ions in a different way in order to better understand the reactions. This method will be described next.

SYMBOLS AND EQUATIONS WITH LIGANDS

The ions of non-transition metals tend to have larger radii than those of transition metals in the same period of the Periodic Table. For example, the ionic radius of K^+ is 0.133 nm, and that of Fe^{2+} is 0.076 nm. The relatively small size of transition metal ions enables them to attract electron-rich species more strongly, including the water molecules present in aqueous solutions. These water molecules are attracted to the transition metal ions so strongly that they form a specific number of bonds, usually six, giving a structure that can be represented in diagrams such as these:

(i) $[Fe(:OH_2)_6]^{2+}$ — Fe with $H_2O:$, $:OH_2$ ligands and arrows

(ii) $[Fe(OH_2)_6]^{2+}$ — Fe with OH_2 ligands shown with wedges

fig B Here are two ways to represent a complex ion: (i) shows one of the lone pairs of electrons on the oxygen atoms and the dative covalent bonds with arrows; (ii) shows the 3D arrangement of bonds and water molecules around the metal ion.

In **fig B** (i):

- the bonds are shown with arrowheads, indicating that they are **dative (coordinate) bonds** – one of the lone pairs of electrons on each oxygen atom is used to form the bond
- the whole structure is shown inside square brackets, and the original charge of the Fe^{2+} ion is shown outside the bracket
- the water molecules are arranged in a regular pattern around the Fe – this arrangement can be explained in terms of the electron pair repulsion theory.

In **fig B** (ii):

- solid wedges represent bonds coming out of the plane of the paper
- striped wedges represent bonds going behind the plane of the paper.

EXAM HINT

Practise drawing these complexes as in **fig B**. Make sure you show which part of the ligand forms the dative bond with the central metal ion. For example, you can represent water as $:OH_2$, to show that the oxygen part of the water molecule is responsible for the bond.

Although it is often useful to think about diagrams like this, when writing equations it is more usual to abbreviate them. In an equation, the example above would be shown as $[Fe(H_2O)_6]^{2+}$. Notice that the square brackets and the position of the charge are the same in this abbreviated form.

These water molecules, and other electron-rich species that can form dative bonds in the same way, are called **ligands**. The complete formulae are called **complexes**. As most complexes have charges, they are often called **complex ions**, but where there is no overall charge, it is better to use the term 'complex'. The total number of dative bonds around the metal ion is called the **coordination number**.

EXAMPLES OF LIGANDS

Many species can act as ligands. However, those that we will meet most often are shown in **table A**, which also shows how the name of the ligand changes when used in the name of a complex.

LIGAND	FORMULA	CHARGE	NAME IN COMPLEX
water	H_2O	0	aqua
hydroxide	^-OH	−1	hydroxo
ammonia	NH_3	0	ammine
chloride	Cl^-	−1	chloro

table A

Note these points about the names used in complexes:

- ligands with a negative charge end in -o
- ammine should not be confused with the organic term amine.

NAMING COMPLEXES

Table B shows some examples to illustrate the naming system used.

COMPLEX	NAME	NOTES
$[Fe(H_2O)_6]^{2+}$	hexaaquairon(II)	In order, there is: • the number of ligands • the name of the ligand • the name of the metal ion • the oxidation number of the metal ion.
$[FeCl_4]^-$	tetrachloroferrate(III)	The overall charge is 1− because the ion is formed from one Fe^{3+} ion and four Cl^- ions. '-ate' is added to the metal name to show that it is now part of a negatively charged complex ion. As it is negatively charged, a Latin name for the metal is used (in Latin, iron is ferrum).
$[Cu(NH_3)_4(H_2O)_2]^{2+}$	tetraamminediaquacopper(II)	With two different ligands, they appear in alphabetical order (so ammine before aqua), ignoring the tetra- and di- parts of the name.

table B

EXAM HINT

In an exam question you may be asked to **name** complex ions.

LEARNING TIP

You can work out the charge on the metal ion in a complex by considering the overall charge on the complex and knowing the charges on the ligands. For example, $[FeCl_4]^{2-}$ has an overall charge of 2−, and there are four chloride ions each with a charge of 1−, so the charge on the metal ion must be 2+.

CHECKPOINT

SKILLS PROBLEM-SOLVING

1. Explain why the methane molecule cannot act as a ligand.
2. An equation for a reaction involving a transition metal complex is:

$$[Cu(H_2O)_6]^{2+} + 2OH^- \rightarrow [Cu(H_2O)_4(OH)_2] + 2H_2O$$

Explain why the first product is a complex but not a complex ion.

SUBJECT VOCABULARY

dative (coordinate) bond a covalent bond formed between the central metal atom or ion and a ligand, in which both of the bonding electrons are supplied by the ligand
ligand a species that uses a lone pair of electrons to form a dative bond with a metal ion
complex a species containing a metal ion joined to ligands
complex ion a complex with an overall positive or negative charge
coordination number the number of dative (coordinate) bonds in the complex

17A 3 THE ORIGIN OF COLOUR IN COMPLEXES

SPECIFICATION REFERENCE 17.7 17.8 17.9

LEARNING OBJECTIVES

- Know that aqueous solutions of transition metal ions are usually coloured.
- Understand that the colour of aqueous ions, and other complex ions, is a consequence of the splitting of the energy levels of the d-orbitals by ligands.
- Understand why there is a lack of colour in some aqueous ions and other complex ions.

A COMPLEX EXPLANATION!

Colour is very important in our world – think of traffic lights, paint charts, clothing and the colours of flowers. There are scientific explanations for colour in plants, paints and dyes. However, these explanations are often specific to the type of chemical substance being considered.

As far as transition metal complexes are concerned, a full explanation includes three main concepts:

- the electromagnetic spectrum and the colour wheel
- the connection between colour, energy and wavelength
- the electronic configurations of transition metal ions.

THE ELECTROMAGNETIC SPECTRUM

We first need to find out about the **electromagnetic spectrum** (**fig A**). You are probably familiar with the idea that the visible part of the electromagnetic spectrum (sometimes described as 'white light') is made up of a mixture of colours (sometimes described as the colours of the rainbow).

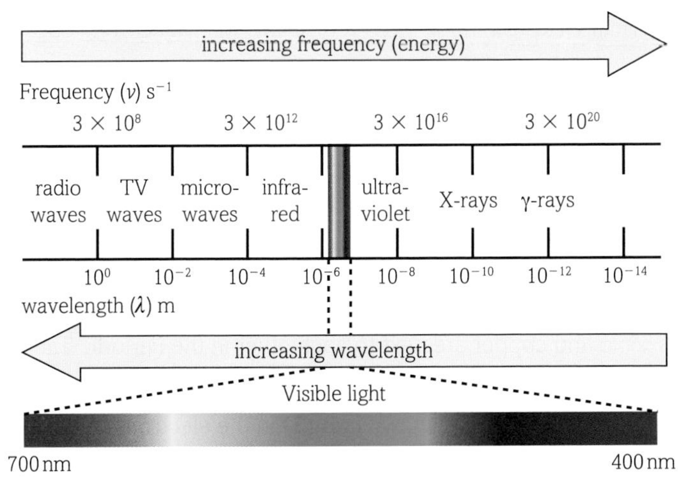

fig A The diagram shows how the visible part of the spectrum fits into the complete electromagnetic spectrum.

So, how many colours make up white light? A traditional answer is seven:

red orange yellow green blue indigo violet

However, seven is a randomly chosen number, and this list of colours does not contain some that are now more familiar, such as cyan and magenta.

The real answer is that visible light is made up of a limitless number of colours! If we take red light to have a wavelength of 700 nm and violet light to have a wavelength of 400 nm, then each precise value within this range (such as 651.0 and 438.4) represents a different wavelength and therefore a different colour. There is clearly no limit to the number of values in this range.

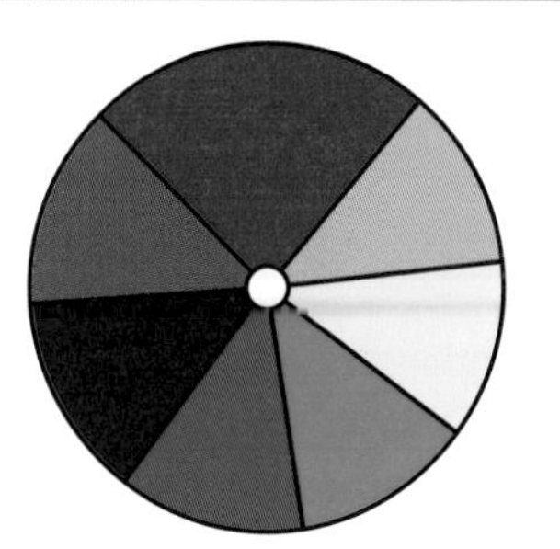

▲ **fig B** This is one example of a colour wheel.

THE COLOUR WHEEL

THE COLOURS IN A COLOUR WHEEL

Now consider the colours in the visible spectrum shown as a wheel, with the red and violet ends of the spectrum next to each other. There are many different colour wheels with varying numbers of colours (just try a web search), but **fig B** shows one that contains seven colours corresponding to those listed above. Clockwise from the top, they are shown in order from red to violet.

COMPLEMENTARY COLOURS

In colour wheels, **complementary colours** are the colours shown opposite each other. In this one, red is opposite blue and green. When white light is passed through a solution containing a transition metal complex, some wavelengths of light are absorbed by the complex. The light emerging will therefore contain proportionately more of the complementary colour. So, if a complex absorbs red light, the light emerging will look blue or green (**fig C**).

▲ **fig C** A range of transition metal solutions. From left to right: Ti^{2+}, V^{3+}, VO^{2+}, Cr^{3+}, $Cr_2O_7^{2-}$, Mn^{2+}, MnO_4^-, Fe^{3+}, Co^{2+}, Ni^{2+} and Cu^{2+}.

COLOUR DEPENDS ON ELECTRONS IN 3d ENERGY LEVELS

A complete explanation of colour, or absence of colour, in aqueous solutions of metal ions is quite complicated. However, the following explanation should help you to understand the basics. We will use an aqueous solution containing copper(II) ions as an example.

An aqueous solution of zinc sulfate is colourless, but an aqueous solution of copper(II) sulfate is blue. As the sulfate ion is present in both solutions we can assume that any difference in colour is caused by the zinc ions and the copper(II) ions. Zinc and copper are next to each other in the Periodic Table, so why should copper, but not zinc, form coloured ions?

The electronic configurations of the Zn^{2+} and Cu^{2+} ions are shown in **fig D** in the electrons-in-boxes format.

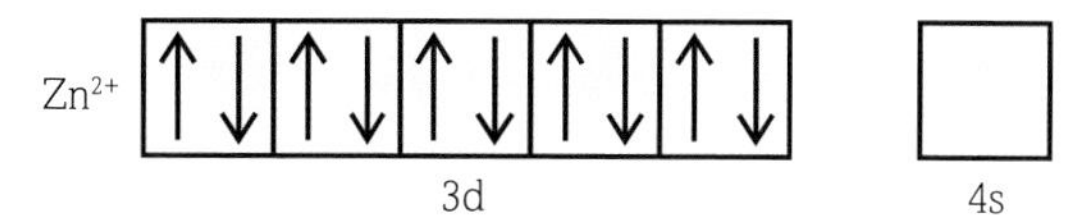

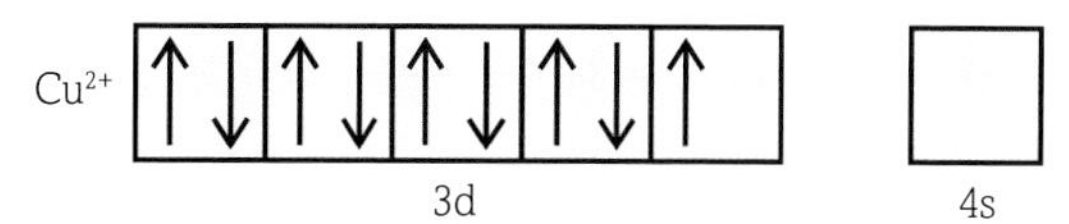

▲ **fig D** The diagram shows the configurations of the 3d and 4s electrons in Zn^{2+} and Cu^{2+}.

Ions that have completely filled 3d energy levels (such as Zn^{2+}) and ions that have no electrons in their 3d energy levels (such as Sc^{3+}) are not coloured. However, the copper(II), or Cu^{2+}, ion has only nine electrons in the 3d energy level, so it is not completely filled. When water ligands are attached to the copper(II) ion, the energy level splits into two levels with slightly different energies (**fig E**). The lower energy level contains six electrons and the higher energy level contains three electrons.

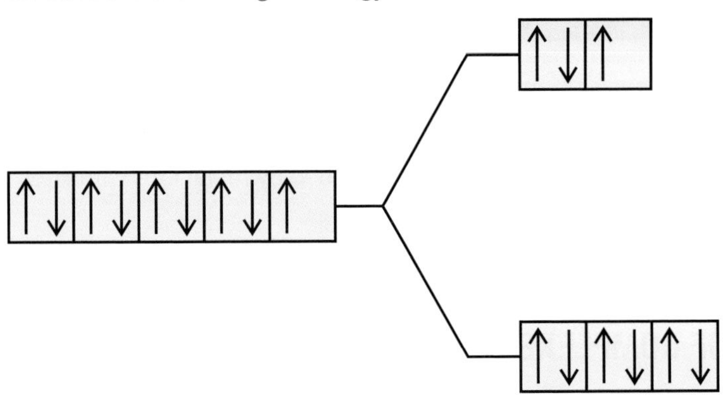

▲ **fig E** In the Cu^{2+} ion, the water ligands split the 3d energy level into lower and higher energy levels.

DID YOU KNOW?

As with a number of theories suggested at advanced levels of the study of chemistry, the concept that colour in transition metal complexes is the result solely of d–d transitions is too simple. The colours of a number of transition metal ions, such as the purple colour of MnO_4^-, are a result of charge-transfer transitions. These transitions occur when an electron moves from a ligand-based orbital to a metal-based orbital, or vice versa. If you are interested, you might like to do some research into this theory.

If one of the electrons in the lower energy level absorbs energy from the visible spectrum, it can move to the higher energy level (**fig F**). Movement from a lower energy level to a higher energy level is called **promotion** or **excitation**. When an electron moves to a higher energy level, the amount of energy it absorbs depends on the difference in energy between the two levels – the bigger the energy difference, the more energy the electron absorbs. It is important to know that the amount of energy gained by the electron is directly proportional to the frequency of the absorbed light (so the energy gained increases as the frequency increases), and inversely proportional to the wavelength of the light (so the energy gained increases as the wavelength decreases). In the case of the Cu^{2+} ion, the small difference in energy levels means that low frequency (or high wavelength) radiation is absorbed from the red end of the spectrum. Therefore, blue light is transmitted.

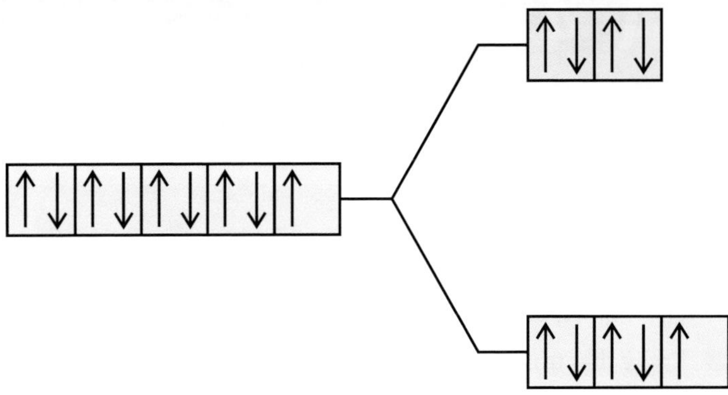

▲ **fig F** Note that one electron has moved from the lower energy level to the higher energy level.

LEARNING TIP

Try to find a way to remember that red light has a longer wavelength than violet light. The Red Sea (between Africa and Asia) is a long (and narrow) sea. Perhaps you can find a better way!

EXAM HINT

Changing the type of ligand in a complex (for example, from water to ammonia) can change the gap between the split d-orbitals. This results in a change in the colour of the complex.

As an example you could think about the difference in colour between $[Cu(H_2O)_6]^{2+}$ and $[Cu(NH_3)_4(H_2O)_2]^{2+}$ in **Section 17B.1**. Use the colour wheel to decide which colours are being absorbed by each complex.

CHECKPOINT

1. Solutions containing Fe^{3+}(aq) ions are yellow. Explain which part of the electromagnetic spectrum is absorbed by these ions.
2. State why solutions containing Al^{3+}(aq) ions are colourless.

SUBJECT VOCABULARY

electromagnetic spectrum the range of all wavelengths and frequencies of all the types of radiation
complementary colours colours opposite each other on a colour wheel
promotion (or **excitation**) when an electron moves from a lower energy level to a higher energy level

17A 4 COMMON SHAPES OF COMPLEXES

LEARNING OBJECTIVES

- Understand why complexes with six-fold coordination have an octahedral shape, such as those formed by metal ions with H_2O, OH^- and NH_3 as ligands.
- Know that transition metal ions may form tetrahedral complexes with relatively large ions such as Cl^-.

PREDICTING THE SHAPES OF COMPLEXES

USING ELECTRON PAIR REPULSION THEORY

You already know from **Topic 3 (Book 1: IAS)** about the use of the electron pair repulsion theory to predict and explain the shapes of simple molecules and ions. It is quite easy to extend this idea to predicting and explaining the shapes of complexes. A ligand bonds to the central metal ion by donating a pair of electrons to form a dative bond. The only differences are that you should ignore the 3d electrons in the transition metal ion and the overall charge on the complex – just count the number of electron pairs donated by the ligands. **Table A** shows how this theory can be applied in exactly the same way to predict the shape of a transition metal complex and the shape of a simple molecule.

NUMBER OF LIGANDS	ELECTRON PAIRS DONATED	SHAPE	BOND ANGLE	EXAMPLE OF A COMPLEX ION	EXAMPLE OF A SIMPLE MOLECULE
6	6	octahedral	90°	$[Co(NH_3)_6]^{2+}$	SF_6
4	4	tetrahedral	109.5°	$[CuCl_4]^{2-}$	CH_4
2	2	linear	180°	$[Ag(NH_3)_2]^+$	$BeCl_2$

table A The electron pair repulsion theory can predict the shapes of simple molecules and complex ions.

OCTAHEDRAL COMPLEXES

The most common ligands in most octahedral complexes are water, ammonia and the hydroxide ion. These ligands have different numbers of lone pairs of electrons (ammonia has one, water has two and the hydroxide ion has three). However each ligand uses only one lone pair to form a coordinate bond with the transition metal ion. As they contain six ligands, the complexes are sometimes described as having **six-fold coordination**. Note that the electron pair donor in each of these ligands is an element in Period 2 of the Periodic Table, so all three ligands are of approximately equal size. **Table B** shows some common examples.

ABBREVIATED FORMULA	NAME	COLOUR
$[Mn(H_2O)_6]^{2+}$	hexaaquamanganese(II)	very pale pink (usually described as colourless)
$[Fe(H_2O)_4(OH)_2]$	tetraaquadihydroxoiron(II)	pale green
$[Cr(OH)_6]^{3-}$	hexahydroxochromate(III)	green

table B Examples of octahedral complexes.

Fig A shows the shapes of these complexes.

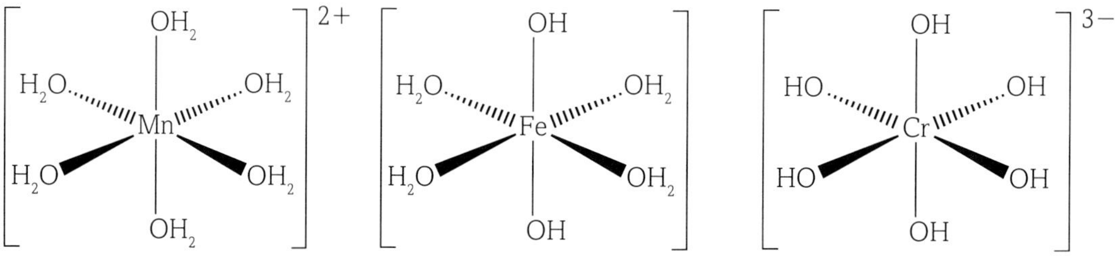

fig A The structures of some octahedral complexes.

Note that:

- Solid and striped wedge bonds are used to indicate the shapes, although the 3D shape can be shown in other ways.
- The overall charges are all different – this depends on the charge on the original transition metal ion and on how many negatively charged ligands there are.
- All three structures use square brackets, although these are sometimes left out when the complex is neutral.
- The water ligands are shown as both H_2O and OH_2 – the important point is that the bond must be shown to come from oxygen because it supplies the lone pair of electrons for the dative bond.

TETRAHEDRAL AND LINEAR COMPLEXES

These are much less common than octahedral complexes. The only tetrahedral complexes that you are likely to see in your course are those in which chloride ions act as ligands, such as in the $[CuCl_4]^{2-}$ ion. Note that because chlorine is a Period 3 element, its ions are much bigger than water molecules, ammonia molecules and hydroxide ions. So, there is usually not enough room around the central metal ion for six chloride ions to act as ligands.

The only linear complex you are likely to see in your course is the reactive ion present in Tollens' reagent (sometimes called ammoniacal silver nitrate). An explanation of why the Ag^+ ion has only two ligands, and not six, is beyond the aims of this book, but note that silver is a transition metal in Period 5 (not Period 4) of the Periodic Table. It therefore behaves differently from the transition metals in Period 4.

Fig B shows the shapes of these two complexes.

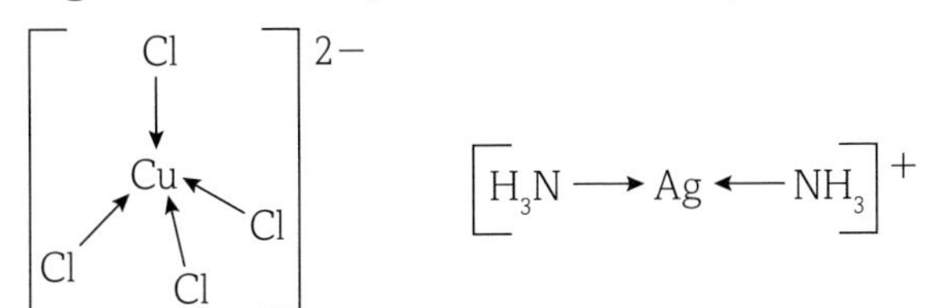

▲ **fig B** The structures of a tetrahedral complex and a linear complex.

LEARNING TIP

You can usually assume that a complex with six ligands is octahedral, one with four ligands is tetrahedral and one with two ligands is linear.

CHECKPOINT

SKILLS REASONING

1. A complex that forms between cobalt ions and nitrite ions (NO_2^-) has the abbreviated formula $[Co(NO_2)_6]^{3-}$. Predict its shape and the oxidation number of the transition metal ion.
2. A complex forms between nickel(II) ions and chloride ions. Predict its shape, name and formula.

SUBJECT VOCABULARY

six-fold coordination complexes in which there are six ligands forming coordinate bonds with the transition metal ion

17A 5 SQUARE PLANAR COMPLEXES

SPECIFICATION REFERENCE 17.15

LEARNING OBJECTIVES

- Know that square planar complexes are also formed by transition metal ions and that *cis*-platin is an example of such a complex, which is used in cancer treatment where it is supplied as a single isomer and not in a mixture with the *trans* form.

SQUARE PLANAR MOLECULES

We know that the electron pair repulsion theory can be applied to predict the shapes of simple molecules and ions. However, you may not have seen an example of a **square planar** shape. One example is xenon tetrafluoride, XeF_4. Although xenon is a noble gas, it does form some stable compounds.

An atom of xenon has eight electrons in its outermost energy level, and each fluorine atom uses one of its electrons to form a covalent bond. The outer energy shell now contains 12 electrons. These are arranged in six pairs, forming an octahedral arrangement. Two of these pairs are lone pairs, which repel each other and are therefore located opposite each other. So, the four bonding pairs are in a plane, with the four fluorine atoms at the corners of a square and an F–Xe–F bond angle of 90°, as shown in **fig A**.

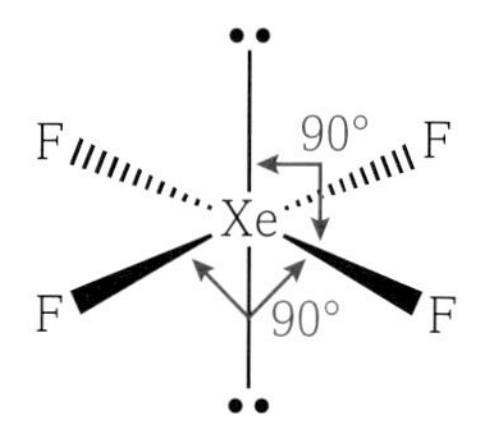

fig A The square planar shape of the xenon tetrafluoride molecule.

cis-PLATIN

Square planar complexes are much less common than octahedral and tetrahedral complexes. One particular complex of this type, *cis*-platin, has become well known in recent years because of its use as an effective treatment for some types of cancer, especially testicular cancer.

An explanation of why the four ligands in this complex form a square planar shape, and not a tetrahedral shape, is beyond the aims of this book. It is not easily explained by the electron pair repulsion theory. Part of the explanation is that platinum is a transition metal in Period 6, and so behaves differently from the transition metals in Period 4.

CIS-TRANS ISOMERS

You have learned about *E–Z* and *cis-trans* isomerism in **Topic 5 (Book 1: IAS)**. Your understanding of this type of isomerism can now be applied to inorganic compounds such as *cis*-platin and its isomer, *trans*-platin. These consist of a platinum(II) ion, two ammonia ligands and two chloride ion ligands.

Fig B shows the structures of these isomers and the relationship between them.

- The *cis*- prefix indicates that identical ligands are next to each other.
- The *trans*- prefix indicates that they are opposite each other.

H_3N, NH_3, Pt, Cl, Cl — *Cis*

H_3N, Cl, Pt, Cl, NH_3 — *Trans*

fig B Structures of *cis*-platin and *trans*-platin.

ANTI-CANCER ACTION

A full understanding of how *cis*-platin kills cancer cells is beyond the aims of this book. Put simply:

- All cells, including cancer cells, contain deoxyribonucleic acid (DNA).
- During cell division, the two strands of DNA must separate from each other to form more DNA.
- The structure of *cis*-platin enables it to form a bond between the two strands of DNA (**fig C**), which prevents them from separating and so prevents the cancer cells from dividing.

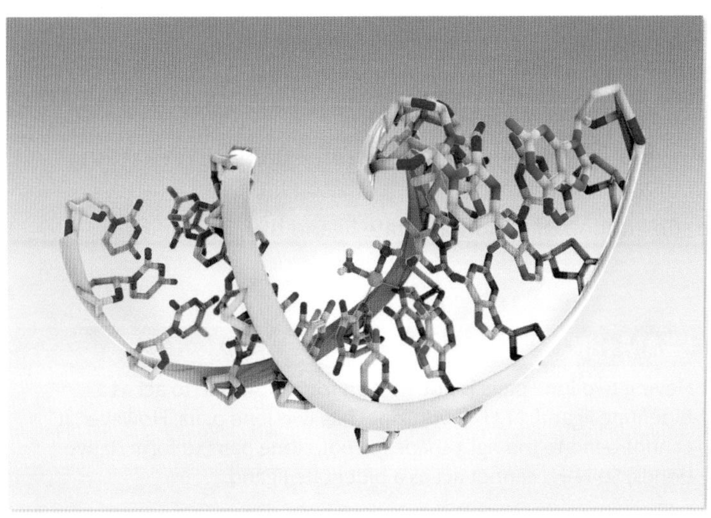

▲ **fig C** DNA strands are shown in grey and *cis*-platin in pink.

trans-PLATIN

The isomers *trans*-platin and *cis*-platin have different structures. The difference in structure means that *trans*-platin is a much less effective cancer treatment than *cis*-platin. It is also more toxic. So, it is important to use only the *cis*- isomer in the cancer treatment.

LEARNING TIP

You should know the structures of *cis*-platin and *trans*-platin, but you do not need to know exactly why *cis*-platin is effective at treating cancer but *trans*-platin is not.

CHECKPOINT

1. Explain why you would expect the ICl_4^- ion to have a square planar shape.
2. (a) Draw the structures of the two *cis-trans* isomers of the $IF_2Cl_2^-$ ion.
   (b) Why are there no *cis-trans* isomers of the $IFCl_3^-$ ion?

SUBJECT VOCABULARY

square planar this shape contains a central atom or ion surrounded by four atoms or ligands in the same plane and with bond angles of 90°

17A 6 MULTIDENTATE LIGANDS

SPECIFICATION REFERENCE 17.12 17.16 17.17 17.25

LEARNING OBJECTIVES

- Understand that H_2O, Cl^- and NH_3 act as monodentate ligands
- Understand the terms 'bidentate' and 'hexadentate' in relation to ligands, and be able to identify examples such as $NH_2CH_2CH_2NH_2$ and $EDTA^{4-}$
- Know that haemoglobin is an iron(II) complex containing a polydentate ligand and that ligand exchange occurs when an oxygen molecule bound to haemoglobin is replaced by a carbon monoxide molecule.
- Understand, in terms of the positive increase in ΔS_{system}, that the substitution of a monodentate ligand by a bidentate or hexadentate ligand leads to a more stable complex ion.

DENTICITY

Denticity is a rarely used English word, but it is a property of ligands that you should be aware of. It comes from the Latin word *dentis* (meaning 'tooth'), from which we get the familiar word 'dentist'. There is a connection!

When ligands were introduced earlier in this book we could have described them as **monodentate ligands**. You can think of monodentate as meaning 'one tooth' or 'one bite', which means that the ligand uses one lone pair of electrons on one atom to form the dative bond with the metal ion. H_2O, OH^- and NH_3 all act as monodentate ligands because they use only one pair of electrons to form a dative (coordinate) bond in complexes.

Now for **bidentate ligands.** *Bi* means two, so a bidentate ligand has two atoms, each of which can use a lone pair of electrons to form a dative bond with the metal ion. You can imagine that a bidentate ligand has two atoms that can 'bite' onto the metal ion.

You can work out for yourself what a **multidentate ligand** is – a ligand with several atoms, each of which uses a lone pair of electrons to form a dative bond with the metal ion. The most common example you are likely to meet has six such atoms, so it could be described as a **hexadentate** ligand.

BIDENTATE LIGANDS

The most likely bidentate ligand you will meet in your chemistry course is the organic compound with the structural formula $NH_2CH_2CH_2NH_2$. Its correct name is 1,2-diaminoethane, although for a reason we will soon see, it is sometimes called ethylenediamine – the ethylene comes from the CH_2CH_2 part, and it is correctly described as a diamine because there are two amino groups. Its structure is:

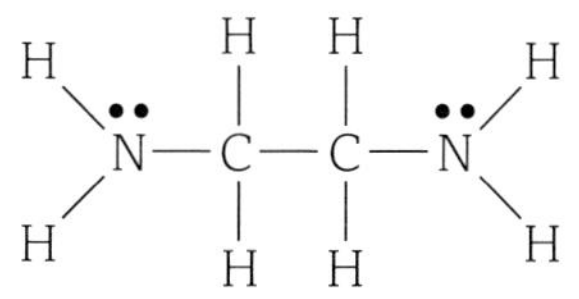

fig A The structure of 1,2-diaminoethane.

EXAM HINT

Having two lone pairs is not enough for a molecule to act as a bidentate ligand. N_2H_4 (hydrazine) has two lone pairs. However, it cannot bend to the right shape for both lone pairs to form dative bonds. So N_2H_4 cannot act as a bidentate ligand.

Occasionally it is abbreviated to 'en', especially when used in equations representing its reactions as a ligand.

When it acts as a bidentate ligand, it uses the lone pair of electrons on each nitrogen atom to attach to the metal ion. **Fig B** shows an Ni^{2+} ion joined to three molecules of 1,2-diaminoethane.

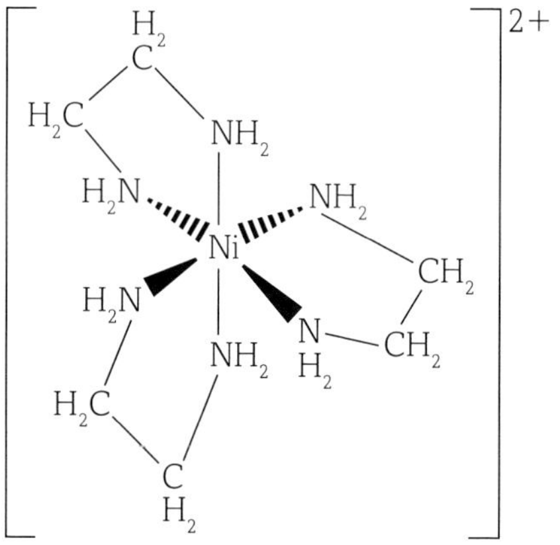

fig B The structure of the complex formed when Ni^{2+} reacts with three molecules of 1,2-diaminoethane.

If this complex were formed by the reaction between a hexaaqua metal ion and 1,2-diaminoethane, this abbreviated equation could be written:

$$[Ni(H_2O)_6]^{2+} + 3en \rightarrow [Ni(en)_3]^{2+} + 6H_2O$$

MULTIDENTATE LIGANDS

The most likely multidentate ligand you will see in your chemistry course is an organic ion with a rather complicated structure.

First, consider the structure of 1,2-diaminoethane. Next, imagine that each of the four hydrogen atoms on the two nitrogens are replaced by $-CH_2COOH$ (this is ethanoic acid bonded to the nitrogens via the CH_3 group), as shown in **fig C**.

fig C The structure of the EDTA molecule.

You may know that the old name for ethanoic acid is acetic acid. Four of these molecules have been used to form this structure. Now, remembering the alternative name for 1,2-diaminoethane, you should be able to see that one name for this structure could be ethylenediaminetetraacetic acid. This is difficult to say, so pick out four key letters from the name to get the abbreviation that is normally used for the compound. EthyleneDiamineTetraAcetic acid now becomes 'EDTA'. We will not attempt the IUPAC name!

The final step is to consider the ion formed when each of the ethanoic acid groups loses its H^+ ion – this gives an ion with four negative charges. It is a hexadentate ligand referred to as $EDTA^{4-}$ and has this structure:

fig D The structure of the $EDTA^{4-}$ ion.

The advantage of displaying the $EDTA^{4-}$ ion in this way is that it clearly shows the six lone pairs of electrons that are used to form the dative bonds when the ion acts as a ligand (two pairs on the nitrogen atoms and four pairs on the oxygen atoms). **Fig E** uses a skeletal formula to show the complex formed between one M^{2+} ion and one $EDTA^{4-}$ ion.

fig E The structure of the complex formed between M^{2+} and $EDTA^{4-}$.

The six dative bonds formed by $EDTA^{4-}$ are shown as dashed lines.

THE STABILITY OF COMPLEXES

Most complexes are stable, in the sense that they do not decompose easily. In this heading, 'stability' does not refer to complexes containing transition metal ions with unstable oxidation numbers, such as +2 in scandium (Sc^{2+}). Instead, it refers to a comparison of the stabilities of two complexes in which the number of ligands has changed. For a discussion of 'stability' see **Topic 12**. Consider this ligand exchange reaction in which a monodentate ligand is replaced by a bidentate ligand:

$$[Cu(H_2O)_6]^{2+} + 3en \rightarrow [Cu(en)_3]^{2+} + 6H_2O$$

Six water ligands are replaced by three 1,2-diaminoethane ligands, so the total number of species has increased from four to seven. This means that the system is more disordered, and so there is an increase in ΔS_{system}. Ligand exchange reactions of this sort lead to an increase in stability of the products compared to the reactants, and so formation of the products is favoured.

When a monodentate ligand is replaced by a bidentate or hexadentate ligand, the increase in stability is even greater, as in this example with $EDTA^{4-}$:

$$[Cu(H_2O)_6]^{2+} + EDTA^{4-} \rightarrow [Cu(EDTA)]^{2-} + 6H_2O$$

Here, six water ligands are replaced by one $EDTA^{4-}$ ligand, so the total number of species has increased from two to seven.

HAEMOGLOBIN AND OXYGEN TRANSPORT

You may know that haemoglobin is a protein in red blood cells that plays a vital role in transporting oxygen through the bloodstream in humans and other animals. Here is a simplified explanation of the role of haemoglobin in oxygen transport, without including details of its structure and of the haem group that it contains:

- Haemoglobin consists of three main parts, the largest of which is protein (the 'globin' part).
- Within the protein, there are four haem groups that are made up mostly of carbon and hydrogen atoms.
- Inside each haem group, there are four nitrogen atoms that hold an Fe^{2+} ion by forming dative bonds with it in a square planar structure.
- There is a fifth dative bond from the protein to the Fe^{2+} ion.
- When blood passes through the lungs, haemoglobin collects oxygen molecules and transports them to cells, where it is released.
- When haemoglobin collects oxygen, the oxygen molecule acts as a ligand by using one of its lone pairs of electrons to form a dative bond with one of the Fe^{2+} ions.

Fig F is a very simplified diagram showing the six dative bonds in oxyhaemoglobin.

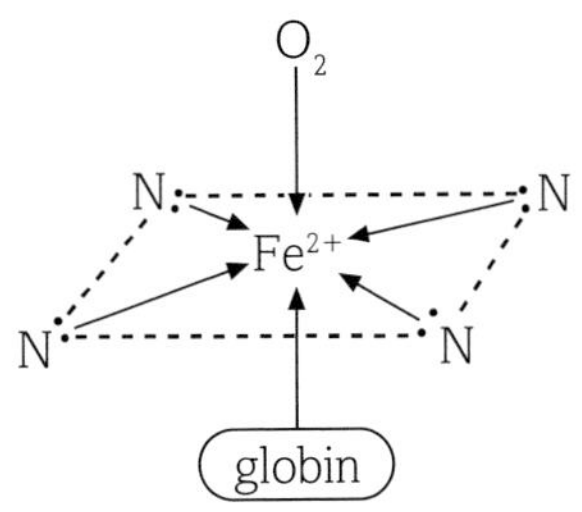

fig F Part of the structure of oxyhaemoglobin, showing the dative covalent bonds to the central metal ion.

HAEMOGLOBIN AND CARBON MONOXIDE

A carbon monoxide molecule has a lone pair of electrons on its carbon atom that enables it to act as a ligand. The strength of the dative bond between oxygen and haemoglobin is not particularly strong. However, the advantage of this is that the oxygen molecule is relatively easily released when it is needed.

Unfortunately, a much stronger dative bond forms between carbon monoxide and haemoglobin. This means that any carbon monoxide breathed in is very likely to replace the oxygen already bound to haemoglobin – this is a ligand substitution reaction. Once the carbon monoxide has formed carboxyhaemoglobin, the dative bond is so strong that it does not break easily. This means that if enough haemoglobin molecules are converted to carboxyhaemoglobin, there may be too little oxygen transported to support life.

DID YOU KNOW?

Carbon monoxide poisoning can be treated by breathing in pure oxygen to reverse the process of absorption of carbon monoxide. In severe cases of carbon monoxide poisoning, a complete blood transfusion is required.

Put very simply, this reaction is reversible:

$$\text{haemoglobin} + \text{oxygen} \rightleftharpoons \text{oxyhaemoglobin}$$

but this reaction is not easily reversible

$$\text{haemoglobin} + \text{carbon monoxide} \rightarrow \text{carboxyhaemoglobin}$$

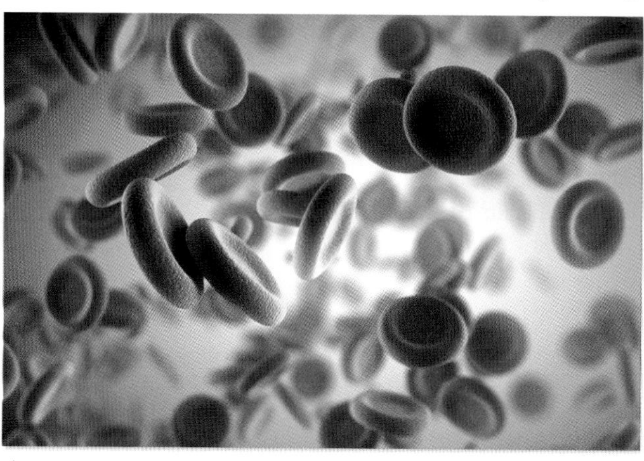

▲ **fig G** These red blood cells contain haemoglobin that transports oxygen from the lungs to the cells in the body.

LEARNING TIP

In 1,2-diaminoethane, there are only two lone pairs of electrons in the molecule and both are used when it acts as a ligand. In $EDTA^{4-}$ the lone pairs of electrons on the O of the C=O are not used. Only those on the N atoms and the negatively charged O atoms are used.

CHECKPOINT

1. The water molecule contains three atoms and has two lone pairs of electrons. Explain why it can only act as a monodentate ligand, and not as a bidentate ligand.
2. Write an equation for the ligand substitution reaction that occurs when four of the ligands in the hexaaquairon(III) ion are replaced by 1,2-diaminoethane (en).

SUBJECT VOCABULARY

monodentate ligand a molecule or ion that forms one dative bond with a metal ion
bidentate ligand a molecule or ion that forms two dative bonds with a metal ion
multidentate ligand a molecule or ion that forms several dative bonds with a metal ion
hexadentate ligand a molecule or ion that forms six dative bonds with a metal ion

17B 1 DIFFERENT TYPES OF REACTIONS

SPECIFICATION REFERENCE
17.11 | 17.24(i) | 17.24(ii) | CP14

LEARNING OBJECTIVES

- Understand that colour changes in transition metal ions may arise as a result of changes in:
  (i) oxidation number of the ion
  (ii) ligand
  (iii) coordination number of the complex.
- Understand that ligand exchange, and an accompanying colour change, occurs in the formation of:
  (i) $[Cu(NH_3)_4(H_2O)_2]^{2+}$ from $[Cu(H_2O)_6]^{2+}$ via $Cu(OH)_2(H_2O)_4$
  (ii) $[CuCl_4]^{2-}$ from $[Cu(H_2O)_6]^{2+}$.
- Be able to write ionic equations to show the meaning of amphoteric behaviour, deprotonation and ligand exchange reactions.

TYPES OF REACTIONS

So far, we have considered the origin of colour in transition metal ions. We can now consider why there are often colour changes when transition metal ions take part in reactions. Four main types of reactions can occur:

- redox – the oxidation number of the transition metal ion changes
- **deprotonation** – one or more of the ligands gains or loses a hydrogen ion (proton)
- **ligand exchange** – one or more of the ligands around the transition metal ion is replaced by a different ligand
- coordination number change – the number of ligands changes.

Any one of these types of reactions can cause a change in the colour of the complex. Some reactions involve more than one of these types of reactions.

CHANGE IN OXIDATION NUMBER

An aqueous solution containing $Fe^{2+}(aq)$ ions is pale green, but when it is exposed to air it gradually turns yellow or brown, as the oxidation number of iron increases from +2 to +3. The type and number of ligands remain unchanged in this oxidation reaction, so the formulae of the two complexes are $[Fe(H_2O)_6]^{2+}$ and $[Fe(H_2O)_6]^{3+}$. Colour changes such as the one in this reaction are best illustrated using solid samples containing the ions (see **fig A**).

Equations are not usually written for oxidation reactions in which the only change is the oxidation number of the transition metal ion.

Iron(II) sulphate ($FeSO^4$)

Iron(III) sulphate ($Fe2(SO^4)^3$)

fig A Solid samples clearly show colour differences between ions.

FORMATION OF $[Cu(NH_3)_4(H_2O)_2]^{2+}$: DEPROTONATION AND LIGAND EXCHANGE REACTIONS

Consider the reaction that occurs when aqueous sodium hydroxide is added to copper(II) sulfate solution. The observation is that a pale blue solution forms a blue precipitate. The equation for this reaction is:

$$[Cu(H_2O)_6]^{2+} + 2OH^- \rightarrow [Cu(H_2O)_4(OH)_2] + 2H_2O$$

You might think that this is a ligand substitution reaction – that two hydroxide ions have replaced two water molecules. In fact, it is a deprotonation reaction – the two hydroxide ions have removed hydrogen ions from two of the water ligands and converted them into water molecules. The two water ligands that have lost hydrogen ions are now hydroxide ligands.

PRACTICAL SKILLS — CP14

Tetramminecopper(II) sulfate-1-water, $[Cu(NH_3)_4.SO_4].H_2O$, can be prepared by adding aqueous ammonia to an aqueous solution of copper(II) sulfate.

The overall equation for the reaction is:

$$[Cu(H_2O)_6]^{2+} + 4NH_3 + SO_4^{2-} \rightarrow [Cu(NH_3)_4].SO_4.H_2O + 5H_2O$$

This is an example of a ligand exchange reaction. This is the reaction you might investigate in **CP14: The preparation of a transition metal complex**.

Exactly the same observations can be made during the careful addition of aqueous ammonia instead of aqueous sodium hydroxide. The equation for this reaction is:

$$[Cu(H_2O)_6]^{2+} + 2NH_3 \rightarrow [Cu(H_2O)_4(OH)_2] + 2NH_4^+$$

DID YOU KNOW?

The formula for the copper(II)-ammine complex in aqueous solution is sometimes given as $[Cu(NH_3)_4]^{2+}$. This is *not* correct. The correct formula is shown in the text. The confusion arose because the bonds from the Cu^{2+} ion to the water ligands are longer than the bonds from the Cu^{2+} ion to the ammonia ligands. This is the result of something called the Jahn–Teller effect. The explanation for this effect is beyond the aims of this book.

▲ **fig B** The pale blue solution contains $[Cu(H_2O)_6]^{2+}$, the pale blue precipitate contains $[Cu(H_2O)_4(OH)_2]$ and the deep blue solution contains $[Cu(NH_3)_4(H_2O)_2]^{2+}$.

▲ **fig C** The pale blue solution contains $[Cu(H_2O)_6]^{2+}$, the yellow solution on the right contains $[CuCl_4]^{2-}$ and the green solution in the middle contains a mixture of the two.

SUBJECT VOCABULARY

deprotonation the removal of one or more hydrogen ions (protons) from a complex ion
ligand exchange when one ligand in a complex ion is replaced by a different ligand
amphoteric (substance) a substance that can act both as an acid and as a base
amphoteric behaviour the ability of a species to react with both acids and bases

It is again easy to see that this equation involves a deprotonation – two of the water ligands transfer a hydrogen ion to the ammonia molecules.

When aqueous ammonia is added to the blue precipitate formed, it dissolves to form a deep blue solution. The equation for this reaction is:

$$[Cu(H_2O)_4(OH)_2] + 4NH_3 \rightarrow [Cu(NH_3)_4(H_2O)_2]^{2+} + 2H_2O + 2OH^-$$

This is a ligand exchange reaction – four ammonia molecules replace two water molecules and two hydroxide ions. The solutions at the start and end of the reaction and the intermediate precipitate are shown in **fig B**.

FORMATION OF $[CuCl_4]^{2-}$: CHANGE IN COORDINATION NUMBER

These reactions always involve a change of ligand as well as a change in coordination number. A good example is the reaction between copper(II) sulfate solution and concentrated hydrochloric acid. When the acid is added slowly and continuously, the colour gradually changes from blue to green and finally to yellow. This is the equation for the reaction:

$$[Cu(H_2O)_6]^{2+} + 4Cl^- \rightleftharpoons [CuCl_4]^{2-} + 6H_2O$$

The state symbols, all (aq), have been left out for clarity.

You can see that all six water ligands have been replaced by four chloride ions. This reaction is also an example of a change in coordination number, from 6 to 4. Note that although the charge on the complex has changed from 2+ to 2−, there has been no change in oxidation number.

The arrow shows that the reaction is reversible, which helps to explain the colour change observed. The hexaaquacopper(II) ion is blue and the tetrachlorocuprate(II) ion is yellow, so the green colour is due to a mixture of the blue and yellow complex ions. The solutions at the start and end of the reaction and the intermediate mixture are shown in **fig C**.

LEARNING TIP

You should remember the colours of the species in these reactions, but you do not need to understand why the species have the actual colours.

AMPHOTERIC HYDROXIDES

We have already seen that hydrated transition metal ions can be deprotonated by adding a base such as aqueous sodium hydroxide to form a precipitate of the metal hydroxide. For example, chromium(III) hydroxide is precipitated from an aqueous solution containing Cr^{3+} ions:

$$\underset{\text{green solution}}{[Cr(H_2O)_6]^{3+}(aq)} + 3OH^-(aq) \rightarrow \underset{\text{green precipitate}}{[Cr(H_2O)_3(OH)_3](s)} + 3H_2O(l)$$

When an excess of sodium hydroxide is added to this precipitate, further deprotonation can take place:

$$[Cr(H_2O)_3(OH)_3](s) + 3OH^-(aq) \rightarrow \underset{\text{green solution}}{[Cr(OH)_6]^{3-}(aq)} + 3H_2O(l)$$

In this reaction, chromium(III) hydroxide is acting as an acid, since it is reacting with a base.

Chromium(III) hydroxide is also a base, because it can react with acids:

$$[Cr(H_2O)_3(OH)_3](s) + 3H^+(aq) \rightarrow [Cr(H_2O)_6]^{3+}(aq)$$

A metal hydroxide that can act as both an acid and a base is called an **amphoteric** hydroxide. This is an example of **amphoteric behaviour**.

CHECKPOINT

1. The equation for a reaction of a transition metal ion is:

   $$[Fe(H_2O)_6]^{3+} + SCN^- \rightarrow [Fe(SCN)(H_2O)_5]^{2+} + H_2O$$

   Explain which of the four main types of reactions are illustrated by this equation.

2. The complex $[Co(H_2O)_6]^{2+}$ is converted into $[Co(NH_3)_6]^{3+}$.

   Explain which types of reactions occur in this conversion.

17B 2 REACTIONS OF COBALT AND IRON COMPLEXES

LEARNING OBJECTIVES

- Be able to record observations and write suitable equations for the reactions of Fe^{2+}(aq), Fe^{3+}(aq) and Co^{2+}(aq) with aqueous sodium hydroxide and aqueous ammonia, including in excess.
- Understand that ligand exchange, and an accompanying colour change, occurs in the formation of $[CoCl_4]^{2-}$ from $[Co(H_2O)_6]^{2+}$.

REACTIONS INVOLVING COBALT COMPLEXES

fig A The bright colour of these glass bottles is caused by blue cobalt compounds.

REACTION WITH ALKALIS

Consider the reaction that occurs when aqueous sodium hydroxide is added to a solution containing the hexaaquacobalt(II) ion until no further change is seen. The observation is that a pink solution forms a blue precipitate. The equation for this reaction is:

$$[Co(H_2O)_6]^{2+} + 2OH^- \rightarrow [Co(H_2O)_4(OH)_2] + 2H_2O$$

As with copper, this is a deprotonation reaction – the two hydroxide ions have removed hydrogen ions from two of the water ligands and converted them into water molecules. The two water ligands that have lost hydrogen ions are now hydroxide ligands. Upon standing, the colour of the precipitate gradually changes to pink.

The same observations can be made when aqueous ammonia is used as the alkali. However, when aqueous ammonia is added to excess, there is an observation not made with aqueous sodium hydroxide – the precipitate dissolves to form a pale yellow solution. The equation for the deprotonation reaction is:

$$[Co(H_2O)_6]^{2+} + 2NH_3 \rightarrow [Co(H_2O)_4(OH)_2] + 2NH_4^+$$

The equation for the ligand exchange reaction forming the pale yellow solution is:

$$[Co(H_2O)_4(OH)_2] + 6NH_3 \rightarrow [Co(NH_3)_6]^{2+} + 4H_2O + 2OH^-$$

Some features of transition metal chemistry are impossible to explain without going well beyond the aims of this book. One of these is why six ammonia ligands are involved in the reaction with the cobalt complex, while in the corresponding reaction with copper only four ammonia ligands are involved.

Upon standing, this pale yellow solution changes colour because of oxidation by the oxygen in the atmosphere. The oxidation number of cobalt increases from +2 to +3, and the darker yellow $[Co(NH_3)_6]^{3+}$ ion forms. Unfortunately, the resulting solution usually looks brown rather than yellow. This is because other products are formed, from ligand exchange reactions with both water molecules and any negative ions present.

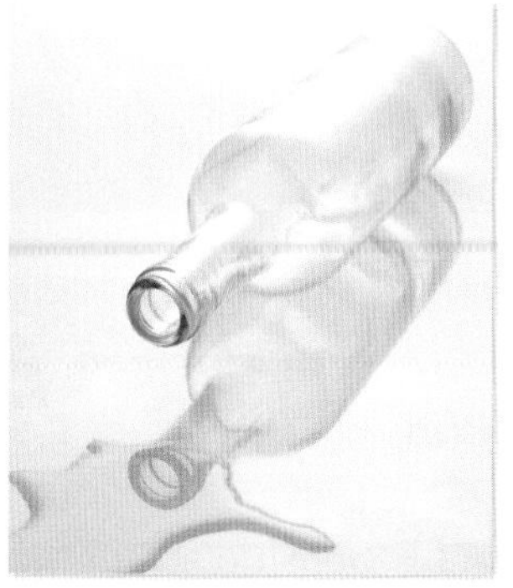

fig B Ordinary glass is often very pale green. This is more noticeable if you look through a considerable thickness of glass. The green colour is caused by the presence of impurities containing iron.

REACTION WITH CONCENTRATED HYDROCHLORIC ACID

This reaction is very like the one with the $[Cu(H_2O)_6]^{2+}$ ion. When concentrated hydrochloric acid is slowly added to a solution containing the hexaaquacobalt(II) ion, the pink solution gradually changes to blue. This is the equation for the reaction:

$$[Co(H_2O)_6]^{2+} + 4Cl^- \rightarrow [CoCl_4]^{2-} + 6H_2O$$

You can see that all six water ligands have been replaced by four chloride ions. This reaction is also an example of a change in coordination number, from 6 to 4. Note that although the charge on the complex has changed from 2+ to 2−, there has been no change in oxidation number.

REACTIONS INVOLVING IRON COMPLEXES

REACTION OF IRON(II) COMPLEXES WITH ALKALIS

Consider the reaction that occurs when aqueous sodium hydroxide is added to a solution containing the hexaaquairon(II) ion until no further change is seen. The observation is that a pale green solution forms a green precipitate. The equation for this reaction is:

$$[Fe(H_2O)_6]^{2+} + 2OH^- \rightarrow [Fe(H_2O)_4(OH)_2] + 2H_2O$$

As with copper and cobalt, this is a deprotonation reaction – the two hydroxide ions have removed hydrogen ions from two of the water ligands and converted them into water molecules. The two water ligands that have lost hydrogen ions are now hydroxide ligands.

The same observations can be made when aqueous ammonia is used as the alkali. The equation for the deprotonation reaction is:

$$[Fe(H_2O)_6]^{2+} + 2NH_3 \rightarrow [Fe(H_2O)_4(OH)_2] + 2NH_4^+$$

Upon standing, the colour of the green precipitate gradually changes to brown as oxygen from the atmosphere causes oxidation, forming $[Fe(H_2O)_3(OH)_3]$ – this is the triaquatrihydroxoiron(III) complex.

EXAM HINT

You will always be asked questions on transition metal complexes in your exam. To answer these questions well you must learn the reactions, remember the colours and make sure you use the correct notation.

REACTION OF IRON(III) COMPLEXES WITH ALKALIS

Consider the reaction that occurs when aqueous sodium hydroxide is added to a solution containing the hexaaquairon(III) ion until no further change is seen. The observation is that a yellow-brown solution forms a brown precipitate. The equation for this reaction is:

$$[Fe(H_2O)_6]^{3+} + 3OH^- \rightarrow [Fe(H_2O)_3(OH)_3] + 3H_2O$$

As with iron(II), this is a deprotonation reaction – the three hydroxide ions have removed hydrogen ions from three of the water ligands and converted them into water molecules. The three water ligands that have lost hydrogen ions are now hydroxide ligands.

The same observations can be made when aqueous ammonia is used as the alkali. The equation for the deprotonation reaction is:

$$[Fe(H_2O)_6]^{3+} + 3NH_3 \rightarrow [Fe(H_2O)_3(OH)_3] + 3NH_4^+$$

There are no further reactions when an excess of either aqueous sodium hydroxide or aqueous ammonia is added and no further changes upon standing.

LEARNING TIP

Note that atmospheric oxidation occurs with complexes containing a transition metal ion with a +2 oxidation number, but not for those with a +3 oxidation number.

CHECKPOINT

1. When aqueous ammonia is added in excess to a solution containing $[Co(H_2O)_6]^{2+}$ ions, an acid-base reaction occurs, followed by a ligand exchange reaction. Write an equation to describe this reaction.

2. A student wrote this equation to explain the formation of the brown precipitate when aqueous ammonia was added to a solution containing Fe^{3+} ions:

$$[Fe(H_2O)_6]^{3+} + 3OH^- \rightarrow [Fe(H_2O)_3(OH)_3] + 3H_2O$$

Why is the above reaction less likely to occur than the one shown below?

$$[Fe(H_2O)_6]^{3+} + 3NH_3 \rightarrow [Fe(H_2O)_3(OH)_3] + 3NH_4^+$$

17B 3 THE CHEMISTRY OF CHROMIUM

SPECIFICATION REFERENCE

17.20(i) 17.20(ii) 17.21 17.22 PART

LEARNING OBJECTIVES

- Be able to record observations and write suitable equations for the reactions of Cr^{3+}(aq) with aqueous sodium hydroxide and aqueous ammonia, including in excess.
- Understand, in terms of the relevant *E* values, that the dichromate(VI) ion, $Cr_2O_7^{2-}$:
  (i) can be reduced to Cr^{3+} and Cr^{2+} ions using zinc in acidic conditions
  (ii) can be produced by the oxidation of Cr^{3+} ions using hydrogen peroxide in alkaline conditions (followed by acidification).
- Know that the dichromate(VI) ion, $Cr_2O_7^{2-}$, can be converted into chromate(VI) ions as a result of the equilibrium
  $Cr_2O_7^{2-} + H_2O \rightleftharpoons 2CrO_4^{2-} + 2H^+$

INTRODUCTION

Some of the reactions of chromium are similar to those we have already met for the ions of copper, cobalt, iron(II) and iron(III). However, chromium has ions in which the metal has an oxidation number of +6, so reactions involving these ions must be considered in a different way.

A WORD OF WARNING

We have already discussed that there are an infinite number of colours in the spectrum of visible light, although we sometimes refer to the seven colours of the rainbow. The perception of colours by humans varies a lot. A specific colour might be described as pink or purple by different people, and a colour described as blue-green by one person might be described as blue, green or turquoise by others.

The situation with the colours used to describe chromium compounds is further complicated by these factors:

- Some compounds have different colours as solids and aqueous solutions.
- The colour of a solution depends on concentration.
- The presence of dissolved oxygen in an aqueous solution can affect the colour observed.

This means that the colours used in this book may not always correspond to those in other books, and may be different from those observed in test tube reactions that you see.

REACTIONS INVOLVING CHROMIUM(III) COMPLEXES

REACTION OF CHROMIUM(III) COMPLEXES WITH ALKALIS

Consider the reaction that occurs when aqueous sodium hydroxide is added to a solution containing the hexaaquachromium(III) ion until no further change is seen. The observation is that a green solution forms a green precipitate. The equation for this reaction is:

$$[Cr(H_2O)_6]^{3+} + 3OH^- \rightarrow [Cr(H_2O)_3(OH)_3] + 3H_2O$$

LEARNING TIP

The colour of the $[Cr(NH_3)_6]^{3+}$ ion in a solid, such as chrome alum, is violet. However, when this solid is dissolved in water the resulting solution is green, not violet. This is because of ligand exchange reactions that occur with the negative ions present.

As with iron(III), this is a deprotonation reaction – the three hydroxide ions have removed hydrogen ions from three of the water ligands and converted them into water molecules. The three water ligands that have lost hydrogen ions are now hydroxide ligands.

The same observations can be made when aqueous ammonia is used as the alkali. The equation for the deprotonation reaction is:

$$[Cr(H_2O)_6]^{3+} + 3NH_3 \rightarrow [Cr(H_2O)_3(OH)_3] + 3NH_4^+$$

EXAM HINT

In reactions with transition metal complexes, make sure you know *when* and *how* to show ammonia acting as a base and ammonia acting as a ligand.

When an excess of aqueous sodium hydroxide is added, the green precipitate dissolves to form a green solution. The equation for this deprotonation reaction can be represented as:

$$[Cr(H_2O)_3(OH)_3] + OH^- \rightarrow [Cr(H_2O)_2(OH)_4]^- + H_2O$$

If the aqueous sodium hydroxide is more concentrated, further deprotonation reactions occur, such as

$$[Cr(H_2O)_2(OH)_4]^- + 2OH^- \rightarrow [Cr(OH)_6]^{3-} + 2H_2O$$

although there is no further change in colour.

The reactions involving hydroxide ions can be reversed by the addition of acid, illustrating the amphoteric nature of the neutral complex.

When an excess of aqueous ammonia is added to the green precipitate, the precipitate is slow to dissolve, but eventually a violet or purple solution forms:

$$[Cr(H_2O)_3(OH)_3] + 6NH_3 \rightarrow [Cr(NH_3)_6]^{3+} + 3H_2O + 3OH^-$$

In all of the reactions so far, the oxidation number of chromium has remained unchanged at +3. However, provided that the solutions are alkaline, oxidation is easily achieved by the addition of the oxidising agent hydrogen peroxide. In this reaction the solution changes from green to yellow as the chromate(VI) ion, with oxidation number +6, is formed. The equation for this oxidation is:

$$2[Cr(OH)_6]^{3-} + 3H_2O_2 \rightarrow 2CrO_4^{2-} + 2OH^- + 8H_2O$$

Note that although the chromate(VI) ion is a complex, it is not enclosed in square brackets.

CHROMATE(VI) AND DICHROMATE(VI) IONS

Chromate(VI) ions are stable in alkaline solution, but in acidic conditions the dichromate(VI) ion is more stable. So, if acid is added there is a colour change from yellow to orange as the following reaction occurs:

$$2CrO_4^{2-} + 2H^+ \rightleftharpoons Cr_2O_7^{2-} + H_2O$$

This reaction is easily reversed by adding alkali. When considering redox reactions, it is often easier to use simplified formulae by leaving out square brackets and ligands that do not undergo redox reactions, especially water.

REDUCTION OF DICHROMATE(VI) IONS

When zinc metal is added to an acidic solution containing dichromate(VI) ions, reduction reactions occur in which the oxidation number of chromium decreases first to +3 and then to +2.

The first stage of the reduction involves a colour change from orange to green, as this reaction occurs:

$$Cr_2O_7^{2-} + 14H^+ + 3Zn \rightarrow 2Cr^{3+} + 7H_2O + 3Zn^{2+}$$

The second stage of the reduction involves a colour change from green to blue, as this reaction occurs:

$$2Cr^{3+} + Zn \rightarrow 2Cr^{2+} + Zn^{2+}$$

EXPLANATION OF REDOX REACTIONS USING $E^\ominus$ VALUES

It is often helpful to explain why redox reactions occur by considering the standard electrode potentials of the different redox systems involved. For chromium, we need to consider these values:

1	$Zn^{2+} + 2e^- \rightleftharpoons Zn$	$E^\ominus = -0.76\,V$
2	$Cr^{3+} + e^- \rightleftharpoons Cr^{2+}$	$E^\ominus = -0.41\,V$
3	$CrO_4^{2-} + 4H_2O + 3e^- \rightleftharpoons Cr(OH)_3 + 5OH^-$	$E^\ominus = -0.13\,V$
4	$H_2O_2 + 2e^- \rightleftharpoons 2OH^-$	$E^\ominus = +1.24\,V$
5	$Cr_2O_7^{2-} + 14H^+ + 6e^- \rightleftharpoons 2Cr^{3+} + 7H_2O$	$E^\ominus = +1.33\,V$

If you have already learned about standard electrode potentials earlier in this book you may remember that when half-equations are arranged from high negative $E^\ominus$ values at the top to high positive values at the bottom, then the best reducing agent is at the top and on the right, and the best oxidising agent is at the bottom and on the left.

EXPLAINING OXIDATION FROM +3 TO +6

The simplified equation for the oxidation of chromium(III) by hydrogen peroxide in alkaline conditions is:

$$2Cr(OH)_3 + 3H_2O_2 + 4OH^- \rightarrow 2CrO_4^{2-} + 8H_2O$$

How can this equation be obtained from the half-equations?

The relevant half-equations are 3 and 4 (see previous page). In these equations, the best reducing agent is $Cr(OH)_3$, and the best oxidising agent is H_2O_2. For these two species to react together, we need to add half-equation 3 (reversed) to half-equation 4. When adding half-equations, they may need to be multiplied so that the number of electrons is the same in both. In this example, we need to multiply equation 3 by 2, and equation 4 by 3, so as to obtain $6e^-$ in both:

$$2Cr(OH)_3 + 10OH^- \rightleftharpoons 2CrO_4^{2-} + 8H_2O + 6e^-$$

$$3H_2O_2 + 6e^- \rightleftharpoons 6OH^-$$

Adding the half-equations and cancelling identical species on both sides gives:

$$2Cr(OH)_3 + 3H_2O_2 + 4OH^- \rightarrow 2CrO_4^{2-} + 8H_2O$$

Since the $E^\ominus$ value for half-equation 3 is more negative than the $E^\ominus$ value for half-equation 4, $Cr(OH)_3$ is electron releasing with respect to H_2O_2, and therefore the reaction is thermodynamically feasible. If the solution is then acidified by adding dilute sulfuric acid, the chromate(VI) ion is converted into the dichromate(VI) ion:

$$2CrO_4^{2-} + 2H^+ \rightarrow Cr_2O_7^{2-} + H_2O$$

EXPLAINING REDUCTION FROM 6+ TO +3

The simplified equation for the reduction of chromium(VI) by zinc in acidic conditions is:

$$Cr_2O_7^{2-} + 14H^+ + 3Zn \rightarrow 2Cr^{3+} + 7H_2O + 3Zn^{2+}$$

How can this equation be obtained from the half-equations?

The relevant half-equations are 1 and 5 (see previous page). In these equations, the best reducing agent is Zn and the best oxidising agent is $Cr_2O_7^{2-}$. For these two species to react together, we need to add half-equation 1 (reversed) to half-equation 6, bearing in mind multiplying half-equations:

$$3Zn \rightleftharpoons 3Zn^{2+} + 6e^-$$

$$Cr_2O_7^{2-} + 14H^+ + 6e^- \rightleftharpoons 2Cr^{3+} + 7H_2O$$

As in the previous example, adding gives:

$$Cr_2O_7^{2-} + 14H^+ + 3Zn \rightarrow 2Cr^{3+} + 7H_2O + 3Zn^{2+}$$

Since the $E^\ominus$ value for half-equation 1 is more negative than the $E^\ominus$ value for half-equation 5, Zn is electron releasing with respect to $Cr_2O_7^{2-}$, and therefore the reaction is thermodynamically feasible.

EXPLAINING REDUCTION FROM +3 TO +2

The simplified equation for the reduction of chromium(III) by zinc in acidic conditions is:

$$2Cr^{3+} + Zn \rightarrow 2Cr^{2+} + Zn^{2+}$$

How can this equation be obtained from the half-equations?

The relevant half-equations are 1 and 2. In these equations, the best reducing agent is Zn and the best oxidising agent is Cr^{3+}. For these two species to react together, we need to add half-equation 1 (reversed) to half-equation 2, bearing in mind multiplying half-equations:

$$Zn \rightleftharpoons Zn^{2+} + 2e^-$$

$$2Cr^{3+} + 2e^- \rightleftharpoons 2Cr^{2+}$$

As in the previous example, adding gives:

$$Cr^{3+} + Zn \rightarrow 2Cr^{2+} + Zn^{2+}$$

Since the $E^\ominus$ value for half-equation 1 is more negative than the $E^\ominus$ value for half-equation 2, Zn is electron releasing with respect to Cr^{3+}, and therefore the reaction is thermodynamically feasible.

CHROMIUM CHEMISTRY SUMMARY

Table A summarises the important reactions of chromium complexes in four sequences. Some of the coloured solutions are shown in **fig A**.

START	REAGENT	INTERMEDIATE	REAGENT	FINISH
$[Cr(H_2O)_6]^{3+}$ green solution	add NaOH(aq)	$[Cr(H_2O)_3(OH)_3]$ green precipitate	add NaOH(aq)	$[Cr(OH)_6]^{3-}$ green solution
$[Cr(H_2O)_6]^{3+}$ green solution	add NH_3(aq)	$[Cr(H_2O)_3(OH)_3]$ green precipitate	add NH_3(aq)	$[Cr(NH_3)_6]^{3+}$ violet solution
$[Cr(OH)_6]^{3-}$ green solution	add H_2O_2/OH^-	CrO_4^{2-} yellow solution	add H^+	$Cr_2O_7^{2-}$ orange solution
CrO_4^{2-} yellow solution	add Zn/H^+	$[Cr(H_2O)_6]^{3+}$ green solution	add Zn/H^+	$[Cr(H_2O)_6]^{2+}$ blue solution

table A

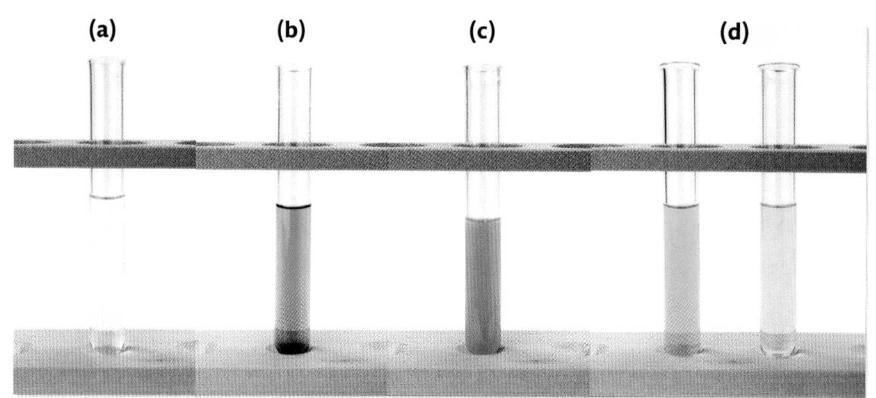

fig A Tube (a) contains the very pale blue $[Cr(H_2O)_6]^{2+}$ ion, produced by reduction using zinc. Tube (b) contains the $[Cr(H_2O)_6]^{3+}$ ion, which is normally green, but can appear violet in the presence of other ions, as here. Tube (c) contains the $[Cr(OH)_6]^{3-}$ ion, which is also green, present in alkaline solutions. The tubes in (d) contain chromium with oxidation number +6. The orange solution contains the $Cr_2O_7^{2-}$ ion and the yellow solution contains the CrO_4^{2-} ion.

LEARNING TIP

Understanding chromium chemistry involves more effort than for some other transition metals because of the number of different ions. As well as all of the complexes involving Cr^{2+} and Cr^{3+}, there are two complexes with oxidation number +6 (CrO_4^{2-} and $Cr_2O_7^{2-}$).

CHECKPOINT

1. Excess acid is added to a green solution containing $[Cr(OH)_6]^{3-}$ ions. Write an equation for the reaction in which a different green chromium-containing solution is formed.
2. The standard electrode potential for the redox system $Mn^{2+} + 2e^- \rightleftharpoons Mn$ is −1.18 V.
   Explain whether or not zinc can reduce Mn^{2+} ions to the element Mn.

17B 4 REACTIONS OF MANGANESE COMPLEXES

SPECIFICATION REFERENCE

17.22 PART 17.23 PART

LEARNING OBJECTIVES

- Be able to record observations and write suitable equations for the reactions of $Mn^{2+}(aq)$ with aqueous sodium hydroxide and aqueous ammonia, including in excess.
- Be able to write ionic equations to show the meaning of deprotonation and ligand exchange in the reactions of manganese ions.

REACTIONS OF MANGANESE(II) COMPLEXES WITH ALKALIS

Consider the reaction that occurs when aqueous sodium hydroxide is added to a solution containing the hexaaquamanganese(II) ion until no further change is seen. The observation is that a pale pink solution forms a pale brown precipitate. The equation for the reaction is:

$$[Mn(H_2O)_6]^{2+} + 2OH^- \rightarrow [Mn(H_2O)_4(OH)_2] + 2H_2O$$

This is a deprotonation reaction – the two hydroxide ions have removed hydrogen ions from two of the water ligands and converted them into water molecules. The two water ligands that have lost hydrogen ions are now hydroxide ligands. The pale brown precipitate turns darker brown on standing in air as it is oxidised to $[Mn(H_2O)_3(OH)_3]$, and then turns very dark brown, forming hydrated manganese(IV) oxide, $MnO_2.xH_2O$.

The precipitate does not dissolve in excess aqueous sodium hydroxide.

The same observations are made when aqueous ammonia is used as the alkali. The equation for this deprotonation reaction, forming the pale brown precipitate, is:

$$[Mn(H_2O)_6]^{2+} + 2NH_3 \rightarrow [Mn(H_2O)_4(OH)_2] + 2NH_4^+$$

LEARNING TIP

There are two important points you need to be aware of for the reaction between manganese(II) ions and aqueous sodium hydroxide. The first is that dilute solutions containing $[Mn(H_2O)_6]^{2+}$ are very likely to appear colourless, not pale pink. The second is that the true colour of $[Mn(H_2O)_4(OH)_2]$ is white, but the white colour may not be seen as the precipitate rapidly turns pale brown.

CHECKPOINT

SKILLS REASONING/ ARGUMENTATION

1. An aqueous solution of an inorganic solid, **Q**, forms a pale brown precipitate, **R**, when aqueous sodium hydroxide is added. **R** darkens on standing in air to form a dark brown solid, **S**. When solid **S** is heated it forms a dark brown metal oxide, $\mathbf{T}O_2$.
   (a) $\mathbf{T}O_2$ contains 36.82% by mass of oxygen. Use this information to calculate the molar mass of the metal **T**, and use this to identify **T**.
   (b) Identify the cation present in the aqueous solution of **Q**.
   (c) Identify **R** and **S**.
2. For the reactions above:
   (a) Write an ionic equation for the formation of **R** from the solution of **Q**. Include state symbols.
   (b) Write an equation for the action of heat on solid **S**.

17B 5 THE CHEMISTRY OF VANADIUM

SPECIFICATION REFERENCE 17.18 17.19

LEARNING OBJECTIVES

- Know the colours of the oxidation states of vanadium (+5, +4, +3 and +2) in its compounds.
- Understand redox reactions for the interconversion of the oxidation states of vanadium (+5, +4, +3 and +2), in terms of the relevant $E^\ominus$ values.

EXAM HINT

You can easily lose marks in an exam question if you are careless when showing the difference between VO_2^+ and VO^{2+} in your answer. Make sure you write your answer clearly.

REDOX REACTIONS

Like chromium, vanadium is a transition metal that forms ions with several oxidation numbers. **Table A** shows the important ones.

OXIDATION NUMBER	FORMULA	NAME	COLOUR OF AQUEOUS SOLUTION
+2	V^{2+}	vanadium(II)	purple
+3	V^{3+}	vanadium(III)	green
+4	VO^{2+}	oxovanadium(IV)	blue
+5	VO_2^+	dioxovanadium(V)	yellow

table A

Unlike with chromium and the other transition elements, the focus here with vanadium is purely on redox reactions. Vanadium is a good choice for this because there is one distinct colour for each of the main oxidation numbers found in its compounds. So, it is relatively easy to demonstrate the change in oxidation number by observing the change in colour during the reaction. As with chromium, we will consider the feasibility of these redox reactions in terms of $E^\ominus$ values. Although most of the species involved are complexes, and some contain water ligands, the square brackets and water ligands are left out for clarity.

REDUCING VANADIUM FROM +5 TO +2

The usual source of vanadium with oxidation number +5 is the compound ammonium vanadate(V), NH_4VO_3. An acidic solution of this compound contains the dioxovanadium(V) ion, VO_2^+. When zinc is added to this solution, reduction begins and there is a gradual colour change from yellow, through blue and green, to purple as the oxidation number decreases from +5 to +2. All of these colours are of solutions – there are no precipitates involved. The test tubes in **fig A** show the colours.

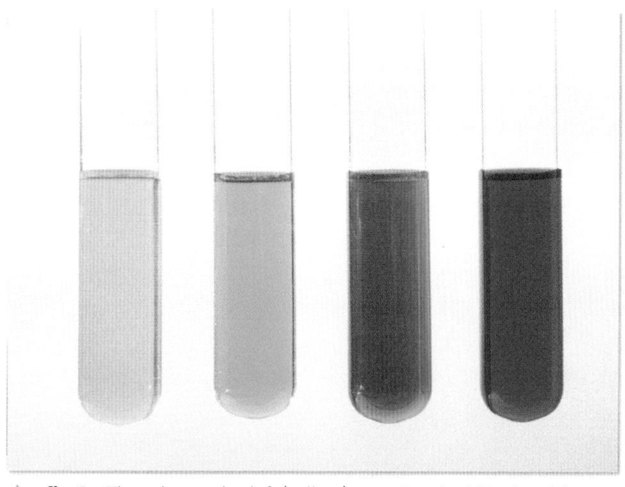

▲ **fig A** The tube on the left (yellow) contains the VO_2^+ ion. The second tube (blue) contains the VO^{2+} ion. The third tube (green) contains the V^{3+} ion. The tube on the right (purple) contains the V^{2+} ion.

EXPLANATION USING $E^{\ominus}$ VALUES

For vanadium, we need to consider these values:

1	$V^{2+} + 2e^- \rightleftharpoons V$	$E^{\ominus} = -1.18\,V$
2	$Zn^{2+} + 2e^- \rightleftharpoons Zn$	$E^{\ominus} = -0.76\,V$
3	$V^{3+} + e^- \rightleftharpoons V^{2+}$	$E^{\ominus} = -0.26\,V$
4	$VO_2^+ + 2H^+ + e^- \rightleftharpoons V^{3+} + H_2O$	$E^{\ominus} = +0.34\,V$
5	$VO_2^+ + 2H^+ + e^- \rightleftharpoons VO^{2+} + H_2O$	$E^{\ominus} = +1.00\,V$

If you followed the explanations for the redox reactions of chromium in the previous topic, you should be able to understand a shortened explanation for each stage that just shows the two relevant half-equations and then the overall equation.

REDUCTION FROM +5 TO +4

$$VO_2^+ + 2H^+ + e^- \rightarrow VO^{2+} + H_2O$$

$$Zn \rightarrow Zn^{2+} + 2e^-$$

Overall reaction:

$$2VO_2^+ + 4H^+ + Zn \rightarrow 2VO^{2+} + Zn^{2+} + 2H_2O$$

Since the $E^{\ominus}$ value for half-equation 2 is more negative than the $E^{\ominus}$ value for half-equation 5, Zn is electron releasing with respect to VO_2^+. Therefore, the reaction is thermodynamically feasible.

REDUCTION FROM +4 TO +3

$$VO^{2+} + 2H^+ + e^- \rightarrow V^{3+} + H_2O$$

$$Zn \rightarrow Zn^{2+} + 2e^-$$

Overall reaction:

$$2VO^{2+} + 4H^+ + Zn \rightarrow 2V^{3+} + Zn^{2+} + 2H_2O$$

Since the $E^{\ominus}$ value for half-equation 2 is more negative than the $E^{\ominus}$ value for half-equation 4, Zn is electron releasing with respect to VO^{2+}. Therefore, the reaction is thermodynamically feasible.

REDUCTION FROM +3 TO +2

$$V^{3+} + e^- \rightarrow V^{2+}$$

$$Zn \rightarrow Zn^{2+} + 2e^-$$

Overall reaction:

$$2V^{3+} + Zn \rightarrow 2V^{2+} + Zn^{2+}$$

Since the $E^{\ominus}$ value for half-equation 2 is more negative than the $E^{\ominus}$ value for half-equation 3, Zn is electron releasing with respect to V^{3+}. Therefore, the reaction is thermodynamically feasible.

REDUCTION FROM +2 TO 0

$$V^{2+} + 2e^- \rightarrow V$$

$$Zn \rightarrow Zn^{2+} + 2e^-$$

Overall reaction:

$$2V^{2+} + Zn \rightarrow 2V + Zn^{2+}$$

Since the $E^{\ominus}$ value for half-equation 2 is less negative than the $E^{\ominus}$ value for half-equation 1, Zn is not electron releasing with respect to V^{2+}. Therefore, the reaction is not thermodynamically feasible.

PREDICTING OXIDATION REACTIONS

A similar method can be used to predict whether a given oxidising agent will oxidise a vanadium species to one with a higher oxidation number. You can see how to do this in Question 2 in this topic.

LEARNING TIP

It is important that you do not get confused between the two oxo ions of vanadium: VO_2^+ and VO^{2+}.

CHECKPOINT

SKILLS PROBLEM-SOLVING

1. The standard electrode potential for the redox system $Sn^{2+} + 2e^- \rightleftharpoons Sn$ is −0.14 V.

Explain why the use of tin as a reducing agent will not obtain V^{2+} from a solution containing VO_2^+ ions.

2. The standard electrode potential for the redox system $Cu^{2+} + 2e^- \rightleftharpoons Cu$ is +0.34 V.

Explain whether copper(II) ions can be used to oxidise VO^{2+} ions to VO_2^+ ions.

17B 6 REACTIONS OF NICKEL AND ZINC COMPLEXES

SPECIFICATION REFERENCE
17.22 PART | 17.23

LEARNING OBJECTIVES

- Be able to record observations and write suitable equations for the reactions of Ni^{2+}(aq) and Zn^{2+}(aq) with aqueous sodium hydroxide and aqueous ammonia, including in excess.
- Be able to write ionic equations to show the meaning of amphoteric behaviour, deprotonation and ligand exchange in the reactions of nickel ions and zinc ions.

REACTIONS OF NICKEL COMPLEXES WITH ALKALIS

Consider the reaction that occurs when aqueous sodium hydroxide is added to a solution containing the hexaaquanickel(II) ion until no further change is seen. The observation is that a green solution forms a green precipitate. The equation for the reaction is:

$$[Ni(H_2O)_6]^{2+} + 2OH^- \rightarrow [Ni(H_2O)_4(OH)_2] + 2H_2O$$

This is a deprotonation reaction – the two hydroxide ions have removed hydrogen ions from two of the water ligands and converted them into water molecules. The two water ligands that have lost hydrogen ions are now hydroxide ligands.

The same observations are made when aqueous ammonia is used as the alkali. However, when aqueous ammonia is added in excess, there is an observation not made with aqueous sodium hydroxide – the precipitate dissolves to form a deep blue solution. The equation for the deprotonation reaction forming the green precipitate is:

$$[Ni(H_2O)_6]^{2+} + 2NH_3 \rightarrow [Ni(H_2O)_4(OH)_2] + 2NH_4^+$$

The equation for the ligand exchange reaction forming the blue solution is:

$$[Ni(H_2O)_4(OH)_2] + 6NH_3 \rightarrow [Ni(NH_3)_6]^{2+} + 4H_2O + 2OH^-$$

REACTIONS OF ZINC COMPLEXES WITH ALKALIS

Consider the reaction that occurs when aqueous sodium hydroxide is added to a solution containing the hexaaquazinc(II) ion until no further change is seen. The observation is that a colourless solution forms a white precipitate. The equation for the reaction is:

$$[Zn(H_2O)_6]^{2+} + 2OH^- \rightarrow [Zn(H_2O)_4(OH)_2] + 2H_2O$$

This is a deprotonation reaction – the two hydroxide ions have removed hydrogen ions from two of the water ligands and converted them into water molecules. The two water ligands that have lost hydrogen ions are now hydroxide ligands.

When an excess of aqueous sodium hydroxide is added, the white precipitate dissolves to form a colourless solution. The equation for this deprotonation reaction is:

$$[Zn(H_2O)_4(OH)_2] + 2OH^- \rightarrow [Zn(H_2O)_2(OH)_4]^{2-} + 2H_2O$$

Zinc hydroxide also reacts with acid, showing its amphoteric behaviour:

$$[Zn(H_2O)_4(OH)_2] + 2H^+ \rightarrow [Zn(H_2O)_6]^{2+}$$

Similar observations are made when aqueous ammonia is used as the alkali. The equation for the deprotonation reaction forming the white precipitate is:

$$[Zn(H_2O)_6]^{2+} + 2NH_3 \rightarrow [Zn(H_2O)_4(OH)_2] + 2NH_4^+$$

The equation for the ligand exchange reaction forming the colourless solution is:

$$[Zn(H_2O)_4(OH)_2] + 4NH_3 \rightarrow [Zn(NH_3)_4]^{2+} + 4H_2O + 2OH^-$$

LEARNING TIP

Some texts quote the formula of the ammine complex as $[Zn(NH_3)_6]^{2+}$ rather than $[Zn(NH_3)_4]^{2+}$.

A research article in the 1993 edition of *Chemical Physics Letters* suggests that the coordination number can be either 4 or 6, but that $[Zn(NH_3)_4]^{2+}$ is favoured as it has a lower energy. Therefore, it is reasonable to say that both formulae are correct.

SUMMARY

Table A summarises the reactions of the hydrated d-block metal ions with aqueous sodium hydroxide.

ION IN SOLUTION	COLOUR OF SOLUTION	COLOUR OF PRECIPITATE	EFFECT OF ADDING EXCESS NaOH(aq)
$[Cr(H_2O)_6]^{3+}$	green	green	forms green solution
$[Mn(H_2O)_6]^{2+}$	pale pink	pale brown (turns darker brown on exposure to air)	none
$[Fe(H_2O)_6]^{2+}$	pale green	green (turns brown on exposure in air)	none
$[Fe(H_2O)_6]^{3+}$	yellow-brown	brown	none
$[Co(H_2O)_6]^{2+}$	pink	blue (turns pink on exposure to air)	none
$[Ni(H_2O)_6]^{2+}$	green	green	none
$[Cu(H_2O)_6]^{2+}$	blue	blue	none
$[Zn(H_2O)_6]^{2+}$	colourless	white	forms colourless solution

table A

Table B summarises the reactions of the hydrated d-block metal ions with aqueous ammonia.

ION IN SOLUTION	COLOUR OF SOLUTION	COLOUR OF PRECIPITATE	EFFECT OF ADDING EXCESS $NH_3(aq)$
$[Cr(H_2O)_6]^{3+}$	green	green	forms purple solution
$[Mn(H_2O)_6]^{2+}$	pale pink	pale brown (turns darker brown on exposure to air)	none
$[Fe(H_2O)_6]^{2+}$	pale green	green (turns brown on exposure in air)	none
$[Fe(H_2O)_6]^{3+}$	yellow-brown	brown	none
$[Co(H_2O)_6]^{2+}$	pink	blue (turns pink on exposure to air)	forms pale yellow solution (turns brown on exposure to air)
$[Ni(H_2O)_6]^{2+}$	green	green	forms deep blue solution
$[Cu(H_2O)_6]^{2+}$	blue	blue	forms deep blue solution
$[Zn(H_2O)_6]^{2+}$	colourless	white	forms colourless solution

table B

CHECKPOINT

1. The electronic configurations of nickel and zinc are:
   Ni [Ar] $3d^8\ 4s^2$
   Zn [Ar] $3d^{10}\ 4s^2$
   (a) State why both elements are classified as d-block elements.
   (b) State why nickel is classified as a transition metal, but zinc is not.
2. The EDTA4– ion forms a complex ion with the Ni2+ ion.
   (a) Complete the electronic configuration of the Ni^{2+} ion.
   $1s^2\ 2s^2\ 2p^6\ 3s^2\ 3p^6$........................
   (b) The visible spectrum of the complex ion formed between Ni2+ and EDTA4– is shown below.

SKILLS ANALYSIS

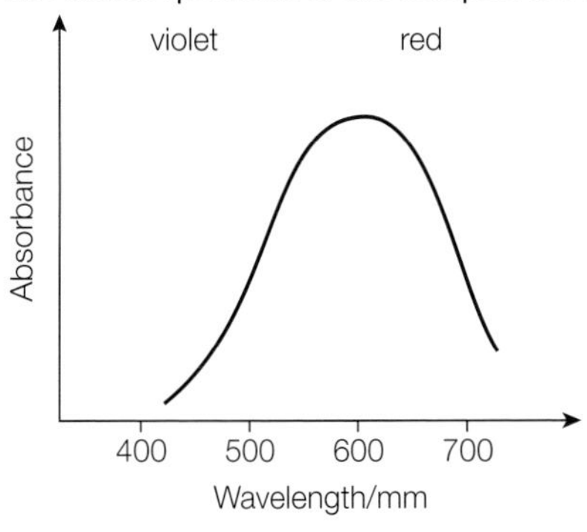

   Explain the colour of this complex ion.
   (c) The $EDTA^{4-}$ ion has the following structure.

   $(^-OOC-CH_2)_2N-CH_2-CH_2-N(CH_2-COO^-)_2$

   (i) Draw a ring around two different types of atoms in the $EDTA^{4-}$ ion that are capable of forming a dative (coordinate) bond with the Ni^{2+} ion.
   (ii) What feature of these atoms allows them to form a bond with Ni^{2+}?

17C 1 HETEROGENEOUS CATALYSIS

SPECIFICATION REFERENCE

17.26 PART | 17.27 | 17.28 | 17.29(i) | 17.29(ii)

LEARNING OBJECTIVES

- Know that transition metals and their compounds can act as heterogeneous catalysts.
- Know that a heterogeneous catalyst is in a different phase from the reactants and that the reaction occurs at the surface of the catalyst.
- Understand, in terms of oxidation number, how V_2O_5 acts as a catalyst in the Contact process.
- Understand how a catalytic converter decreases carbon monoxide and nitrogen monoxide emissions from internal combustion engines by:
  (i) adsorption of CO and NO molecules onto the surface of the catalyst, resulting in the weakening of bonds and chemical reaction
  (ii) desorption of CO_2 and N_2 product molecules from the surface of the catalyst.

DIFFERENT WAYS TO UNDERSTAND CATALYSIS

When you first learned about catalysts you probably recognised them as substances that increased the rate of a reaction and remained chemically unchanged at the end of the reaction. This description is fine as an introduction to catalysis, but it does not tell you anything about how catalysts work.

In **Topic 9 (Book 1: IAS)** you learned about catalysis in terms of its effect on the activation energy of a reaction. You also learned how to represent catalytic action using Maxwell–Boltzmann energy distributions and energy profiles.

In this section and the next one, we will consider catalysis in more detail by looking at the two main types of action: heterogeneous and homogeneous.

TRANSITION METALS AS HETEROGENEOUS CATALYSTS

A **heterogeneous catalyst** is one that is in a different phase from that of the reactants. You can probably remember an example of this type of catalysis from many years ago. Oxygen gas can be made in the laboratory by decomposing a hydrogen peroxide solution. The catalyst used in this reaction is usually manganese(IV) oxide – note that this is a compound of a transition metal. However, what makes this an example of heterogeneous catalysis is that manganese(IV) oxide is a solid and hydrogen peroxide solution is a liquid.

Many transition metals, and their compounds, are used as solid catalysts. Their action can be explained in terms of what happens on the surface of the catalyst. The fact that the action takes place only on the surface explains why many of them are used in a finely divided form, as small particles (including powder) rather than as large lumps. Sometimes, they are used instead as a thin coating on an inert support material.

EXAM HINT

Industrial chemists often use heterogeneous catalysts rather than homogeneous catalysts. This is because it is easier to separate the reaction products from the catalyst if they are in different phases.

THE CONTACT PROCESS

Sulfuric acid is one of the most widely used chemicals, both in terms of the quantity produced and the number of industries that use it. Its biggest single use is in the manufacture of fertilisers. A complete description of the Contact process used to manufacture sulfuric acid is beyond the aims of this book, but the key reaction in the process is the conversion of sulfur dioxide to sulfur trioxide in this reaction:

$$2SO_2 + O_2 \rightleftharpoons 2SO_3$$

At the temperatures and pressures used in the process, all of the substances are in the gas phase, and the mixture of reactants is passed over a catalyst of vanadium(V) oxide, V_2O_5, usually known in industry as vanadium pentoxide (**fig A**).

▲ **fig A** Sample of vanadium(V) oxide, as used as the catalyst in the Contact process.

SURFACE ADSORPTION THEORY

This theory is often used to explain the way that a heterogeneous catalyst works. It is usually considered as having three steps:

1. **Adsorption**, in which one or more reactants become attached to the surface of the catalyst.
2. Reaction, following the weakening of bonds in the adsorbed reactants.
3. **Desorption**, in which the reaction product becomes detached from the surface of the catalyst.

In the Contact process, the reaction step has two parts:

Part 1: sulfur dioxide adsorbs onto the vanadium(V) oxide and a redox reaction occurs:

$$V_2O_5 + SO_2 \rightarrow V_2O_4 + SO_3$$

Note that the oxidation number of vanadium decreases from +5 to +4. The sulfur trioxide then desorbs.

Part 2: oxygen reacts with the V_2O_4 on the surface of the catalyst and another redox reaction occurs:

$$V_2O_4 + \frac{1}{2}O_2 \rightarrow V_2O_5$$

Note that the original catalyst is regenerated as the oxidation number increases from +4 to +5.

CATALYTIC CONVERTERS

THE PROBLEMS

One of the major problems associated with increased road vehicle usage over several decades is the increased amount of pollution from vehicle exhaust gases. In some cities, the quality of air breathed by humans is below acceptable levels. Without catalytic converters, the situation would now be much worse.

Although many different pollutants come from vehicle exhausts, two of the most significant are carbon monoxide and nitrogen monoxide.

- Carbon monoxide is a toxic gas that interferes with oxygen transport from the lungs through the bloodstream to vital organs in the body (see **Section 17A.6**).
- Nitrogen monoxide is easily oxidised in the atmosphere to nitrogen dioxide. It can act as a respiratory irritant and contribute to the formation of acid rain.

Carbon monoxide forms through the incomplete combustion of hydrocarbon fuels. Nitrogen monoxide forms through the reaction between nitrogen and oxygen at the high temperatures that exist in an internal combustion engine.

THE SOLUTIONS

For over 20 years, cars sold worldwide have been fitted with catalytic converters in an attempt to reduce the effect of vehicle emissions on air quality. The main transition metals used in catalytic converters are platinum and rhodium, and sometimes palladium (**fig B**). The method of action can be explained by the surface adsorption theory described above. In one type of catalyst, molecules of carbon monoxide and nitrogen monoxide are adsorbed onto the surface. Then, because their bonds are weakened, they react together to form carbon dioxide and nitrogen, which are then desorbed from the surface of the catalyst. The overall reaction can be represented by this equation:

$$2CO + 2NO \rightarrow 2CO_2 + N_2$$

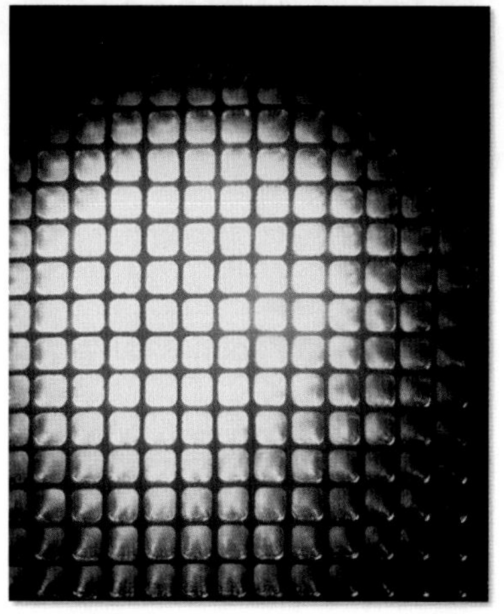

▲ **fig B** View through the element of a catalytic converter from a car exhaust. The inner surface is coated with an alloy containing platinum, rhodium and palladium.

LEARNING TIP

Do not confuse absorb/absorption with adsorb/adsorption. Very simply, absorption involves one substance becoming distributed throughout another (for example, water in a sponge), while adsorption only happens at the surface of a substance.

CHECKPOINT

SKILLS CREATIVITY, ETHICS

1. How does the description of the catalytic role of vanadium(V) oxide show that this statement is not true?

   'A catalyst increases the rate of a chemical reaction but does not take part in the reaction.'

2. Write an equation to show how nitrogen monoxide can form acid rain.

SUBJECT VOCABULARY

heterogeneous catalyst a catalyst that is in a different phase from the reactants

adsorption the process that occurs when reactants form weak bonds with a solid catalyst

desorption the process that occurs when products leave the surface of a solid catalyst

17C 2 HOMOGENEOUS CATALYSIS

SPECIFICATION REFERENCE
17.26 PART | 17.30 | 17.31 | 17.32

LEARNING OBJECTIVES

- Know that transition metals and their compounds can act as homogeneous catalysts.
- Know that a homogeneous catalyst is in the same phase as the reactants and appreciate that the catalysed reaction will proceed via an intermediate species.
- Understand the role of Fe^{2+} ions in catalysing the reaction between I^- and $S_2O_8^{2-}$ ions.
- Know the role of Mn^{2+} ions in autocatalysing the reaction between MnO_4^- and $C_2O_4^{2-}$ ions.

TRANSITION METALS AS HOMOGENEOUS CATALYSTS

A **homogeneous catalyst** is one that is in the same phase as the reactants. This means that they are either all gases, or more often, all in aqueous solution. You have not seen many examples of this type of catalyst, compared with those that take part in heterogeneous catalysis. Homogeneous catalysis is also much less common in industry.

The key feature of homogeneous catalysis is the formation of an intermediate species for which a specific formula can be written.

A REACTION OF THE $S_2O_8^{2-}$ ION

The ion with this formula has several names, the simplest of which is the 'persulfate ion', although it is often known as the 'peroxydisulfate ion'. We will not use the IUPAC name, which is very complicated!

It acts as an oxidising agent in its reaction with iodide ions, the equation for which is:

$$S_2O_8^{2-} + 2I^- \rightarrow 2SO_4^{2-} + I_2$$

One reason why this reaction is slow at room temperature is that the two reactant ions are both negatively charged and so repel each other. The reaction is much faster in the presence of Fe^{2+} ions, which act as a catalyst. All of the reactants and products, and the catalyst, are in the aqueous phase, so this reaction is an example of homogeneous catalysis.

STEPS IN THE CATALYSED REACTION

Step 1: The Fe^{2+} ions are not repelled by the $S_2O_8^{2-}$ ions because they have the opposite charge. They react together as follows:

$$S_2O_8^{2-} + 2Fe^{2+} \rightarrow 2SO_4^{2-} + 2Fe^{3+}$$

Step 2: The Fe^{3+} ions formed in Step 1 now react with the I^- ions (these also have opposite charges) as follows:

$$2Fe^{3+} + 2I^- \rightarrow 2Fe^{2+} + I_2$$

The iron(II) ions are used in Step 1 and regenerated in Step 2, so the two steps can repeat continuously.

ALTERNATIVE MECHANISM

The reaction is also catalysed by Fe^{3+} ions, for which the following mechanism can be written.

Step 1: $2Fe^{3+} + 2I^- \rightarrow 2Fe^{2+} + I_2$

Step 2: $S_2O_8^{2-} + 2Fe^{2+} \rightarrow 2SO_4^{2-} + 2Fe^{3+}$

The same two reactions are involved, but they occur in a different order.

OXIDATION OF ETHANEDIOATE IONS

You may know about titrations in which potassium manganate(VII) in acidic conditions acts as an oxidising agent. For these titrations to work well enough, it is important that the reactions are fast, so that the end point of the reaction can be accurately observed (**fig A**).

One example of a redox reaction that can be used in titrations is the oxidation of ethanedioate ions by potassium manganate(VII). The equation for this reaction is:

$$2MnO_4^- + 5C_2O_4^{2-} + 16H^+ \rightarrow 2Mn^{2+} + 5CO_2 + 8H_2O$$

As with the $S_2O_8^{2-}/I^-$ reaction, the reacting species are both negatively charged. Therefore, the reaction is slow. However, as more potassium manganate(VII) solution is added, the reaction rate increases.

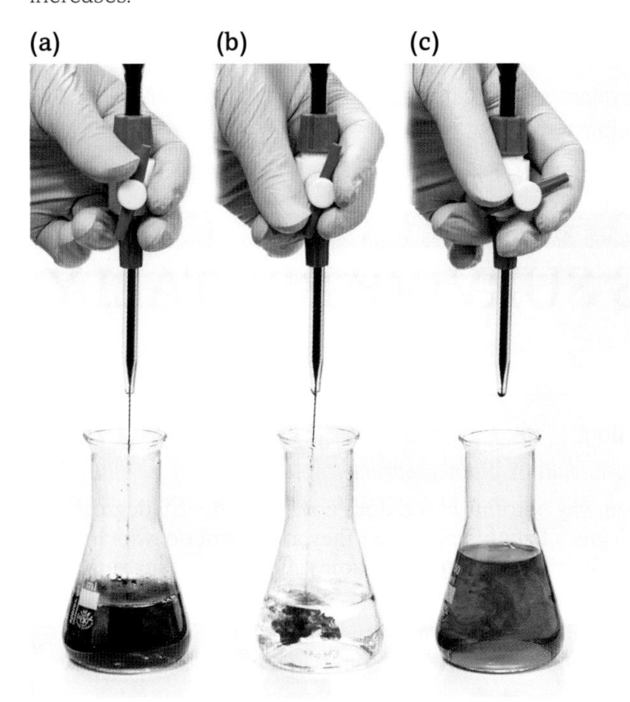

▲ **fig A** (a) At the start of the titration, the colour of the potassium manganate(VII) solution takes some time to disappear; (b) the colour of the potassium manganate(VII) solution now disappears more quickly because there are Mn^{2+} ions to catalyse the reaction; (c) at the end of the titration, there are no more ethanedioate ions left to react, so the remaining potassium manganate(VII) solution gives the mixture a pink colour.

Note that one of the reaction products is the Mn^{2+} ion. This can act as a catalyst in the reaction, which explains why the reaction rate increases as the titration proceeds. The formation of a reaction product that increases the rate of the reaction is sometimes referred to as **autocatalysis**. This effect is sometimes represented using a graph (**fig B**).

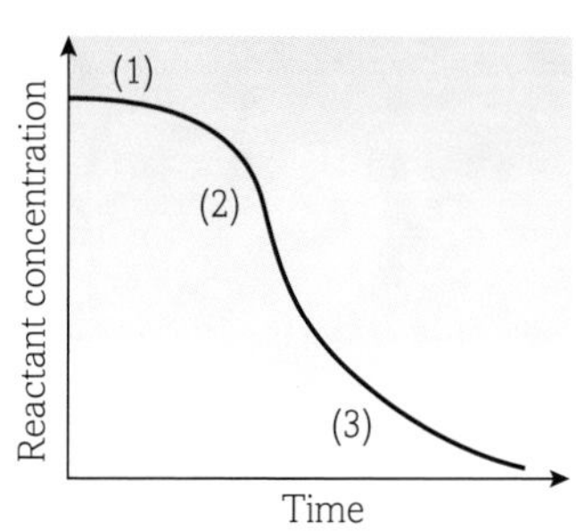

▲ **fig B** (1) The slow decrease in the reactant concentration is because the reaction is uncatalysed at first. (2) The reactant concentration decreases more rapidly as the reaction rate increases because of catalysis. (3) The reactant concentration then decreases more slowly because there is little reactant left.

LEARNING TIP

Note that in **Section 17C.1**, all of the transition metal catalysts were solids, but that in this section they are all ions in aqueous solution.

CHECKPOINT

1. State why Mg^{2+} ions do not act as a catalyst in the reaction between $S_2O_8^{2-}$ and I^- ions.
2. State why autocatalysis of the reaction between MnO_4^- and $C_2O_4^{2-}$ ions does not occur under alkaline conditions.

SUBJECT VOCABULARY

homogeneous catalyst a catalyst that is in the same phase as the reactants
autocatalysis when a reaction product acts as a catalyst for the reaction

17 THINKING BIGGER

POWERFUL PIGMENTS

INITIATIVE,
SELF-DIRECTION

Octopuses are complex and intelligent creatures. One of their many fascinating characteristics is the blue colour of their blood. What is the chemical that causes this colour and what is its function? Read the article below to find out.

BLUE BLOOD HELPS OCTOPUS SURVIVE BRUTALLY COLD TEMPERATURES

Posted by Stefan Sirucek in *Weird & Wild* on 10 July 2013

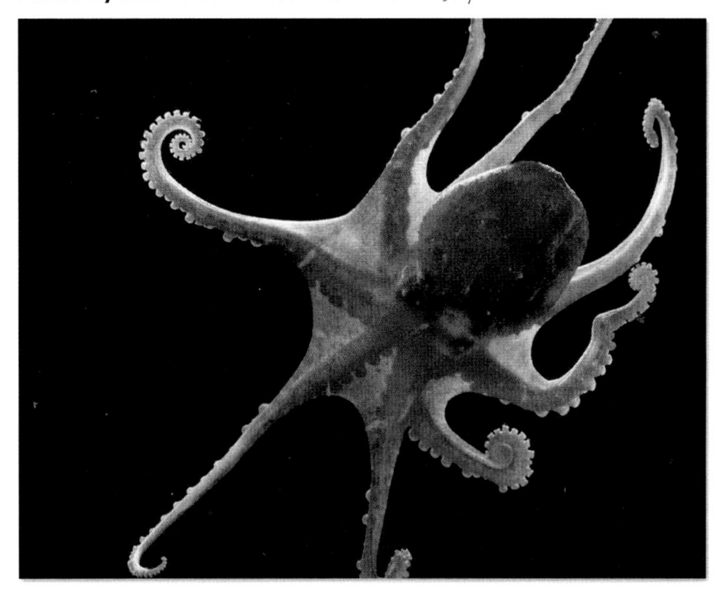

▲ **fig A** A Pareledone octopus photographed near South Georgia Island.

Researchers at the Alfred Wegener Institute for Polar and Marine Research in Germany have found that a specialized pigment in the blood of Antarctic octopods allows them to survive temperatures that often drop below freezing.

It's all down to a respiratory pigment in their blood called haemocyanin that allows the octopus 'to maintain an aerobic lifestyle at sub-zero temperatures,' said Michael Oellermann, who took part in the research. Haemocyanin also has the effect of making the octopods' blood blue.

'Haemocyanin contains copper, which is the responsible ion for binding oxygen,' explained Oellermann in an email. 'Haemoglobins contain iron instead, which gives our blood a red color.'

Blood chilling

It's not a question of blood freezing.

'The reason why octopus blood does not freeze at −1.9 degrees C [28.58 degrees Fahrenheit] is that they are isosmotic, which means that their blood shares the same salinity as the surrounding seawater,' explained Oellermann.

The problem in very cold temperatures is that, without special adaptations, the aerobic process would, in most cases, shut down.

'Cold temperatures increase oxygen affinity to the extent that oxygen cannot be released in the tissue anymore,' explained Oellermann, making survival in waters that can drop to as low as 29 degrees Fahrenheit a challenge to say the least.

Comparing octopods

The researchers looked at a particular species of Antarctic octopod called *Pareledone charcoti* and compared it with temperate and warm-adapted octopods to observe the difference in how cold-water octopods transport oxygen in their blood.

'*Pareledone charcoti* decreased the oxygen affinity of its haemocyanin to counter the adverse effect of temperature on oxygen binding accompanied by changes in their protein sequence, to assure sufficient oxygen supply to its tissues and organs,' explained Oellermann.

Octopods likely developed the adaptation because they require a lot of oxygen compared to other invertebrates, notes Oellermann, and because they are largely non-migratory and must instead adapt to their environment.

In contrast, said Oellermann, Antarctic icefish survive in the same frigid environment by dint of their lower oxygen demand, rather than through the adapted system of oxygen transport employed by their blue-blooded neighbors.

Interestingly, the same adaptation may be what allows octopods to survive temperatures on the other end of the thermometer, such as the 86-degree F heat often found near thermal vents.

From the *National Geographic* magazine website https://blog.nationalgeographic.org/2013/07/10/blue-blood-helps-octopus-survive-brutally-cold-temperatures

SCIENCE COMMUNICATION

1 How might an article like the one opposite be used to highlight the importance of species diversity and environmental conservation?

INTERPRETATION NOTE

Why might it be more effective to focus on one specific example of a fascinating species than to talk about species diversity in general?

CHEMISTRY IN DETAIL

2 Oxygenated haemocyanin is blue in colour, whereas deoxygenated haemocyanin is colourless. Explain this observation using your knowledge of copper chemistry.

3 Describe the type of the bond between copper and nitrogen in oxyhaemocyanin, as shown below in **fig B**.

▲ **fig B** An oxyhaemocyanin molecule.

4 Suggest why zinc is unlikely to be able to form a protein complex that is capable of reversibly binding oxygen.

5 (a) Use the two half-equations below to show that the reaction between oxygen and copper(I) ions under standard conditions is feasible:

$$Cu^{2+}(aq) + e^- \rightleftharpoons Cu^+(aq)$$

$$O_2(g) + 4H^+(aq) + 4e^- \rightleftharpoons 2H_2O(l)$$

(b) Calculate the total entropy change for this reaction under standard conditions (298 K) ($F = 96\,500\ C\,mol^{-1}$)

ACTIVITY

Design an A3 poster for the general public to illustrate the role of transition metals and their compounds as catalysts. You could choose to focus on their role in biological systems (enzymes) or their role in industrial chemistry. In each case, choose a specific example to illustrate the general principles.

WRITING SCIENTIFICALLY

Focusing on a specific example of an enzyme or industrial catalyst can be helpful if you want to illustrate general principles.

17 EXAM PRACTICE

1 A precipitate of copper(II) hydroxide dissolves in aqueous ammonia. What is the formula of the complex ion formed?

A $[Cu(NH_3)_2]^{2+}$

B $[Cu(NH_3)_4]^{2+}$

C $[Cu(NH_3)_6]^{2+}$

D $[Cu(NH_3)_4(H_2O)_2]^{2+}$ [1]

(Total for Question 1 = 1 mark)

2 What is the electronic configuration of the iron cation that can form the complex ion $[Fe(CN)_6]^{4-}$? [The cyanide ligand has the formula CN^-.]

A [Ar] $3d^4$ $4s^2$

B [Ar] $3d^5$ $4s^0$

C [Ar] $3d^6$ $4s^0$

D [Ar] $3d^6$ $4s^2$ [1]

(Total for Question 2 = 1 mark)

3 The anti-cancer drug *cis*-platin has the general formula $Pt(NH_3)_2Cl_2$. In the human body, one of the chloride ions of *cis*-platin is replaced by one water molecule:

$$Pt(NH_3)_2Cl_2 + H_2O \rightarrow [Pt(NH_3)_2(H_2O)Cl]^+ + Cl^-$$

What are the oxidation numbers of platinum in these substances?

	Oxidation number in *cis*-platin	Oxidation number in the aqua complex
A	+2	+1
B	+2	+2
C	+4	+3
D	+4	+4

[1]

(Total for Question 3 = 1 mark)

4 A compound of chromium with the general formula $CrCl_3.6H_2O$ is dissolved in water. When an excess of aqueous silver nitrate is added, only one-third of the total chloride present is precipitated as AgCl.

What is the formula of the chromium complex present in $CrCl_3.6H_2O$?

A $[Cr(H_2O)_6]^{3+}$

B $[Cr(H_2O)_5Cl]^{2+}$

C $[Cr(H_2O)_4Cl_2]^+$

D $[Cr(H_2O)_3Cl_3]$ [1]

(Total for Question 4 = 1 mark)

5 Titanium has the electronic configuration of $1s^2$ $2s^2$ $3s^2$ $3p^6$ $3d^2$ $4s^2$.

Which of the following compounds is *unlikely* to exist?

A TiO

B TiO_2

C K_3TiF_6

D K_2TiO_4 [1]

(Total for Question 5 = 1 mark)

6 What are the colours of aqueous solutions containing $Cr_2O_7^{2-}$, CrO_4^{2-}, $[Cr(H_2O)_6]^{3+}$ and $[Cr(H_2O)_6]^{2+}$?

	$Cr_2O_7^{2-}$	CrO_4^{2-}	$[Cr(H_2O)_6]^{3+}$	$[Cr(H_2O)_6]^{2+}$
A	yellow	orange	green	blue
B	orange	yellow	green	blue
C	orange	yellow	blue	green
D	orange	green	yellow	blue

[1]

(Total for Question 6 = 1 mark)

7 Why is the hexaaquacopper(II) ion, $[Cu(H_2O)_6]^{2+}$, blue?

A The water ligands split the p-orbital energies and p–p electron transitions emit blue light.

B The water ligands split the d-orbital energies and d–d electron transitions absorb all but blue light.

C The water ligands split the p-orbital energies and p–p electron transitions absorb all but blue light.

D The water ligands split the d-orbital energies and d–d electron transitions emit blue light. [1]

(Total for Question 7 = 1 mark)

8 Scandium and titanium are both d-block elements, but only titanium is a transition metal.

(a) Complete the electronic configuration of each of these ions.

Sc^{3+} $1s^22s^22p^6$ ....

Ti^{3+} $1s^22s^22p^6$ .... [2]

(b) Explain why titanium is a transition metal, but scandium is not. [2]

(c) What is the oxidation number of the d-block element in each of these compounds?

$Sc(OH)_3$ ....

$CaTiO_3$ .... [2]

(Total for Question 8 = 6 marks)

9 (a) What is the meaning of the term 'ligand'? [2]

(b) (i) Give the formula of the complex with the name tetraaquadihydroxoiron(II). [1]

(ii) Give the name of the complex with the formula $[CoCl_4]^{2-}$. [1]

(c) The arrangement of d-electrons in the complex with the formula $[Cr(H_2O)_6]^{3+}$ can be represented using this diagram.

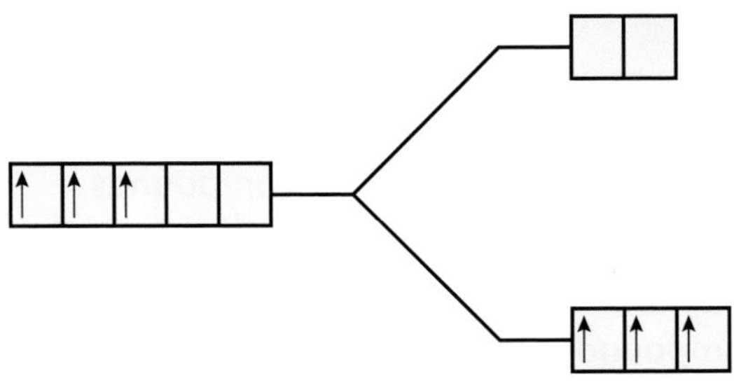

Explain, with reference to this diagram, why a solution containing this complex is coloured. [5]

(Total for Question 9 = 9 marks)

10 Most transition metal complexes have octahedral or tetrahedral shapes.

(a) Explain why the complex formed between one copper(II) ion and six water molecules has an octahedral shape. [2]

(b) Concentrated hydrochloric acid is added to a solution containing the complex in part (a). During the reaction that occurs, there is a colour change from blue to yellow.

(i) Explain why the number of ligands around the metal ion decreases during this reaction. [2]

(ii) State one other feature of the complex that changes during this reaction, and one feature that does *not* change. [2]

(c) A complex with a linear shape has the formula $[Ag(NH_3)_2]^+$, and is used in a test to distinguish between aldehydes and ketones.

(i) Name this complex and the reagent that contains it. [2]

(ii) State the type of change that this complex undergoes when it reacts with an aldehyde, and the name of the metal species formed. [2]

(d) A complex with a different shape, used in cancer treatment, is *cis*-platin, and has the formula $[Pt(NH_3)_2Cl_2]$.

Draw the structure of *cis*-platin and state the name of its shape. [2]

(e) A multidentate ligand has a formula abbreviated to $EDTA^{4-}$. Its structure is:

$$\left[\begin{array}{ccc} OOCCH_2 & & CH_2COO \\ & \diagdown\; NCH_2CH_2N \;\diagup & \\ OOCCH_2 & & CH_2COO \end{array}\right]^{4-}$$

(i) Explain why $EDTA^{4-}$ is described as a multidentate ligand. [1]

(ii) Draw the structure of $EDTA^{4-}$ showing the lone pairs of electrons used in the formation of bonds with a transition metal ion. [1]

(iii) Using the abbreviation EDTA, give the formula of the complex formed when $[Fe(H_2O)_6]^{3+}$ reacts with $EDTA^{4-}$. [1]

(Total for Question 10 = 15 marks)

11 The following sequence summarises some reactions, 1, 2, 3 and 4, of chromium compounds.

$$[Cr(H_2O)_6]^{3+} \xrightarrow{1} [Cr(H_2O)_3(OH)_3] \xrightarrow{2} [Cr(OH)_6]^{3-} \xrightarrow{3} CrO_4^{2-} \xrightarrow{4} Cr_2O_7^{2-}$$

(a) State the appearance of the reactant and product in reaction 1. [2]

(b) State the reagent used in both reactions 1 and 2. [1]

(c) With reference to oxidation numbers, identify which reaction is a redox reaction. [2]

(d) Describe the colour change in reaction 4. [1]

(e) Write an equation for reaction 4. [1]

(Total for Question 11 = 7 marks)

12 The heterogeneous catalyst used in the Contact process to manufacture sulfuric acid has the formula V_2O_5.

Explain, with the help of suitable equations and with reference to the oxidation states of vanadium, the steps in this catalysed reaction. [6]

(Total for Question 12 = 6 marks)

TOPIC 18 ORGANIC CHEMISTRY: ARENES

A ARENES: BENZENE

Your study of organic chemistry so far has been limited to aliphatic compounds. These are compounds containing just unbranched or branched chains of carbon atoms.

In this topic you will extend your knowledge of organic chemistry by looking at aromatic compounds. These are compounds that contain one or more benzene rings. These compounds are also called arenes or aromatic compounds.

Aromatic compounds play a very important role in helping to keep us healthy. For some people, pharmaceuticals mean a better quality of life. For others, they mean the difference between life and death. In both cases, aromatic compounds play an essential part. Phenol, for example, is used as a starting material to make 2,4,6-trichlorophenol (TCP) and also paracetamol. Salicylic acid is used to make aspirin.

Modern transport depends on the products manufactured from aromatic compounds. If all of the components of a car, lorry or aircraft made from aromatic-based products were removed, all of those might cease to exist! Car body parts, bumpers, dashboards, seats and upholstery are generally all made from products derived from aromatic compounds. Synthetic rubbers, also derived from aromatics, give tyres better road-hugging ability, especially on wet roads, and also increase tyre mileage.

Today we want our clothes to be warm, but not thick and itchy. Or we want them to keep us cool, especially when we are exercising. We want them to be colourful, but we do not want the colour to fade after a few washes. We do not want them to crease after a few hours' wear, but we do want them to be supple and soft on the skin. We want them to be wear-resistant, too.

All this has been made possible by the new fibres that have been created thanks to the input of the aromatics industry. Have a look at the labels on your clothes: acrylic fibres, polyester, nylon. These substances are very often added to luxurious fibres such as linen, silk and cashmere wool to give them more resistance to wear.

Aromatic compounds are used today in most of our sports equipment, from polyurethane footballs to nylon parachutes, from light running shoes to polyester swimwear.

The world as we know it today would be very different if there were no aromatics industry.

What prior knowledge do I need?

Topics 4, 5, 10 (IAS: Book 1), 15

- How to use different kinds of formulae to represent organic compounds
- Using IUPAC rules to name organic compounds
- Recognising different types of isomerism, including geometrical isomerism
- How to convert one organic compound into another
- How to write reaction mechanisms

What will I study in this topic?

- How aromatic compounds are different from aliphatic compounds
- Why benzene undergoes substitution reactions rather than addition reactions
- The mechanism of electrophilic substitution reactions into a benzene ring

What will I study later?

Topic 19

- Aromatic amines
- Benzenediazonium ions and azo dyes

18A 1 BENZENE: A MOLECULE WITH TWO MODELS

SPECIFICATION REFERENCE 18.1 18.2 18.3

LEARNING OBJECTIVES

- Be able to use thermochemical, X-ray diffraction and infrared data as evidence for the structure and stability of the benzene ring.
- Understand that the delocalised model for the structure of benzene involves overlap of p-orbitals to form π-bonds.
- Understand why benzene is resistant to bromination, compared to alkenes, in terms of delocalisation of π-bonds in benzene compared to the localised electron density of the π-bond in alkenes.

WHAT ARE AROMATIC COMPOUNDS?

The sections in **Topic 18** deal with compounds containing a benzene ring. You have already seen benzene rings in previous sections. However, no detail was given about them. For example, in the section of **Topic 15** on carbonyl compounds, there was a reaction involving 2,4-dinitrophenylhydrazine, which contains a benzene ring. In the section of **Topic 15** on carboxylic acids, one of the monomers used to make polyesters was terephthalic acid, which also contains a benzene ring.

▲ **fig A** Many brightly coloured dyes are made from aromatic compounds.

ORIGIN OF THE TERM

The word '**aromatic**' in everyday language refers to smells, usually pleasant. Some herbs used in cooking are often described as aromatic.

ALIPHATIC AND AROMATIC

With a few exceptions, almost all the organic compounds you have seen so far in this course could have been described as 'aliphatic', although this is a term we did not use to describe them because there was no point. You can regard aliphatic compounds as all those that are not aromatic!

BENZENE

The compound benzene is at the heart of every aromatic compound. So, the first thing we must do is understand what is special about benzene.

Benzene can be described as an arene, as can many of its derivatives. You already know that *-ene* indicates the presence of a C=C double bond in a molecule (as in ethene). Therefore, you would expect that an arene also contains C=C double bonds. The answer is – yes and no. This obviously needs careful explanation!

PHYSICAL PROPERTIES OF BENZENE

Benzene is a colourless liquid with a boiling temperature of 80 °C, and it is insoluble in water. It is present in crude oil and the fuels obtained from it, so the petrol tank of a car contains some benzene. The one place you will not find it is in a school or college laboratory, because of its toxic nature; in particular, it is a carcinogen (i.e. it can cause cancer). It was first isolated by the English chemist Michael Faraday in 1825. Within a few years, the compound became known as benzene and its molecular formula was established.

Its molecular formula is C_6H_6, which suggests that it is highly unsaturated, because the alkane with six carbon atoms has 14 hydrogen atoms. Finding a structural formula to fit the molecular formula, and which was supported by studies of its chemical reactions, was a challenge for chemists.

THE KEKULÉ STRUCTURE

One of the better-known stories in the history of chemistry is how the German chemist Friedrich August Kekulé came to suggest the structure of benzene that still bears his name. The story goes like this ... Kekulé was sleeping in an armchair in front of a fire. In the flames he imagined some snake-like molecules dancing, one of which held its own tail and whirled around in the flames. Inspired by this dream about snakes, in 1865 he proposed a cyclic or ring structure for benzene. **Fig B** shows the displayed and skeletal forms of this structure.

Displayed Skeletal

▲ **fig B** Kekulé structure of benzene.

Perhaps Kekulé was also influenced by Josef Loschmidt, who had suggested a ring structure for benzene a few years earlier. Loschmidt's role in developing the structure of benzene is not well remembered. However, you may remember his name from **Book 1** as the source of the symbol L for the Avogadro constant.

PROBLEMS WITH THE KEKULÉ STRUCTURE

We still use the original skeletal structure to represent benzene today. However, since it was first used, there has been mounting experimental evidence that does not fit with the structure. Several problems arose in the 19th century, although more became apparent in the 20th century.

PROBLEM 1

If benzene contains three C=C double bonds, it should readily decolourise bromine water in an addition reaction. However, it does not decolourise bromine water. When bromine does react with benzene, a substitution reaction occurs (see **Section 18A.2**). This evidence suggests that there are no C=C double bonds. **Fig C** shows the results of shaking bromine water with a liquid alkane, benzene and a liquid alkene in separate test tubes.

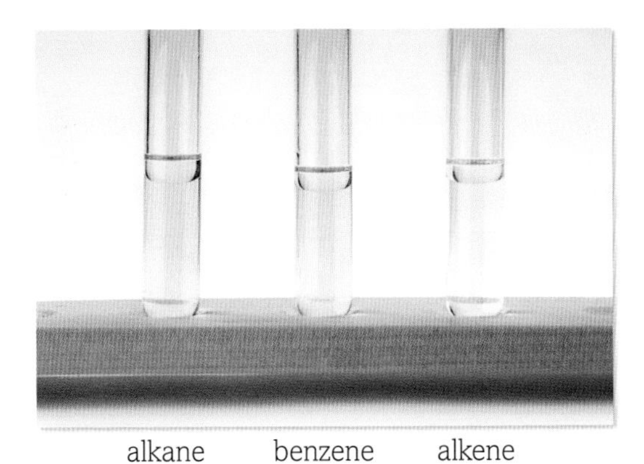

▲ **fig C** The yellow colour is caused by unreacted bromine.

PROBLEM 2

As more compounds containing the benzene ring were discovered, including those such as dibromobenzene ($C_4H_4Br_2$), another problem arose. If the Kekulé structure is correct, there should be four isomers with this molecular formula:

▲ **fig D** These are the four isomers of $C_6H_4Br_2$, assuming that the Kekulé structure is correct.

However, only three were known to exist. So, the two isomers with bromine on adjacent carbon atoms are identical. It is not the case that adjacent atoms in one isomer are joined by C—C and in the other by C=C. This suggests that the bonds between the carbon atoms in the benzene ring are the same, not different.

PROBLEM 3

When data from X-ray diffraction about the lengths of covalent bonds in molecules became available, there was another problem. **Table A** shows the bond lengths in benzene and cyclohexene (a cyclic alkene with five C—C single bonds and one C=C double bond) in picometres.

BOND	BOND LENGTH/pm
C–C in cyclohexene	154
C=C in cyclohexene	134
C–C in benzene	139

table A

This evidence suggests that the carbon–carbon bonds in benzene are all the same, and perhaps also intermediate in character between C—C and C=C bonds.

PROBLEM 4

When thermochemical data about enthalpy changes of hydrogenation became available, there was yet another problem. **Table B** shows these enthalpy changes for three relevant compounds.

COMPOUND	ENTHALPY CHANGE OF HYDROGENATION/kJ mol^{-1}
cyclohexene	-120
cyclohexa-1,4-diene	-239
cyclohexa-1,3,5-triene (theoretical)	-360
benzene (actual)	-208

table B

All of these hydrogenation reactions form the cyclic alkane cyclohexane. This evidence suggests the following.

- The first two values indicate that the enthalpy change for adding 1 mol of H_2 to 1 mol of C=C bonds is around −120 kJ mol^{-1} (the value for cyclohexa-1,4-diene is double that of cyclohexene because there are twice as many C=C bonds).
- The Kekulé structure could be named cyclohexa-1,3,5-triene, for which the value should be treble that of cyclohexene.
- The actual benzene has a value very much lower (152 kJ mol^{-1} lower) than a theoretical structure with three C=C double bonds would have.

The values for cyclohexa-1,3,5-triene (theoretical) and benzene (actual) can be represented on an enthalpy level diagram, as shown in **fig E**.

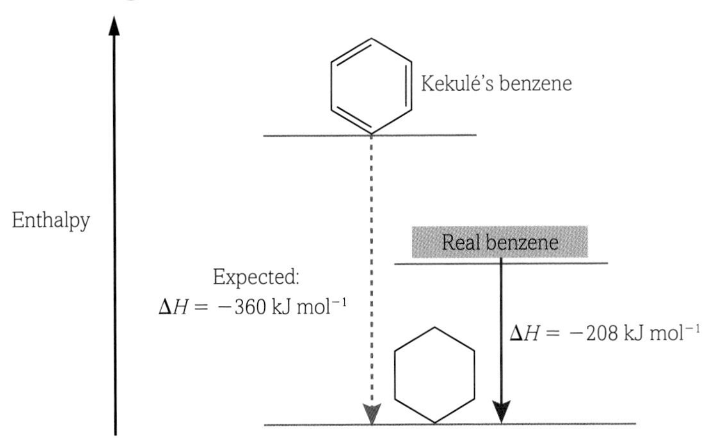

▲ **fig E** This diagram shows the enthalpy changes for Kekulé's benzene and real benzene.

EXAM HINT

When writing an equation for the complete hydrogenation of benzene, make sure you include $3H_2$ to balance the equation (see **Section 18A.2**).

PROBLEM 5

Fig F shows a comparison of the infrared spectra of cyclohexene and benzene.

(a)

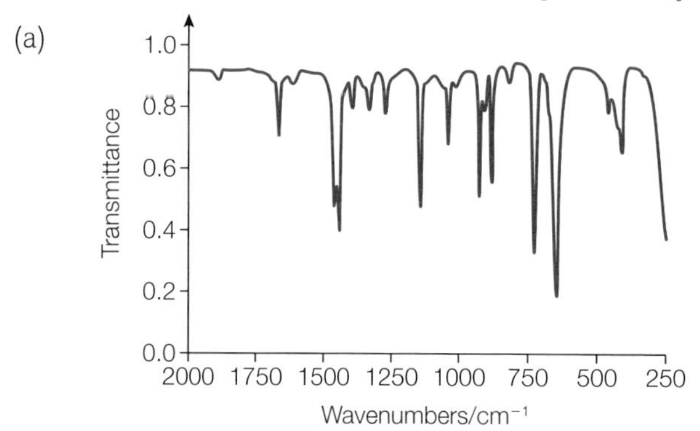

(b)

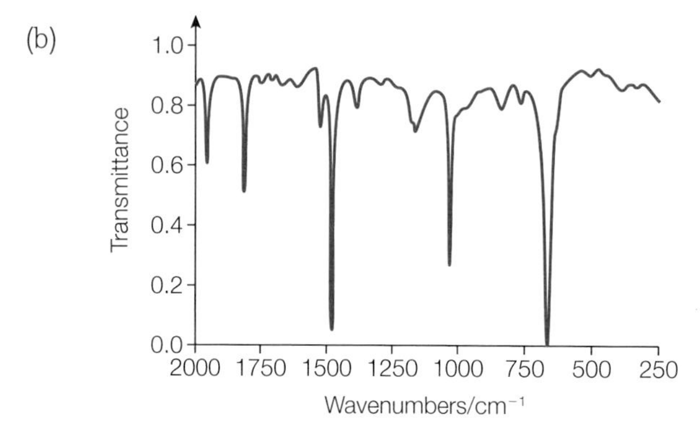

▲ **fig F** Comparison of the infrared spectra of (a) cyclohexene and (b) benzene.

Cyclohexene has an absorption at around 1650 cm^{-1}, which is typical of an isolated alkene C=C stretch (normally between 1680 and 1645 cm^{-1}).

Benzene has a strong absorption at around 1500 cm^{-1}, and also absorptions at 1580 cm^{-1} and 1450 cm^{-1}, which are typical of an aromatic C=C stretch.

A NEW MODEL FOR BENZENE

The weaknesses of the Kekulé structure and growing evidence from other areas of chemistry led to a new model of benzene. In **Book 1**, we explained the C=C double bond in ethene in terms of single sigma C—C bonds and pi bonds formed by the sideways overlap of orbitals.

Now consider the situation in benzene after all the sigma (σ) bonds have formed. Each carbon atom has one electron in a p-orbital. You can imagine the formation of three individual π-bonds, but instead imagine the formation of one large π-bond made up of all six electrons. Three are in a doughnut shape above the atoms and three are in another doughnut shape below the atoms. Together, these six electrons form a delocalised π-bond, as shown in **Fig G**.

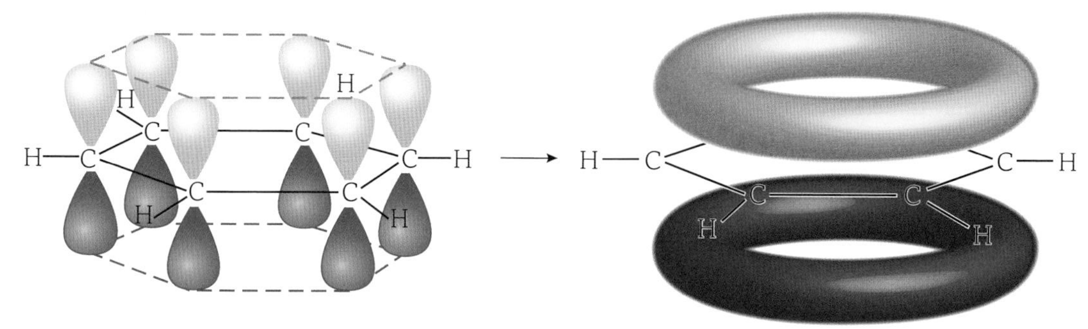

▲ **fig G** The delocalised structure of benzene forms when the p-orbitals overlap sideways forming a π-electron cloud above and below the plane of the carbon atoms. Each carbon atom contributes one electron to the cloud, giving a delocalised system of six electrons in total.

PROBLEMS SOLVED?

Let us see how this new model overcomes the problems associated with the Kekulé structure.

Problem 1 There are no individual C=C bonds, so there is no addition reaction with bromine.

Problem 2 There are three, not four, isomers of $C_6H_4Br_2$ because when the bromine atoms are on adjacent carbon atoms there is no difference in the arrangement of electrons between these atoms.

Problem 3 The carbon–carbon bonds are all the same length because they are identical and are not individual C—C and C=C bonds.

Problem 4 When charge is spread around in a species, there is increased stability, which explains the 152 kJ mol^{-1} greater stability of benzene compared with cyclohexa-1,3,5-triene.

Problem 5 The infrared spectrum of benzene has a very strong absorption at around 1500 cm^{-1}, which is typical of C=C in an aromatic compound. Absorptions for C=C in non-aromatic compounds are between 1680 and 1645 cm^{-1}.

The fact that benzene undergoes a substitution reaction with bromine to form C_6H_5Br, rather than addition to form $C_6H_4Br_2$, can be explained because substitution preserves the stability of the delocalised electrons in the pi bond. The addition reaction would instead produce a compound with two C=C double bonds, and that would lack the extra stability of the product of the substitution reaction.

A NEW MEANING OF 'AROMATIC'

Now we can use the old word '**aromatic**' in a new way – it refers to a hydrocarbon ring containing delocalised electrons. The traditional skeletal formula for benzene can now be replaced by this:

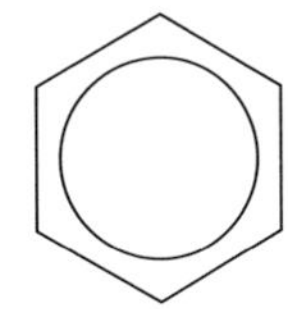

fig H The circle represents the delocalised pi bonds with six electrons in total.

LEARNING TIP

When explaining the bonding in a benzene molecule, be careful to avoid confusion between electrons in p-orbitals and delocalised electrons in π-bonds.

CHECKPOINT

1. Using the Kekulé structure for benzene, draw the isomers of tribromobenzene, $C_6H_3Br_3$.
2. How many electrons in a benzene molecule are involved in:
   (a) σ-bonding
   (b) π-bonding?

SUBJECT VOCABULARY

aromatic the original meaning was a description of the smell of certain organic compounds. The new meaning is a description of the bonding in a compound – delocalised electrons forming pi (π) bonding in a hydrocarbon ring

reactions. [2]

(b) The reagent used in the nitration of benzene is concentrated nitric acid. The reaction is usually carried out using a catalyst of concentrated sulfuric acid.

Write equations to show how the sulfuric acid acts as a catalyst. [2]

(Total for Question 7 = 6 marks)

8 A Kekulé structure of benzene suggests that the molecule consists of alternate single and double carbon to carbon bonds.

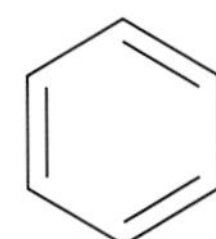

The standard enthalpy change of hydrogenation of a carbon to carbon double bond is $-120\ kJ\ mol^{-1}$.

(a) (i) Calculate the standard enthalpy change of hydrogenation of benzene for the Kekulé structure. [1]

(ii) The actual standard enthalpy change of hydrogenation of benzene is $-208\ kJ\ mol^{-1}$.

Use this information, and your answer to (i), to calculate the difference in stability of benzene and the Kekulé structure.

State what this tells us about the bonding in benzene.

Explain how this influences the type of chemical reactions that benzene undergoes. [4]

(b) The flow chart shows some typical reactions of benzene.

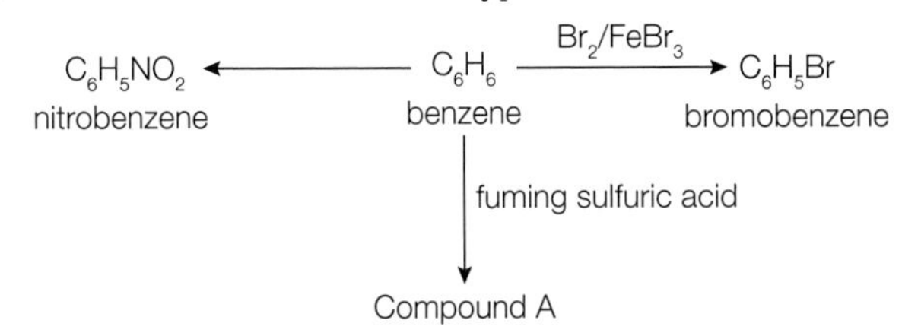

(i) Name the two chemicals needed to make nitrobenzene from benzene. [2]

(ii) Classify the type of reagent required in all three reactions. [1]

(iii) Show, with appropriate diagrams, the mechanism for the reaction of benzene with bromine, in the presence of iron(III) bromide, to form bromobenzene. Include an equation to show formation of the ion attacking the benzene molecule. [4]

(iv) Give the structural formula and name of compound A. [2]

(Total for Question 8 = 14 marks)

9 Benzene reacts with a mixture of concentrated nitric acid and concentrated sulfuric acid to form nitrobenzene.

(a) Give the mechanism for this reaction, including an equation for the formation of the electrophile. [4]

(b) The preparation of nitrobenzene is usually carried out between 50 and 60 °C.

Explain why the temperature used should not be higher or lower. [2]

(Total for Question 9 = 6 marks)

10 Benzene can be represented by either structure A or structure B.

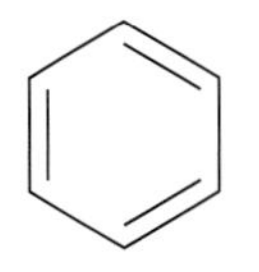

structure A

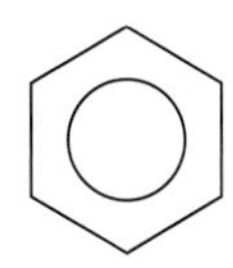

structure B

Explain how the evidence from each of the following supports the view that structure B is the better representation for benzene:

(a) X-ray diffraction [2]

(b) enthalpy change of hydrogenation [2]

(c) the type of reaction that occurs with bromine. [2]

(Total for Question 10 = 6 marks)

TOPIC 19 ORGANIC NITROGEN COMPOUNDS: AMINES, AMIDES, AMINO ACIDS AND PROTEINS

A AMINES, AMIDES, AMINO ACIDS AND PROTEINS

In **Topic 18**, you learned about the chemistry of arenes. In this topic you will extend your understanding of organic chemistry by considering compounds with functional groups such as amines and amides. You will also learn about condensation polymers such as polyesters and polyamides. Amino acids and proteins will be introduced.

MATHS SKILLS FOR THIS TOPIC

- Visualise and represent 2D and 3D forms, including two-dimensional representations of 3D objects
- Understand the symmetry of 2D and 3D shapes

What prior knowledge do I need?

Topics 4, 5, 10 (IAS: Book 1), 15, 18

- How to use different kinds of formulae to represent organic compounds
- Using IUPAC rules to name organic compounds
- Recognising different types of isomerism, including geometrical isomerism
- How to convert one organic compound into another
- How to write reaction mechanisms

What will I study in this topic?

- How to name amines, amides and amino acids
- How to draw structural, displayed and skeletal formulae of amines, amides and amino acids
- The similarities between manufacturing polyamides and the formation of proteins from amino acids

What will I study later?

- Developments involving plastics in the evermore important field of solar cell tecÚology, which may lead to the cost of manufacture falling and efficiency rising, with particular importance for developing countries

Topic 20

- Reaction schemes to convert one organic compound into another
- Methods of preparing and purifying organic compounds

19A 1 AMINES AND THEIR PREPARATIONS

SPECIFICATION REFERENCE 19.1 PART 19.4(i) 19.4(ii) 19.5

LEARNING OBJECTIVES

- Understand the nomenclature of amines and be able to draw their structural, displayed and skeletal formulae.
- Understand, in terms of reagents and general reaction conditions, the preparation of primary aliphatic amines:
  (i) from halogenoalkanes
  (ii) by the reduction of nitriles.
- Know the preparation of aromatic amines by the reduction of aromatic nitro-compounds using tin and concentrated hydrochloric acid.

INTRODUCTION TO AMINES

These nitrogen-containing compounds have some similarities with ammonia. The three bonding pairs of electrons around nitrogen are distributed in a trigonal pyramidal shape. The nitrogen atom has a lone pair of electrons and three bonds to one or more alkyl groups. If there is one alkyl group, the amine is classed as primary. If there are two, then it is secondary. With three it is tertiary. **Fig A** shows the structures of the simplest amine in each of these classes.

fig A These are the simplest examples of primary, secondary and tertiary amines.

In this book we will focus on primary amines. Methylamine is a gas at room temperature, ethylamine has a boiling temperature of 17 °C, and propylamine and butylamine are volatile liquids. This list of names shows the naming system used – the suffix is *-amine* and the usual codes methyl, ethyl, propyl, butyl etc. are used.

We will also meet the simplest aromatic amine, $C_6H_5NH_2$, which is phenylamine.

Amines occur widely in nature, and many drugs (both legal and illegal) are amines, including the group of compounds called amphetamines.

fig B Some cold and flu remedies contain amines.

NOMENCLATURE

We learned about amines in **Topic 10 (Book 1: IAS)**, but only as the product of reactions between halogenoalkanes and ammonia.

Table A shows the structural formulae and names of some examples of amines.

STRUCTURAL FORMULA	NAME
CH_3NH_2	methylamine
$CH_3CH_2NH_2$	ethylamine
$CH_3CH_2CH_2NH_2$	propylamine
$CH_3CH_2CH_2CH_2NH_2$	butylamine
$C_6H_5NH_2$	phenylamine

table A

DISPLAYED AND SKELETAL FORMULAE OF AMINES

Table B shows the displayed and skeletal formulae of the amines in **table A**.

NAME	DISPLAYED FORMULA	SKELETAL FORMULA
methylamine	H, H—C—N, H, H, H	NH_2
ethylamine	H H, H—C—C—N, H H, H, H	NH_2
propylamine	H H H, H—C—C—C—N, H H H, H, H	NH_2
butylamine	H H H H, H—C—C—C—C—N, H H H H, H, H	NH_2
phenylamine	H H, N	NH_2

table B

PREPARATION OF ALIPHATIC AMINES

There are two main ways of making primary aliphatic amines, starting from halogenoalkanes or from nitriles.

PREPARATION FROM HALOGENOALKANES

The method involves heating a halogenoalkane with ammonia. Because ammonia is a gas, this must be done under pressure and in a sealed container. Alternatively, the halogenoalkane can be mixed with concentrated aqueous ammonia. The equation for the preparation of methylamine is:

$$CH_3Cl + NH_3 \rightarrow CH_3NH_2 + HCl$$

The reaction involves nucleophilic attack by the lone pair of electrons of ammonia on the electron-deficient carbon atom in the halogenoalkane. Notice that the amine formed also has a nitrogen atom containing a lone pair of electrons. This means that it could also act as a nucleophile, competing with ammonia in the attack on the halogenoalkane. This would result in the reaction:

$$CH_3Cl + CH_3NH_2 \rightarrow (CH_3)_2NH + HCl$$

The product of this reaction is a secondary amine, in this case dimethylamine.

To prevent such unwanted side-reactions occurring, or at least reduce the chances of them happening, the ammonia is used in excess, so that it outnumbers the molecules of primary amine formed. Some of the excess ammonia reacts with the acidic hydrogen chloride formed, so a better equation to represent the preparation is:

$$CH_3Cl + 2NH_3 \rightarrow CH_3NH_2 + NH_4Cl$$

PREPARATION FROM NITRILES

Nitriles can be reduced to primary amines by reduction, using the reducing agent lithium tetrahydridoaluminate, which you may remember from **Topic 15** can also be used to reduce carbonyl compounds. As before, the reactants are mixed in dry ether, to ensure that there is no water that could affect the reaction. The equation for the reduction of ethanenitrile is:

$$CH_3CN + 4[H] \rightarrow CH_3CH_2NH_2$$

[H] represents hydrogen atoms produced by the reagent in a way that you do not need to know.

EXAM HINT

Although you can use [H] for the reducing agent, you must ensure the equation is balanced by **writing** the appropriate number. In this case it is 4[H].

PREPARATION OF AROMATIC AMINES

A specific method is used to prepare aromatic amines, especially phenylamine, which is made by the reduction of nitrobenzene. In this method, the reducing agent is tin mixed with concentrated hydrochloric acid, and the reaction mixture is heated under reflux (**fig C**). The reduction is achieved partly through the oxidation of tin to tin(II) ions and tin(IV) ions, and partly through the hydrogen produced in the reaction between tin and the acid. To make things simple we will represent the reducing agent by [H]. The equation for the reaction is:

$$C_6H_5NO_2 + 6[H] \rightarrow C_6H_5NH_2 + 2H_2O$$

As with other amines, phenylamine is basic and will react with the acid present to form the phenylammonium ion, but this can easily be converted into phenylamine by adding an alkali such as sodium hydroxide solution:

$$C_6H_5NH_3^+ + OH^- \rightarrow C_6H_5NH_2 + H_2O$$

In the 19th century, phenylamine was called aniline, and was the source of large numbers of dyes used mainly for clothing. It is still used in the manufacture of synthetic indigo, most familiar as the colour of blue denim jeans. Nowadays, much phenylamine is used in the manufacture of polymers and pharmaceuticals.

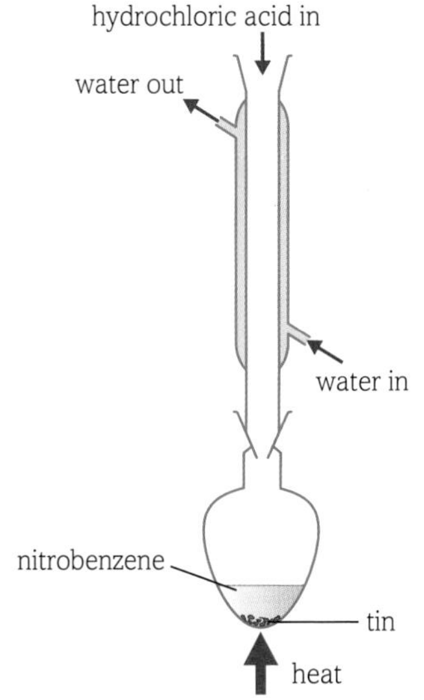

▲ **fig C** Apparatus suitable for converting nitrobenzene into phenylamine.

LEARNING TIP

Note that in the name used for a nitrile, the code indicating the number of carbon atoms includes the C in the CN group, so $CH_3CH_2CH_2CN$ is butanenitrile, not propanenitrile.

CHECKPOINT

SKILLS DECISION-MAKING, PROBLEM-SOLVING

1. Write an equation for the preparation of propylamine starting from:
   (a) a halogenoalkane
   (b) a nitrile.
2. Explain why another organic product would be formed in 1(a) if the two reactants were used in a 1:1 ratio.

19A 2 ACID–BASE REACTIONS OF AMINES

LEARNING OBJECTIVES

- Understand the reactions of primary aliphatic amines (using butylamine as an example) and aromatic amines (using phenylamine as an example) with:
  (i) water to form an alkaline solution
  (ii) acids to form salts.
- Understand that amines are miscible with water as a result of hydrogen bonding, and the reasons for the difference in basicity between ammonia, primary aliphatic amines and primary aromatic amines.

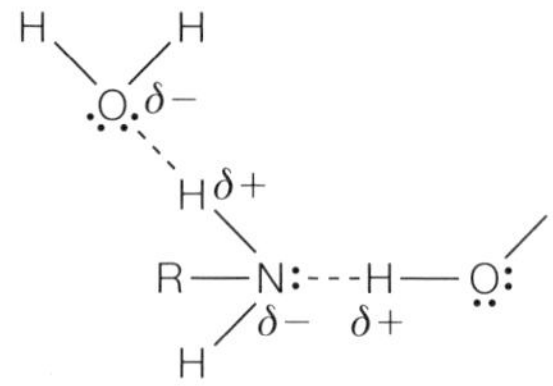

▲ **fig A** The diagram shows the formation of hydrogen bonds between a primary amine and water.

REACTIONS WITH WATER

The first few members of the homologous series of primary aliphatic amines are completely miscible (capable of mixing) with water. However, as the hydrocarbon part of the molecule becomes proportionately larger, the solubility decreases.

They dissolve in water because they can form hydrogen bonds with water molecules (see **fig A**).

Phenylamine is only slightly soluble in water.

As well as dissolving in water, amines also react slightly with water to form alkaline solutions. Compare the equations using methylamine and ammonia:

$$CH_3NH_2 + H_2O \rightleftharpoons CH_3NH_3^+ + OH^-$$

$$NH_3 + H_2O \rightleftharpoons NH_4^+ + OH^-$$

You can see the similarity – this is because the nitrogen atom in both molecules can use its lone pair of electrons to form a dative bond with the hydrogen of a water molecule. The cation in the first equation is the methylammonium ion – think of it as an ammonium ion in which one hydrogen has been replaced by a methyl group. So, how do amines and ammonia compare as bases? What is the position of equilibrium in these reactions?

COMPARING BASICITIES

The **basicity** (basic strength) of a base can be quantified using the constant K_b or the constant pK_b. The pK_b of water at 298 K is 7.00, and any value lower than 7.00 indicates a basicity greater than that of water. Water is equally good as a base and as an acid. **Table A** shows the pK_b values at 298 K for ammonia and some amines.

LEARNING TIP

Note the big difference in basicity between aliphatic amines and aromatic amines and the small differences in basicity between different aliphatic amines.

NAME	FORMULA	pK_b
ammonia	NH_3	4.75
methylamine	CH_3NH_2	3.36
ethylamine	$CH_3CH_2NH_2$	3.27
propylamine	$CH_3CH_2CH_2NH_2$	3.16
phenylamine	$C_6H_5NH_2$	9.38

table A

What these values show is that methylamine is a stronger base than ammonia, and that extending the hydrocarbon chain causes further, but smaller, increases in basicity. Phenylamine is a much weaker base than any of the aliphatic amines, and is a weaker base than even water.

Methylamine is a stronger base than ammonia because the methyl group is electron-releasing, and so has an increased electron density on nitrogen compared with ammonia. The ethyl and propyl groups are only slightly more electron-releasing than methyl, so their effects are very similar. **Fig B** shows the situation.

$CH_3 \rightarrow \overset{\delta-}{NH_2}$ $CH_3CH_2 \rightarrow \overset{\delta-}{NH_2}$ $CH_3CH_2CH_2 \rightarrow \overset{\delta-}{NH_2}$

▲ **fig B** The diagram represents the slight increase in the electron-releasing effect of alkyl groups, causing a slight increase in electron density on the nitrogen atoms.

The equation for the reaction of butylamine with water is

$$CH_3CH_2CH_2CH_2NH_2 + H_2O \rightleftharpoons CH_3CH_2CH_2CH_2NH_3^+ + OH^-$$

The situation with phenylamine is very different. Think back to the diagram in **Section 18A.4** showing that in phenol one of the lone pairs of electrons on oxygen is incorporated into the delocalised electrons in the benzene ring. A similar thing happens with phenylamine, and this makes the nitrogen less electron-rich and the lone pair of electrons less available for donating to the hydrogen of a water molecule. **Fig C** shows how the delocalised electrons now extend above and below the nitrogen. So, as the original lone pair of electrons on nitrogen is now part of the delocalised system, it is less available for bonding to the hydrogen of a water molecule.

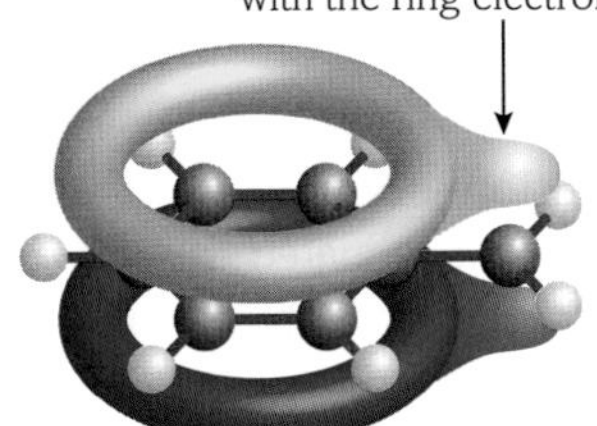

▲ **fig C** The electrons from the lone pair on the nitrogen have joined with the delocalised electrons on the benzene ring.

REACTIONS WITH ACIDS

Even though the amines vary in their basicity, they all react with strong acids to form ionic salts. Sample equations are:

$CH_3NH_2 + HNO_3 \rightarrow CH_3NH_3^+NO_3^-$
methylamine → methylammonium nitrate

$CH_3CH_2CH_2CH_2NH_2 + HCl \rightarrow CH_3CH_2CH_2CH_2NH_3^+Cl^-$
butylamine → butylammonium chloride

$C_6H_5NH_2 + HCl \rightarrow C_6H_5NH_3^+Cl^-$
phenylamine → phenylammonium chloride

CHECKPOINT

1. Write an equation for the reaction of ethylamine with:
   (a) water
   (b) sulfuric acid.
2. Predict, with a reason, whether dimethylamine has a lower or higher basicity than methylamine.

SKILLS CREATIVITY

SUBJECT VOCABULARY

basicity the extent to which a base can donate a lone pair of electrons to the hydrogen atom of a water molecule

19A 3 OTHER REACTIONS OF AMINES

SPECIFICATION REFERENCE
19.2(iii) 19.2(iv)
19.2(v) 19.6

LEARNING OBJECTIVES

- Understand the reactions of primary aliphatic amines (using butylamine as an example) and aromatic amines (using phenylamine as an example) with:
  (i) halogenoalkanes
  (ii) ethanoyl chloride
  (iii) copper(II) ions to form a complex ion.
- Be able to describe the reaction of aromatic amines with nitrous acid to form benzenediazonium ions, followed by a coupling reaction with phenol to form a dye.

LEARNING TIP

Addition-elimination reactions are also called condensation reactions.

REACTIONS WITH ETHANOYL CHLORIDE

You do not need to know the mechanisms of these reactions. However, the name of the mechanism will help you understand how the reactions occur. The reaction type is **addition–elimination**, which means that the two molecules join together, and then a small molecule is eliminated – in these examples, hydrogen chloride. The organic product contains an new functional group – amide – in which a carbonyl group is next to an NH group.

The equation for the reaction of phenylamine with ethanoyl chloride is:

$$CH_3COCl + C_6H_5NH_2 \rightarrow CH_3CONHC_6H_5 + HCl$$

The equation for the reaction between ethanoyl chloride and butylamine is:

$$CH_3{-}C(=O)Cl + H_2N{-}CH_2CH_2CH_2CH_3 \longrightarrow CH_3{-}C(=O){-}N(H){-}CH_2CH_2CH_2CH_3 + HCl$$

We will meet amides in a later section, but we can introduce the naming system here. The most common amide is ethanamide, and the product of the above reaction can be thought of as the result of replacing one of the hydrogens in the NH_2 group by the butyl group. Its name contains two words – the first is the alkyl group from the amine and the second indicates the number of carbon atoms in the original acyl chloride. As the butyl group is attached via a nitrogen atom, *N* is used as the locant. The name is *N*-butylethanamide.

PARACETAMOL

This is one of the most common pharmaceuticals used to relieve the symptoms of fever and pain. It is manufactured in a sequence of reactions, one of which is an addition–elimination reaction. Its structure is:

$$CH_3{-}C(=O){-}N(H){-}C_6H_4{-}OH$$

You can see how paracetamol got its name. The 'para' part indicates that the two groups attached to the benzene ring are at opposite ends of the ring. The 'acetam' part comes from the old name for ethanamide, which used to be called acetamide. Finally, the 'ol' part indicates the presence of a hydroxyl group.

REACTIONS WITH HALOGENOALKANES

You do not need to know the mechanisms of these reactions. However, you can see that the two would react together because a halogenoalkane contains an electron-deficient carbon atom and an amine contains an electron-rich nitrogen atom. Using general formulae, the reaction can be represented as:

$$R'NH_2 + R''X \rightarrow R'NHR'' + HX$$

where R′ is the alkyl group in the amine and R″ is the alkyl group in the halogenoalkane. The reaction is another example of substitution. The organic product is a secondary amine and the inorganic product is a hydrogen halide, often hydrogen chloride.

The equation for the reaction between phenylamine and chloroethane is:

$$C_6H_5NH_2 + CH_3CH_2Cl \rightarrow C_6H_5NHCH_2CH_3 + HCl$$

An example equation using butylamine and chloroethane is:

$$CH_3CH_2CH_2CH_2NH_2 + CH_3CH_2Cl \rightarrow CH_3CH_2CH_2CH_2NHCH_2CH_3 + HCl$$

Notice that the organic product also contains an electron-rich nitrogen atom, so it can also react with chloroethane. The equation for this further reaction is:

$$CH_3CH_2CH_2CH_2NHCH_2CH_3 + CH_3CH_2Cl \rightarrow CH_3CH_2CH_2CH_2N(CH_2CH_3)_2 + HCl$$

The organic product of this reaction is a tertiary amine.

Once again, the tertiary amine contains a nitrogen atom with a lone pair of electrons, which can also react with the halogenoalkanes. The equation for this further reaction is:

$$CH_3CH_2CH_2CH_2N(CH_2CH_3)_2 + CH_3CH_2Cl \rightarrow CH_3CH_2CH_2CH_2N^+(CH_2CH_3)_3Cl^-$$

Note that the equation is different this time – HCl is not formed because this would require the loss of H from the nitrogen of the organic reactant, which a tertiary amine does not have. The product is an ionic compound related to ammonium chloride ($NH_4^+Cl^-$), except that all the hydrogens in the ammonium ion have been replaced by alkyl groups. It is known as a quaternary ammonium salt.

Reaction with halogenoalkanes does not seem to be a good method of preparation, as it is likely that a mixture of similar products will be formed. If there is an excess of the halogenoalkane, then the sequence of all four reactions is more likely to occur. Producing a quaternary ammonium salt in this sequence is what happens in industry. These compounds have many uses, and are usually found in fabric softeners.

REACTIONS WITH COPPER(II) IONS

You will remember from **Topic 17** that ammonia can act as a lone pair donor in its reactions with transition metal ions. This is the overall equation for its reaction with hexaaquacopper(II) ions:

$$[Cu(H_2O)_6]^{2+} + 4NH_3 \rightarrow [Cu(NH_3)_4(H_2O)_2]^{2+} + 4H_2O$$

Amines also have a lone pair of electrons on nitrogen, so can take part in similar reactions. The observations are the same as with ammonia – first of all a pale blue precipitate forms, then with excess butylamine the precipitate dissolves to give a deep blue solution. The equations for these reactions look complicated, but the only difference is the use of $CH_3CH_2CH_2CH_2NH_2$ in place of NH_3.

Formation of the pale blue precipitate:

$$[Cu(H_2O)_6]^{2+} + 2CH_3CH_2CH_2CH_2NH_2 \rightarrow [Cu(H_2O)_4(OH)_2] + 2CH_3CH_2CH_2CH_2NH_3^+$$

Formation of the deep blue solution:

$$[Cu(H_2O)_4(OH)_2] + 4CH_3CH_2CH_2CH_2NH_2 \rightarrow [Cu(CH_3CH_2CH_2CH_2NH_2)_4(H_2O)_2]^{2+} + 2H_2O + 2OH^-$$

With phenylamine, the overall equation to form the complex is:

$$[Cu(H_2O)_6]^{2+} + 4C_6H_5NH_2 \rightarrow [Cu(C_6H_5NH_2)_4(H_2O)_2]^{2+} + 4H_2O$$

LEARNING TIP

Be careful not to confuse amines, ammines and amides, all of which appear in this section.

FORMATION OF AZO DYES

REACTION OF PHENYLAMINE WITH NITROUS ACID

Nitrous acid is a very unstable compound. It exists only in aqueous solution and decomposes at room temperature. This means that any reactions involving nitrous acid have to be carried out under very carefully controlled conditions. The nitrous acid must be prepared in situ by mixing ice-cold solutions of sodium nitrite and dilute hydrochloric acid. The equation for the reaction is:

$$NaNO_2(aq) + HCl(aq) \rightarrow NaCl(aq) + HNO_2(aq)$$

If an aromatic amine, such as phenylamine, is then added to this reaction mixture, a diazonium ion is formed:

$$C_6H_5NH_2(l) + HNO_2(aq) + HCl(aq) \longrightarrow C_6H_5N^+{\equiv}N(aq) + Cl^- + 2H_2O(l)$$

phenylamine → benzendiazonium ion

The reaction vessel must be kept in an ice-water mixture because the reaction must be carried out at 5 °C or lower. If the temperature rises above 5 °C, phenol is formed:

$$C_6H_5NH_2(l) + HNO_2(aq) \longrightarrow C_6H_5OH(aq) + N_2(g) + H_2O(l)$$

phenylamine → phenol

REACTION OF DIAZONIUM IONS WITH PHENOLS

Diazonium ions carry a positive charge and therefore can act as strong electrophiles. For example, the benzenediazonium ion reacts readily in cold, alkaline solution with both aromatic phenols and amines.

The benzenediazonium ion reacts with an alkaline solution of phenol to produce, after acidification with dilute hydrochloric acid, a compound called (4-hydroxyphenyl)azobenzene:

$$C_6H_5{-}N^+{\equiv}N + C_6H_5{-}O^- + OH^- \longrightarrow C_6H_5{-}N{=}N{-}C_6H_4{-}O^- + H_2O$$

benzenediazonium ion, phenoxide ion

$$C_6H_5{-}N{=}N{-}C_6H_4{-}O^- + H^+ \longrightarrow C_6H_5{-}N{=}N{-}C_6H_4{-}OH$$

(4-hydroxyphenyl)azobenzene

This reaction is called a **coupling reaction**. The compound formed is an energetically stable, yellow azo dye. Its stability comes from extensive delocalisation of electrons via the nitrogen-to-nitrogen double bond. If you examine the structure you can see how the double and single bonds alternate. Systems of alternating double and single bonds are called **conjugated systems**. Where extensive conjugation occurs, absorption of light takes place, so the compound is coloured.

The dye (4-hydroxyphenyl)azobenzene is just one of the wide range of dyes that can be made from aromatic amines and other arenes. These are known as azo dyes. They are very stable, so they do not fade. Another example is the acid–base indicator methyl orange, which has the structure:

$$(H_3C)_2N{-}C_6H_4{-}N{=}N{-}C_6H_4{-}S(=O)_2{-}O^-Na^+$$

CHECKPOINT

SKILLS DECISION-MAKING

1. Write an equation for the reaction between propanoyl chloride and propylamine.
2. Draw the structure of the quaternary ammonium salt formed when ethylamine reacts with 1-bromopropane.

SUBJECT VOCABULARY

addition–elimination reaction when two molecules join together, followed by the loss of a small molecule
coupling reaction a reaction in which two organic molecules or ions join together to form one new molecule
conjugated system where single and double bonds alternate, allowing the electrons in the p-orbitals of the atoms to overlap and form a delocalised electron cloud

19A 4 AMIDES AND POLYAMIDES

SPECIFICATION REFERENCE
19.1 | 19.7 | 19.8(i) PART | 19.8(ii) | 19.9 PART | 19.10

LEARNING OBJECTIVES

- Understand the nomenclature of amides and be able to draw their structural, displayed and skeletal formulae.
- Understand that amides can be prepared from acyl chlorides.
- Be able to describe:
  (i) condensation polymerisation for the formation of polyamides such as nylon
  (ii) addition polymerisation, including poly(propenamide) and poly(ethenol).
- Be able to draw the structural formulae of the repeat units of polyamides, poly(propenamide) and poly(ethenol).
- Be able to comment on the physical properties of polyamides and the solubility in water of the addition polymer poly(ethenol) in terms of hydrogen bonding, including soluble laundry bags or liquid-detergent capsules (liquitabs).

AMIDES

We learned about amides in previous sections as the products of reactions. Amides have a functional group consisting of a carbonyl group joined to an amino group as shown.

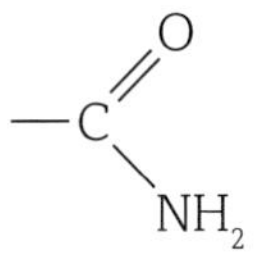

O
—C
NH$_2$

EXAM HINT

You might expect the amide functional group to show basic character, but it does not. The carbonyl group alters the chemical character of the NH_2 group, just like the carbonyl group alters the chemical character of the OH group in carboxylic acids.

As with other compounds in which two functional groups share the same carbon, amides have properties that are different from both carbonyl compounds and amines. They are solids (except for methanamide, which is a liquid). Also, the lower aliphatic amides are soluble in water because they contain two electronegative atoms and polar bonds. This means that they can form hydrogen bonds with water. The carbon atom is very electron-deficient because it is joined to both nitrogen and oxygen.

NOMENCLATURE

Table A shows the structural formulae and names of some amides.

STRUCTURAL FORMULA	NAME
$HCONH_2$	methanamide
CH_3CONH_2	ethanamide
$CH_3CH_2CONH_2$	propanamide
$CH_3CH_2CH_2CONH_2$	butanamide

table A

DISPLAYED AND SKELETAL FORMULAE OF AMIDES

Table B shows the displayed and skeletal formulae of the amides in **table A**.

NAME	DISPLAYED FORMULA	SKELETAL FORMULA
methanamide	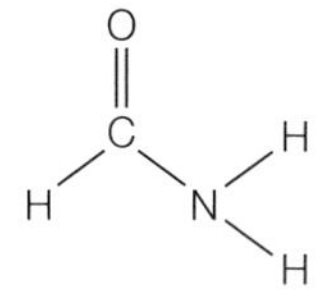	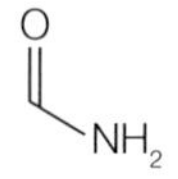
ethanamide	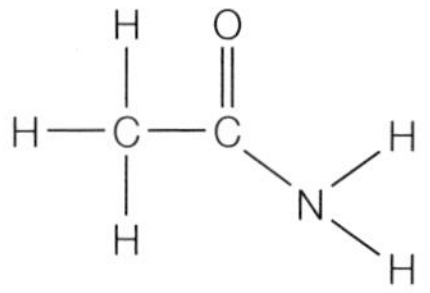	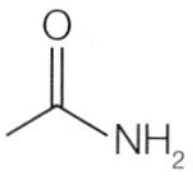
propanamide	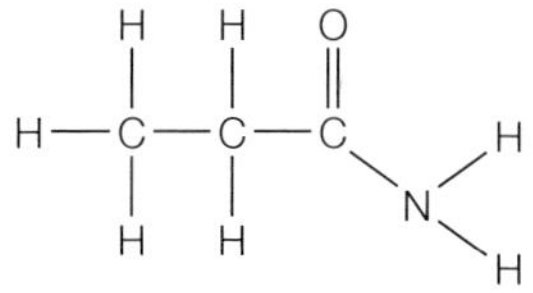	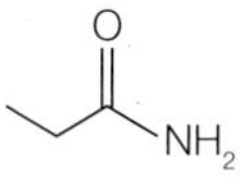
butanamide	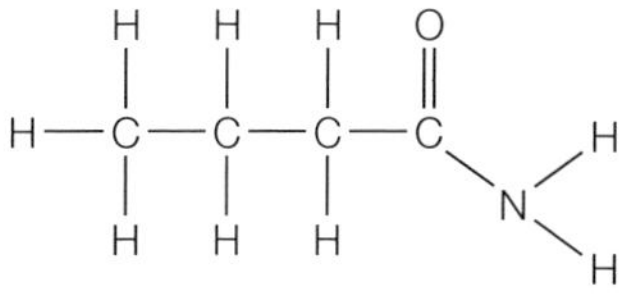	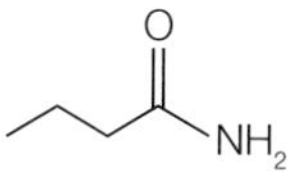

table B

PREPARATION OF AMIDES

A convenient way to prepare amides in the laboratory is by mixing an acyl chloride with concentrated aqueous ammonia. Although you do not need to know the mechanisms of these reactions, it helps to understand the reaction if we refer to the mechanism. The lone pair of electrons on the nitrogen of the ammonia molecule is strongly attracted to the electron-deficient carbon atom of the acyl chloride. The chlorine of acyl chloride combines with one of the hydrogen atoms of ammonia to form hydrogen chloride, which appears as misty fumes. The equation for the reaction between propanoyl chloride and ammonia is:

$$CH_3CH_2COCl + NH_3 \rightarrow CH_3CH_2CONH_2 + HCl$$

Note that ammonia is basic and that the inorganic product is acidic, so there will be a reaction between the two molecules:

$$NH_3 + HCl \rightarrow NH_4Cl$$

So, a better overall equation for the reaction, combining these equations, is:

$$CH_3CH_2COCl + 2NH_3 \rightarrow CH_3CH_2CONH_2 + NH_4Cl$$

POLYAMIDES

▲ **fig A** Balloon fabric is made of nylon.

In **Topic 15** we learned about polyesters as examples of polymers formed by condensation polymerisation reactions. The formation of these polymers needs two monomers – a dicarboxylic acid and a diol. The formation of polyamides also needs two monomers – a dicarboxylic acid and a diamine.

The simplest examples of a dicarboxylic acid and a diamine are HOOCCOOH and $H_2NCH_2NH_2$. You can see how these molecules can react together – one has acidic groups and the other has basic groups, so they can react with each other to form a CONH group with the elimination of H_2O.

NYLON

Nylon is almost certainly the most familiar example of a polyamide. Both of the monomers used in the production of its most common form contain six carbon atoms. Common examples of these monomers are hexanedioic acid (known as adipic acid in the polymer industry) and hexane-1,6-diamine (also known as 1,6-diaminohexane). Both of these monomers have the same number of CH_2 groups with reactive groups at each end of the chain. The formation of this polymer is shown below:

LEARNING TIP

In the formation of a polyamide, the small molecule formed is water (if a dicarboxylic acid is used) or hydrogen chloride (if a dioyl chloride is used).

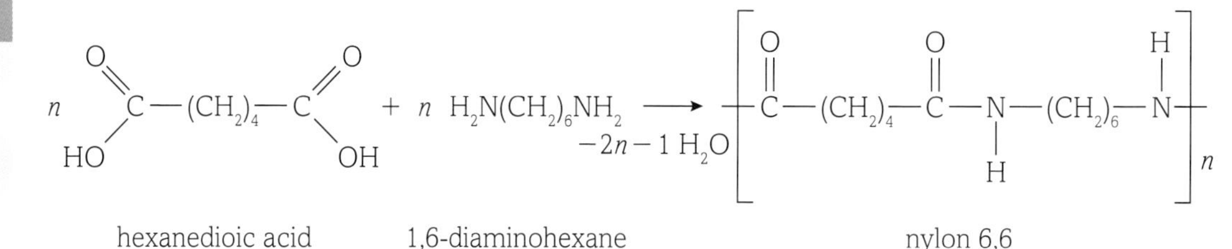

There are different types of nylon, and this example is described as nylon 6,6 because both of the monomers contain six carbon atoms.

KEVLAR®

If benzene rings take the place of the CH_2 groups in the monomers used to make nylon 6,6, and a dioyl chloride is used in place of a dicarboxylic acid, then the monomers and the polymer structure can be shown like this:

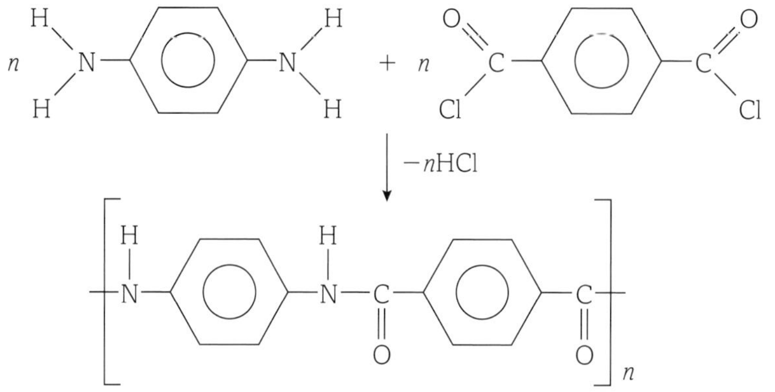

$-n$HCl

▲ **fig B** Kevlar® is used to make bulletproof vests.

This polymer produced is known as Kevlar® and has very many uses, the most familiar of which is as body armour (bulletproof and stabproof vests). However, the list of applications continues to grow, as a web search will reveal.

PROPERTIES OF POLYAMIDES

The majority of polyamides (nylons) tend to be semi-crystalline. They are generally very tough materials with good thermal and chemical resistance. Polyamides tend to absorb moisture from their surroundings. This absorption continues until equilibrium is reached. In general, the impact resistance and flexibility of polyamides tends to increase with moisture content, while the strength and stiffness decrease.

The strong bonds in polyamides make the polyamide chains themselves strong. Polyamides such as nylon can be made into strong fibres for making clothing, fishing lines and carpets. Kevlar® is also a polyamide.

Polyamide film (cling film) is used for food packaging because of its toughness and because gas molecules do not pass through it easily. Polyamide film is also used for 'boil-in-the-bag' food packaging because of its high temperature resistance.

Polyamides tend to provide good resistance to most chemicals. However, they can be attacked by strong acids and alkalis. This is because they contain the amide link (peptide link), which is hydrolysed by acids and alkalis (see **Section 19A.6**).

POLY(PROPENAMIDE)

In **Topic 5 (Book 1: IAS)** you learned about addition polymerisation. This is when many molecules (monomers) with double bonds join together to form a long-chain molecule (polymer). In this reaction, one of the bonds in the double bond is broken and the monomers are linked together by single bonds. The most common example of this is the formation of poly(ethene) from ethene:

$$n\,H_2C{=}CH_2 \longrightarrow \left[\!-CH_2-CH_2-\!\right]_n$$

ethene → poly(ethene)

Poly(propenamide) is an addition polymer. It is formed by polymerising 2-propenamide (sometimes called acrylamide):

$$n\,H_2C{=}CH{-}CONH_2 \longrightarrow \left[\!-H_2C-CH(CONH_2)-\!\right]_n$$

2-propenamide → poly(propenamide)

The polymer is used as a thickener and as a filler in facial surgery. It is also used in water treatment and papermaking.

The polymer chains in poly(propenamide) can be easily cross-linked with other chains. The resulting polymer with cross-chains produces a thick gel that has a larger capacity to absorb water. It can be used for making soft contact lenses (**fig C**).

fig C Soft contact lenses are made of poly(propenamide).

POLY(ETHENOL)

Poly(ethenol) is another addition polymer. It is sometimes called poly(vinyl alcohol). It is not manufactured in the usual way by directly polymerising a monomer. Instead it is made in two stages.

The first stage is the polymerisation of ethenyl ethanoate:

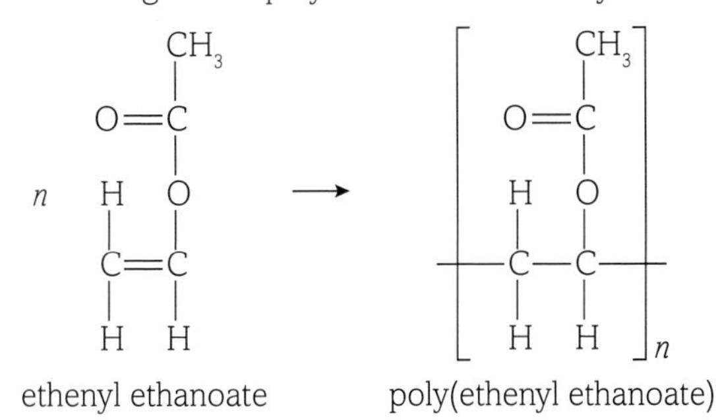

In the second stage, poly(ethenyl ethanoate) is reacted with methanol to form poly(ethenol) and methyl ethanoate, in a process called ester exchange:

$$\left[-\underset{H}{\overset{H}{C}}-\underset{H}{\overset{O-C(=O)CH_3}{C}}- \right]_n + n\,CH_3OH \xrightarrow{\text{ester exchange}} \left[-\underset{H}{\overset{H}{C}}-\underset{H}{\overset{OH}{C}}- \right]_n + n\,CH_3COOCH_3$$

poly(ethenyl ethanoate) poly(ethenol)

The amount of ester exchange can be controlled by altering the temperature.

Poly(ethenol) molecules contain many OH groups. These groups can form hydrogen bonds with water, and so the polymer can be made to be soluble in water. The exact solubility of the final polymer depends on the percentage of ester groups that have been replaced by OH groups. The solubility can range from insoluble, to soluble in hot/warm water, to soluble in cold water.

Poly(ethenol) is used to make disposable laundry bags for use in hospitals. The laundry bags are soluble in water. Dirty bedding and clothes that may be covered in microorganisms from patients are put into the laundry bag. The bag is then placed into a washing machine without the hospital workers touching the dirty fabrics. During washing, the bag dissolves completely and the fabrics are washed clean.

Poly(ethenol) is also used to make liquid-detergent capsules (liquitabs) that contain measured quantities of detergent for use in washing machines and dishwashers. The bags dissolve in the water.

CHECKPOINT

SKILLS INTELLECTUAL INTEREST AND CURIOSITY

1. Write the overall equation for the formation of ethanamide from an acyl chloride.
2. Write an equation for the formation of Kevlar® starting from a diamine and a dicarboxylic acid.

19A 5 AMINO ACIDS

SPECIFICATION REFERENCE
19.1 PART 19.11(i) 19.11(ii)

LEARNING OBJECTIVES

- Understand the nomenclature of amino acids and be able to draw their structural, displayed and skeletal formulae.
- Be able to describe experiments to investigate the characteristic behaviour of amino acids limited to:
  (i) acidity and basicity and the formation of zwitterions
  (ii) effect of aqueous solutions on plane-polarised monochromatic light.

WHAT ARE AMINO ACIDS?

The name suggests the presence of an amino group and a carboxylic acid group. Unlike amides, these two groups are separated by a carbon atom and so retain most of their typical properties. Below is the general displayed formula of an amino acid.

```
H        H
 \       |      O
  N  —   C  — C//
 /       |      \
H        R       O—H
```

When R (the symbol for an alkyl group) is only a hydrogen, the amino acid is the simplest one, glycine. Over 20 amino acids are found in humans. Some of these are synthesised in the body, and others must be included in the diet. Proteins are made from amino acids, and these will be considered in the next section. Although R is normally used to represent an alkyl group, in the case of amino acids it can represent a more complex structure. **Table A** shows information about some important amino acids.

NAME	ABBREVIATION	STRUCTURE	ISOELECTRIC POINT
alanine	ala	$H_2N—CH(CH_3)—COOH$	6.0
cysteine	cys	$H_2N—CH(CH_2—SH)—COOH$	5.1
glutamic acid	glu	$H_2N—CH(CH_2—CH_2—COOH)—COOH$	3.2
glycine	gly	$H_2N—CH_2—COOH$	6.0
lysine	lys	$H_2N—CH(CH_2—CH_2—CH_2—CH_2—NH_2)—COOH$	9.7

table A

There are 20 naturally occurring amino acids. They are often referred to as 2-amino acids because the NH_2 is attached to the second carbon atom in the chain, counting the C in COOH as carbon 1. Note that most amino acids have one NH_2 group and one COOH group, although some have two of one and one of the other.

NOMENCLATURE

Although amino acids are usually referred to by their common names, such as glycine, H_2NCH_2COOH, and alanine, $H_2NCH(CH_3)COOH$, they also have systematic names using the rules defined by IUPAC. They are named as amine derivatives of carboxylic acids, using the prefix *amino* to indicate the presence of the amine group, NH_2. A locant is used to indicate the position of the amine group in the carbon chain of the carboxylic acid.

Table B shows the structural formulae and names of some amino acids. The 3-amino acids are not naturally occurring.

STRUCTURAL FORMULA	NAME
H_2NCH_2COOH	2-aminoethanoic acid
$CH_3CH(NH_2)COOH$	2-aminopropanoic acid
$H_2NCH_2CH_2COOH$	3-aminopropanoic acid
$CH_3CH(NH_2)CH_2COOH$	3-aminobutanoic acid
$C_6H_5CH(NH_2)COOH$	2-aminophenylethanoic acid

table B

DISPLAYED AND SKELETAL FORMULAE OF AMINES

Table C shows the displayed and skeletal formulae of the amines in **table B**.

NAME	DISPLAYED FORMULA	SKELETAL FORMULA
2-aminoethanoic acid	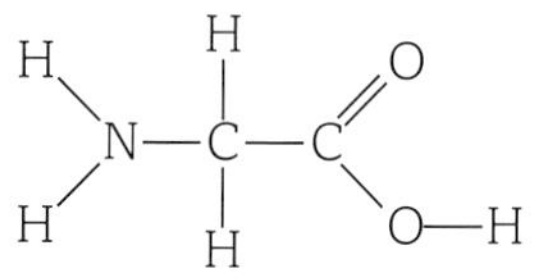	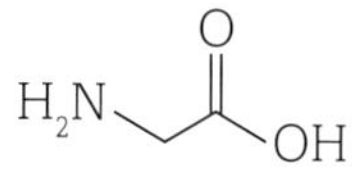
2-aminopropanoic acid	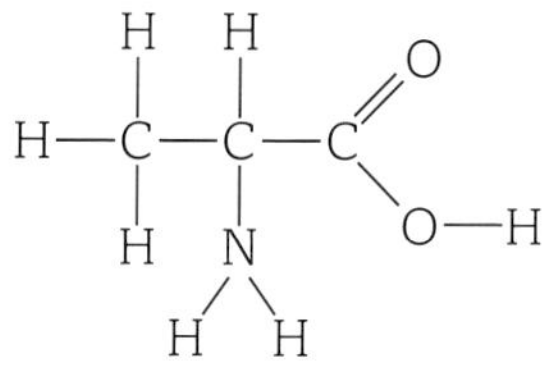	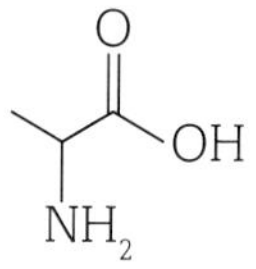
3-aminopropanoic acid	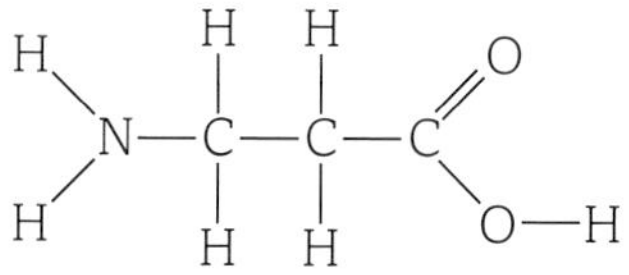	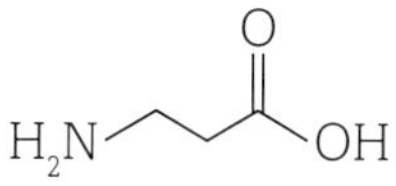
3-aminobutanoic acid	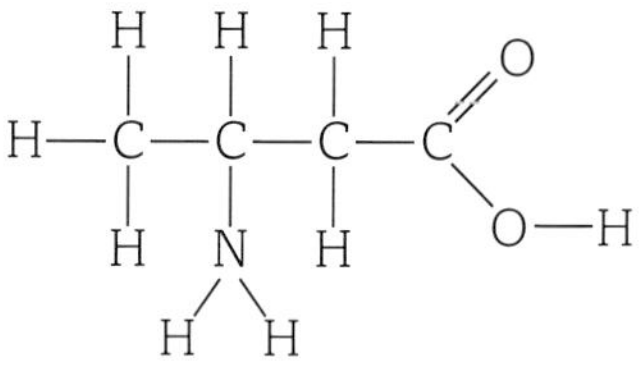	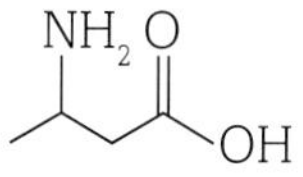
2-aminophenylethanoic acid	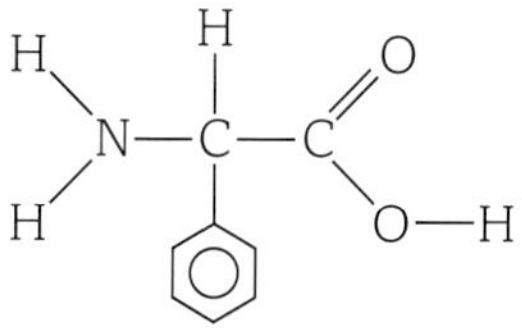	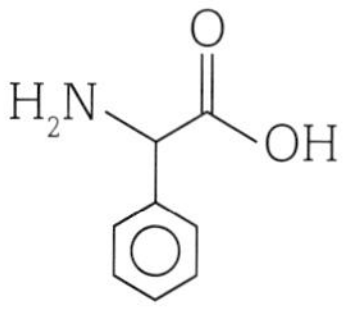

table C

ACIDIC AND BASIC PROPERTIES

The values of the **isoelectric points** of the amino acids have been included to help you understand their acid–base character. They are all soluble in water, and you can imagine that one of two reactions might occur. The molecule could act as a base and form an alkaline solution in a reaction like this:

$$H_2N{-}CHR{-}COOH + H_2O \rightleftharpoons H_3N^+{-}CHR{-}COOH + OH^-$$

Alternatively, it could act as an acid and form an acidic solution in a reaction like this:

$$H_2N—CHR—COOH + H_2O \rightleftharpoons H_2N—CHR—COO^- + H_3O^+$$

There is a third alternativc – an H^+ ion could transfer from the COOH group to the NH_2 group in a reaction like this:

$$H_2N—CHR—COOH \rightleftharpoons H_3N^+—CHR—COO^-$$

The product of this reaction is electrically neutral because it has a positive charge and a negative charge that balance each other. Such species are called **zwitterions**. *Zwitter* is the German word for hybrid, meaning a cross between two things and having the characteristics of both.

Now to the isoelectric point. This is the pH at which the zwitterion exists in aqueous solution. A low isoelectric point indicates that the molecule is predominantly acidic (glutamic acid has a value of 3.2), while a high value indicates that it is predominantly basic (lysine has a value of 9.7). You can see that these values are related to the numbers of NH_2 and COOH groups in the molecule (glutamic acid has an extra carboxylic acid group and lysine has an extra amino group).

EXAM HINT

There is a link here to **Topic 14** on acid-base equilibria. You might like to think about how proteins could be involved in buffering pH.

SALT FORMATION

All amino acids can form salts with acids and bases. For example, alanine can react with acids to form this protonated structure:

$$\begin{array}{l} H_3N^+—CH—COOH \\ \quad\;\; | \\ \quad\; CH_3 \end{array}$$

Glutamic acid can react with sodium hydroxide to form three possible salts. This is because there are two COOH groups, so either of them can react, or both can react. One salt has the structure:

$$\begin{array}{l} H_2N—CH—COO^-Na^+ \\ \qquad\;\; | \\ \qquad CH_2—CH_2—COOH \end{array}$$

You may have heard of this salt – its name is monosodium glutamate and is often used as a flavour enhancer in food.

▲ **fig A** Monosodium glutamate can be used as a flavour enhancer in cooking.

OPTICAL ACTIVITY

Almost all 2-amino acids contain a chiral centre (the C of the CH group), and so are optically active. The exception is glycine, which has a CH_2 group instead. Aqueous solutions of the enantiomers rotate the plane of polarisation of plane-polarised light – some enantiomers are dextrorotatory (+) and others are laevorotatory (−). If an amino acid is synthesised in the laboratory, then a racemic mixture is formed.

LEARNING TIP

Practice writing equations for different amino acids to show how they can react as both acids and bases.

CHECKPOINT

1. Draw the structure of glycine in a solution with:
   (a) a pH of 6.0
   (b) a very low pH
   (c) a very high pH.
2. Draw structures to show how the two enantiomers of cysteine are related to each other.

SUBJECT VOCABULARY

isoelectric point (of an amino acid) the pH of an aqueous solution in which it is neutral
zwitterion a molecule containing positive and negative charges but which has no overall charge

19A 6 PEPTIDES AND PROTEINS

SPECIFICATION REFERENCE

19.9 PART 19.11(iii)

LEARNING OBJECTIVES

- Be able to draw the structural formulae of the repeat units of proteins.
- Be able to describe experiments to investigate the characteristic behaviour of amino acids in the formation of peptide bonds by condensation polymerisation.

WHAT IS A PEPTIDE?

When two amino acid molecules react together, an acid–base reaction occurs. The OH of the COOH group combines with one of the H atoms of the NH_2 group to form water. This is a condensation reaction, in which the two amino acids are joined together by an amide group (CO—NH). The bond that forms is known as a **peptide bond**, and the organic product is a dipeptide.

MORE THAN ONE DIPEPTIDE

When two amino acids combine together to form a dipeptide, there are always two possibilities – sometimes more. For example, when glycine reacts with alanine, the OH could be lost from either molecule, as shown below.

alanine + glycine:

$$H_2N{-}CH(CH_3){-}COOH + H_2N{-}CH_2{-}COOH \longrightarrow H_2N{-}CH(CH_3){-}CO{-}NH{-}CH_2{-}COOH + H_2O$$

glycine + alanine:

$$H_2N{-}CH_2{-}COOH + H_2N{-}CH(CH_3){-}COOH \longrightarrow H_2N{-}CH_2{-}CO{-}NH{-}CH(CH_3){-}COOH + H_2O$$

TRIPEPTIDES

When one molecule of each of three different amino acids reacts together to form a tripeptide, the six possibilities can be summarised using the three-letter abbreviations shown in the table in the previous section. They are:

ala–cys–glu   ala–glu–cys   cys–ala–glu   cys–glu–ala   glu–ala–cys   glu–cys–ala

If you work out the possibilities in which two or more molecules of the same amino acid in these three react to form a tripeptide you will find that there are many more!

POLYPEPTIDES AND PROTEINS

Polypeptides and **proteins** are formed by condensation polymerisation of many amino acids. The main difference between a long-chain polypeptide and a protein is that proteins have further levels to their structures. These are to do with the way in which the polypeptide chains interact with each other in three dimensions, to give secondary, tertiary and quaternary structures, but these are beyond the aims of this book.

ANALYSING PROTEINS

Many proteins have very large molar masses and complex structures. In recent decades, more and more proteins have been analysed. As several million different proteins are thought to exist, this work is still continuing but far from complete. **Table A** shows three examples of proteins, the first two of which have vital roles in the human body.

PROTEIN	WHERE FOUND	APPROXIMATE MOLAR MASS/g mol^{-1}	APPROXIMATE NUMBER OF AMINO ACIDS
insulin	pancreas	5700	51
haemoglobin	blood	66 000	574
urease	soya beans	480 000	4500

table A

The first step in discovering the structure of a protein is to find which amino acids are present. The second is to find the order in which they occur in the polypeptide chains.

HYDROLYSING PROTEINS

The polypeptide chains in a protein can be broken down into their individual amino acids by prolonged heating with concentrated hydrochloric acid. This breaks the peptide bonds between the amino acids, although, because of the strongly acidic conditions, all the amino acids formed will have their NH_2 groups protonated as $^+NH_3$ groups. You can see this process more easily in the hydrolysis of a dipeptide. As an example we will use the dipeptide formed between alanine and glycine, referred to earlier in this section. The equation for this hydrolysis is:

$$H_2N{-}CH(CH_3){-}CO{-}NH{-}CH_2{-}COOH + H_2O + 2H^+ \longrightarrow H_3\overset{+}{N}{-}CH(CH_3){-}COOH + H_3\overset{+}{N}{-}CH_2{-}COOH$$

You can see that this hydrolysis reaction is the reverse of the original dipeptide formation, except that the two NH_2 groups are protonated, which explains the inclusion of the two H^+ ions in the equation.

USING CHROMATOGRAPHY

The different types of chromatography are covered in **Topic 15**, and you already know about simple chromatography from your previous study of chemistry. A mixture of amino acids produced by hydrolysis of a protein can be spotted onto chromatography paper. Using a suitable solvent, the individual amino acids will rise to different heights during the experiment. As amino acids are colourless, the chromatogram can be sprayed with a developing agent so that the positions of the amino acids can be seen. You may read about the use of ninhydrin as a developing agent. Although this works well, ninhydrin is not normally used now because of its toxic nature.

However, once the positions of the amino acids have been established, their R_f values can be calculated, and so the individual amino acids can be identified. Working out how they are all joined together is rather more complicated and beyond the aims of this book.

Insulin is a protein with the molecular formula $C_{256}H_{381}N_{65}O_{79}S_6$. It is made up of a total of 51 amino acids, including 17 different ones. Its structure is shown in abbreviated form in **fig A**.

▲ **fig A** This diagram shows the sequence of amino acids in insulin from a sheep. You can see that there are actually two separate sequences of amino acids in which cysteine molecules are joined by disulfide bonds.

EXAM HINT

Remember that separating amino acids using chromatography depends on the relative solubilities of the amino acids in the mobile and stationary phases. It *does not* depend on the size of the amino acids (see **Topic 15**).

CHECKPOINT

1. Write the structures of the two dipeptides formed when alanine and cysteine react together.
2. Write the structures of the two amino acids formed when this dipeptide is hydrolysed by concentrated hydrochloric acid:

$$H_2N{-}CH(CH_2{-}OH){-}CO{-}NH{-}CH(CH_2{-}COOH){-}COOH$$

SKILLS PROBLEM-SOLVING

SUBJECT VOCABULARY

peptide bond the bond formed by a condensation reaction between the carbonyl group of one amino acid and the amino group of another amino acid
polypeptide a condensation polymer formed from many amino acids
protein a polypeptide that has folded into a specific shape in order to have a specific function

LEARNING TIP

Two possible dipeptides can form between two different amino acids. If one or both of the amino acids has more than one NH_2 or COOH group, then more possible dipeptides can be formed.

19 THINKING BIGGER

PROTEIN MOLECULES: LIFE'S HARDWARE

The science behind protein structures has come a long way since the first protein structure, myoglobin, was solved by Kendrew and Perutz in 1958. Developments in the knowledge of how proteins fold have since led to the development of 'designer' proteins.

DESIGNER PROTEINS

Researchers have successfully designed entirely new proteins based on biological principles.

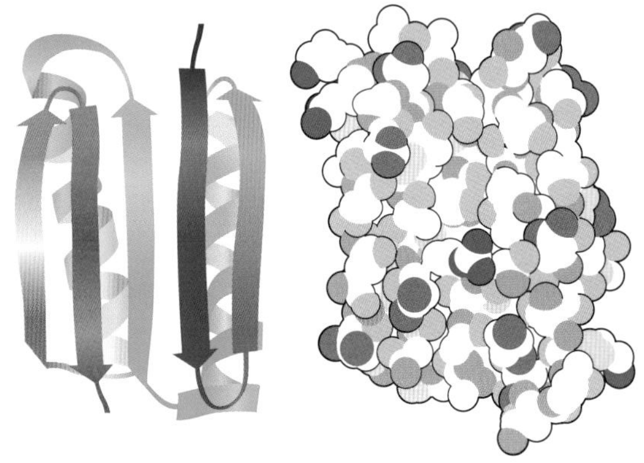

▲ **fig A** Ribbon diagram and space-filling diagram for Top7.

As we learn more and more about proteins and how they work we naturally have the desire to use this knowledge and do some tinkering of our own. Since the early 1980s, scientists have been using the ever-expanding understanding of protein structure and function to redesign existing proteins, and, more recently, to design entirely new proteins.

Designing proteins

As scientists began this quest they quickly found that proteins are more complicated than they might seem. The different types of amino acids, each with their own chemical features, work together to coax a protein chain to fold into a compact stable structure. A collection of carbon-rich amino acids, like leucine and phenylalanine, are usually placed inside the protein, all chosen to lock perfectly together. On the other hand, charged amino acids, such as lysine and aspartic acid, are typically spread across the surface to make the protein soluble in water. Hydrogen-bonding amino acids, such as serine and asparagine, are dotted in strategic places to tie different portions of the chain together. Finally, the odd glycine or proline is added to redirect the chain in the proper direction.

Negative design

This combination of favourable forces lock the protein chain into a stable, compact structure. But this is only the first step in protein design. In order to design a protein that will successfully fold you also need to make sure that the protein has only one stable structure. If there are any other ways to fold the protein they will compete with your desired structure and spoil the construction. So, it is not enough to design a stable protein structure ... you also need to design a protein that is unstable in every other conformation.

Proteins from scratch

Building on decades of protein structure research, scientists at the Fred Hutchinson Cancer Research Center and the University of Washington in Seattle have successfully weighed these positive and negative design elements to create an entirely novel protein. They started with a fold that had never been seen in nature, shown here above. Then, they used a computational method to design a sequence of amino acids that would adopt this fold. When they built this protein, which they call Top7, they found that it folded up and adopted exactly the structure they designed, as seen in PDB entry 1qys.

From the Protein Data Bank's RCSB educational portal: 'Molecule of the Month', PDB 101. http://pdb101.rcsb.org/motm/70

SCIENCE COMMUNICATION

1 There are a number of 'non-scientific' words used in this article, such as 'tinkering' and 'coaxed'. What do you think the author is trying to achieve by using such words?

CHEMISTRY IN DETAIL

2 Suggest why it is important that a protein has "only one stable structure".

3 The IUPAC name for the amino acid known as aspartic acid is 2-amino-butan-1,4-dioic acid. Draw the skeletal structure of this molecule and calculate is relative molecular mass.

4 One of the amino acids mentioned in the text is phenylalanine, whose structure is shown below:

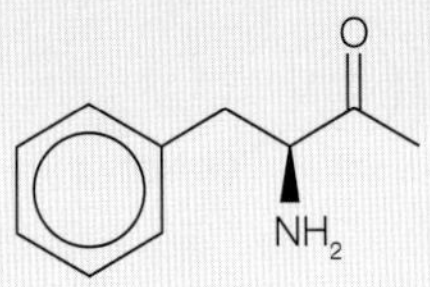

NH_2

(a) Give the molecular formula of phenylalanine.

(b) Suggest why it would be unlikely for this amino acid to be found on the surface of a water-soluble enzyme.

THINKING BIGGER TIP

In this extract you will see how the subjects of chemistry and biology overlap. The Protein Data Bank is an excellent online resource where you can learn a lot more about these 'molecules of life' (https://www.rcsb.org).

CHEMISTRY TIP

Think about which types of functional groups can be classed as hydrophobic or hydrophilic. You can also think about how benzene interacts with water.

ACTIVITY

In the final paragraph the author refers to a protein "fold that had never been seen in nature".

Working in groups if you prefer, answer the following questions:

(i) What is meant by 'bioethics'?

(ii) What are some of the ethical issues scientists might encounter as they use genetic tecÚiques to design new proteins?

19 EXAM PRACTICE

1 After the reduction of nitrobenzene to phenylamine, using tin and hydrochloric acid, an excess of sodium hydroxide is added. What is the purpose of the sodium hydroxide?

A to dry the product

B to liberate phenylamine

C to lower the boiling point for subsequent distillation

D to precipitate tin(II) hydroxide [1]

(Total for Question 1 = 1 mark)

2 The amino acids aspartic acid and glutamic acid can react with each other to form amide linkages.

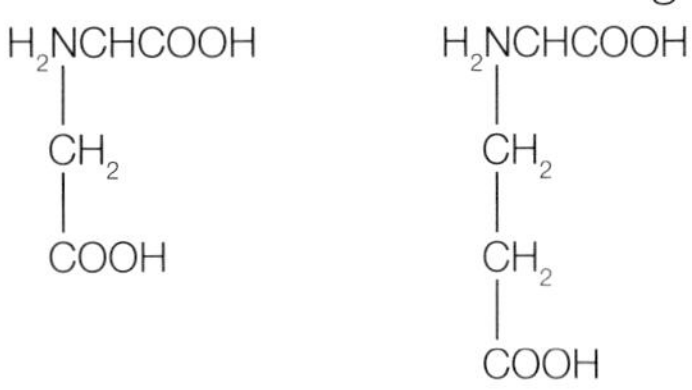

aspartic acid glutamic acid

What is the maximum number of different compounds, each containing only one amide linkage, that can be formed from one molecule of aspartic acid and one molecule of glutamic acid?

A 1 **B** 2 **C** 4 **D** 6 [1]

(Total for Question 2 = 1 mark)

3 What is the formula of the organic product obtained from the coupling reaction of the benzenediazonium ion, $C_6H_5N_2^+$, and the phenoxide ion, $C_6H_5O^-$ in alkaline solution?

A H—N=N—C_6H_4—C_6H_4—O^-

B Cl—N=N—C_6H_4—O—C_6H_5

C C_6H_5—N=N—C_6H_4—O^-

D C_6H_5—N=N—O—C_6H_5

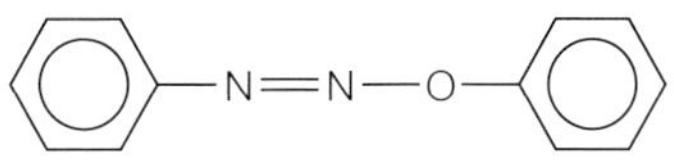

[1]

(Total for Question 3 = 1 mark)

4 The structure of a herbicide called karbutilate may be represented as:

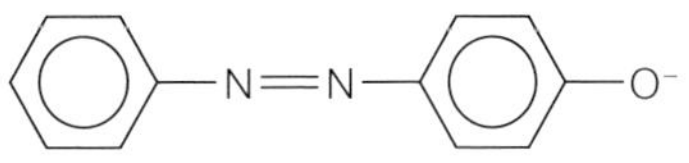

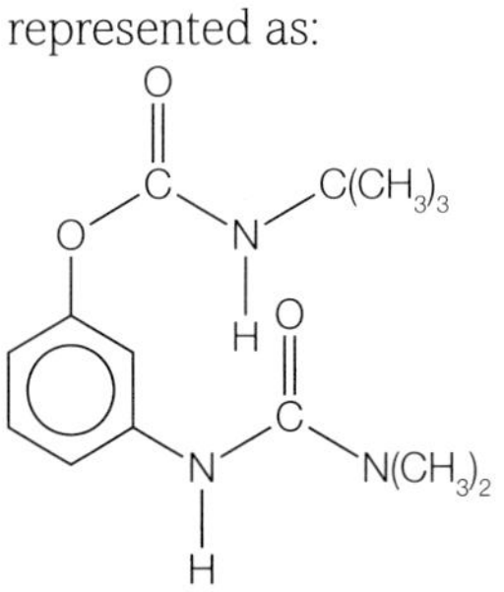

O, C, O, N, $C(CH_3)_3$, H, O, C, N, $N(CH_3)_2$, H

Which compound would be formed by prolonged boiling of karbutilate with aqueous sodium hydroxide?

A O^-Na^+, NH_2

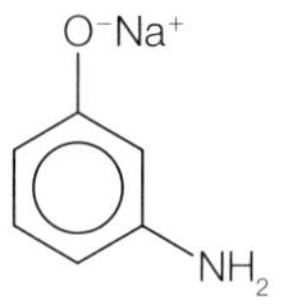

B O^-Na^+, O^-Na^+

C $OCOO^-Na^+$, O^-Na^+

D $OCOO^-Na^+$, NH_2

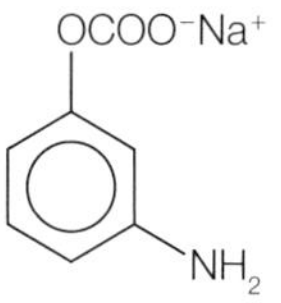

[1]

(Total for Question 4 = 1 mark)

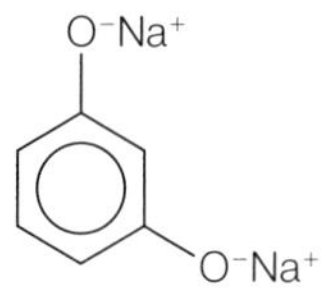

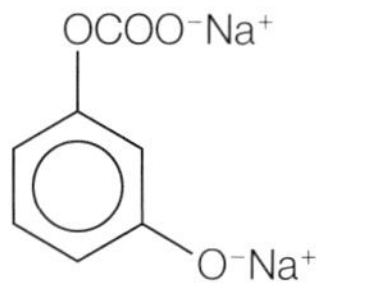

5 Nylon 6 has the following formula:

$$\left[NH(CH_2)_5CO \right]_n$$

What is the product of the acidic hydrolysis of nylon 6?

A $H_2N(CH_2)_5COOH$

B $H_2N(CH_2)_5COO^-$

C $H_3N^+(CH_2)_5COOH$

D $H_3N^+(CH_2)_5COO^-$ [1]

(Total for Question 5 = 1 mark)

6 The addition polymer poly(ethenol) is water-soluble. Which is the repeating unit of poly(ethenol)?

A C(OH)(H)—C(H)(OH)

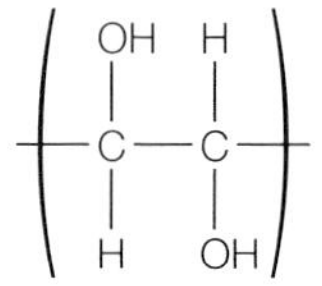

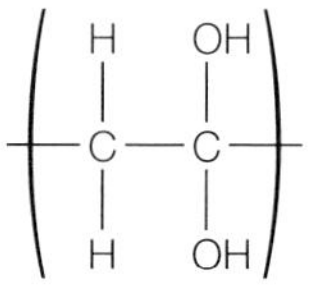

B C(H)(H)—C(H)(OH)

C C(H)(H)—C(OH)(OH)

D C(OH)(OH)—C(OH)(OH)

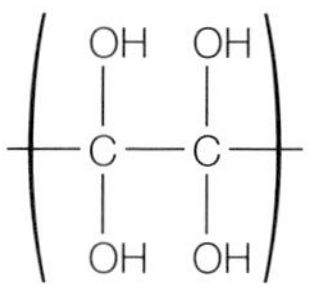

[1]

(Total for Question 6 = 1 mark)

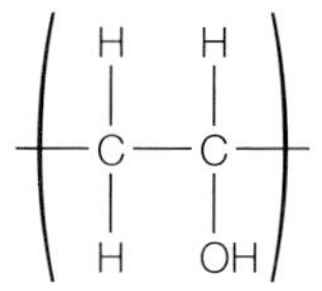

7 A student proposed to make the compound 2,5-dimethylphenylamine using the following scheme:

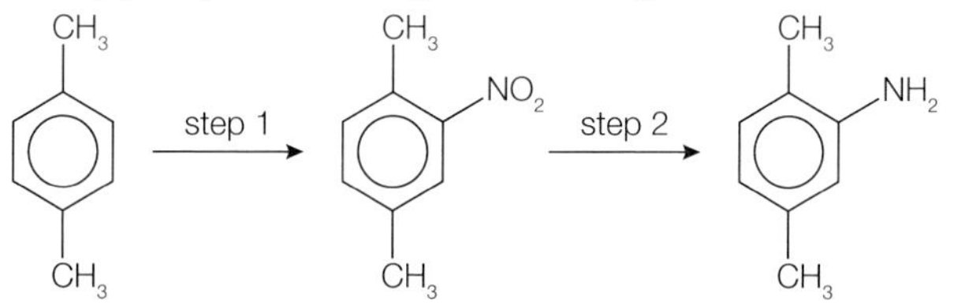

CH_3, CH_3 → step 1 → CH_3, NO_2, CH_3 → step 2 → CH_3, NH_2, CH_3

1,4-dimethylbenzene 2,5-dimethylphenylamine

(a) (i) Name the two reagents needed for Step 1. [2]

(ii) Suggest why 1,4-dimethylbenzene is more reactive than benzene in reactions such as Step 1. [2]

(b) What type of reaction occurs in Step 2? [1]

(c) 2,5-Dimethylphenylamine can be used to make azo-dyes.

State the reagents and conditions needed to make an azo-dye from 2,5-dimethylphenylamine and phenol. Include equations for the organic reactions. [5]

(Total for Question 7= 10 marks)

8 The simplest amino acid is glycine.

NH_2

C H

HOOC H

glycine

(a) Give the systematic (IUPAC) name for glycine. [1]

(b) (i) Give the structural formula for the zwitterion of glycine in the solid state. [1]

(ii) Explain why glycine has a relatively high melting temperature for a molecule with such a low molar mass. [2]

(c) Alanine, $H_2NCH(CH_3)COOH$, is another amino acid. Alanine has optical isomers (enantiomers).

(i) Draw diagrams to show the two optical isomers of alanine. [2]

(ii) Explain how the two optical isomers can be distinguished from each another. [2]

(d) Alanine and glycine react together in a condensation reaction to form two dipeptides.

(i) State why this is described as a condensation reaction. [1]

(ii) Give the structural formula for each of the two dipeptides that can be formed between alanine and glycine. [2]

(Total for Question 8 = 11 marks)

9 Butylamine, $CH_3CH_2CH_2CH_2NH_2$, and phenylamine, $C_6H_5NH_2$, both behave as bases.

(a) (i) State the feature of an amine molecule that allows it to act as a base. [1]

(ii) Explain why phenylamine is a weaker base than butylamine. [2]

(iii) Give the formulae of the salts formed when butylamine reacts with sulfuric acid and with ethanoic acid. [2]

(b) Phenylamine is formed by the reduction of nitrobenzene, $C_6H_5NO_2$.

State the reagents used for this reaction. [2]

(c) When phenylamine is reacted with a mixture of sodium nitrite, $NaNO_2$, and concentrated hydrochloric acid at a temperature between 0 and 5 °C, a diazonium ion is formed.

If phenylamine is added to the cold solution containing the benzenediazonium ion, a coupling reaction takes place similar to that between the benzenediazonium ion and phenol.

(i) Draw the structure of the diazonium ion, clearly displaying the functional group present in the ion. [1]

(ii) Suggest a structure for the organic product of the coupling reaction between the benzenediazonium ion and phenylamine. [1]

(iii) Suggest a use for this organic product. [1]

(Total for Question 9 = 10 marks)

10 Naturally occurring amino acids have the formula $H_2N(RCH)COOH$.

(a) In the amino acid glycine, the formula of the R group is —H.

Write the structural formula of the zwitterion ion formed by glycine. [1]

(b) In the amino acid serine, the formula of the R group is $—CH_2OH$.

Write the structural formula of serine at pH 1.0 and pH 10.0. [2]

(c) In the amino acid alanine, the formula of the R group is $—CH_3$.

Draw a section of the condensation polymer that can be formed from alanine. Show two repeat units of the polymer and display the link between them. [2]

(d) Compound **P** can be converted in three steps into a cyclic compound **T**.

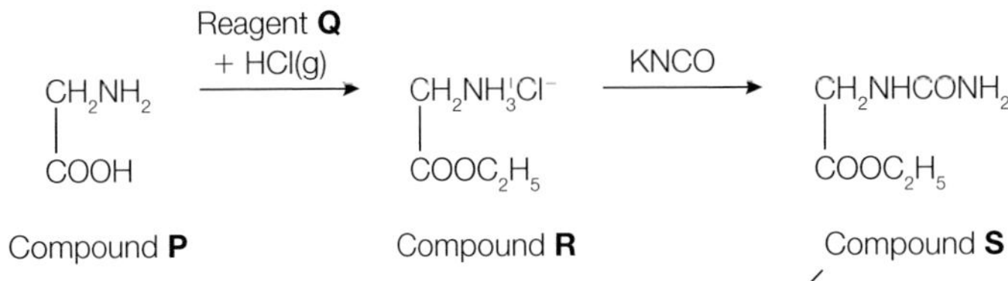

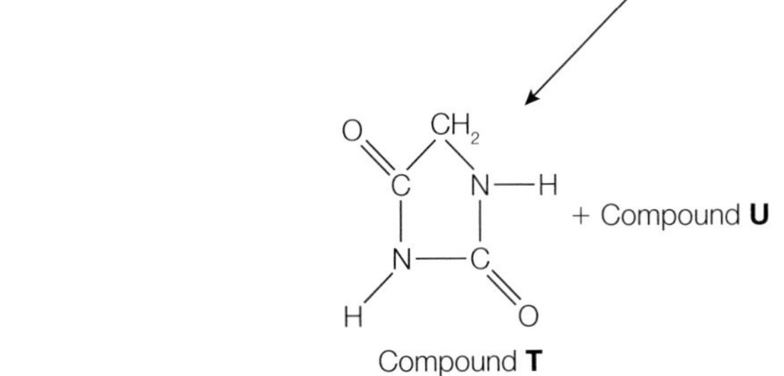

(i) Give the systematic name for compound **P**. [1]

(ii) Name reagent **Q**. [1]

(iii) Name the two types of reactions that are occurring when compound **P** is converted into compound **S**. [2]

(iv) Suggest the identity of compound **U**. [1]

(v) When compound **T** reacts with hot dilute hydrochloric acid, compound **P** is formed, as well as two other products.

Suggest the type of reaction that is occurring. [1]

(vi) Suggest the identity of the two products that also form in the reaction in (d)(v). [2]

(Total for Question 10 = 13 marks)

TOPIC 20 ORGANIC SYNTHESIS

A ORGANIC STRUCTURES

In organic chemistry so far you have learned about the reactions of a number of functional groups, such as alcohols, aldehydes, ketones, carboxylic acids, acyl chlorides, esters, amines and amides.

In this topic you will learn how to use your knowledge of these reactions to produce a reaction synthesis for an organic compound. You will also learn about the various techniques used in the preparation and purification of organic compounds, such as refluxing, solvent extraction, recrystallisation and distillaton.

Perhaps the most important use of organic synthesis is the production of medicines. People have used 'natural' medicines obtained from plants and herbs for thousands of years. During most of that time they had no idea how or why the medicines worked. The effectiveness of the plants or herbs was discovered by trial and error, and sometimes there were disastrous mistakes.

Today's synthetic medicines are increasingly designed to have specific effects, something that is becoming easier as we learn more about the body's chemistry. Aspirin and paracetamol are perhaps two of the best known and earliest medicines to have been synthesised. Today there are thousands of synthesised medicines to help combat all types of illnesses and diseases, ranging from antibiotics, to cure illnesses caused by bacteria, to statins, which help to lower the level of LDL cholesterol in the blood.

Who knows! Maybe one day you will be involved in the synthesis of a life-saving medicine.

MATHS SKILLS FOR THIS TOPIC

- Recognise and use expressions in decimal and ordinary form
- Use ratios, fractions and percentages
- Use an appropriate number of significant figures

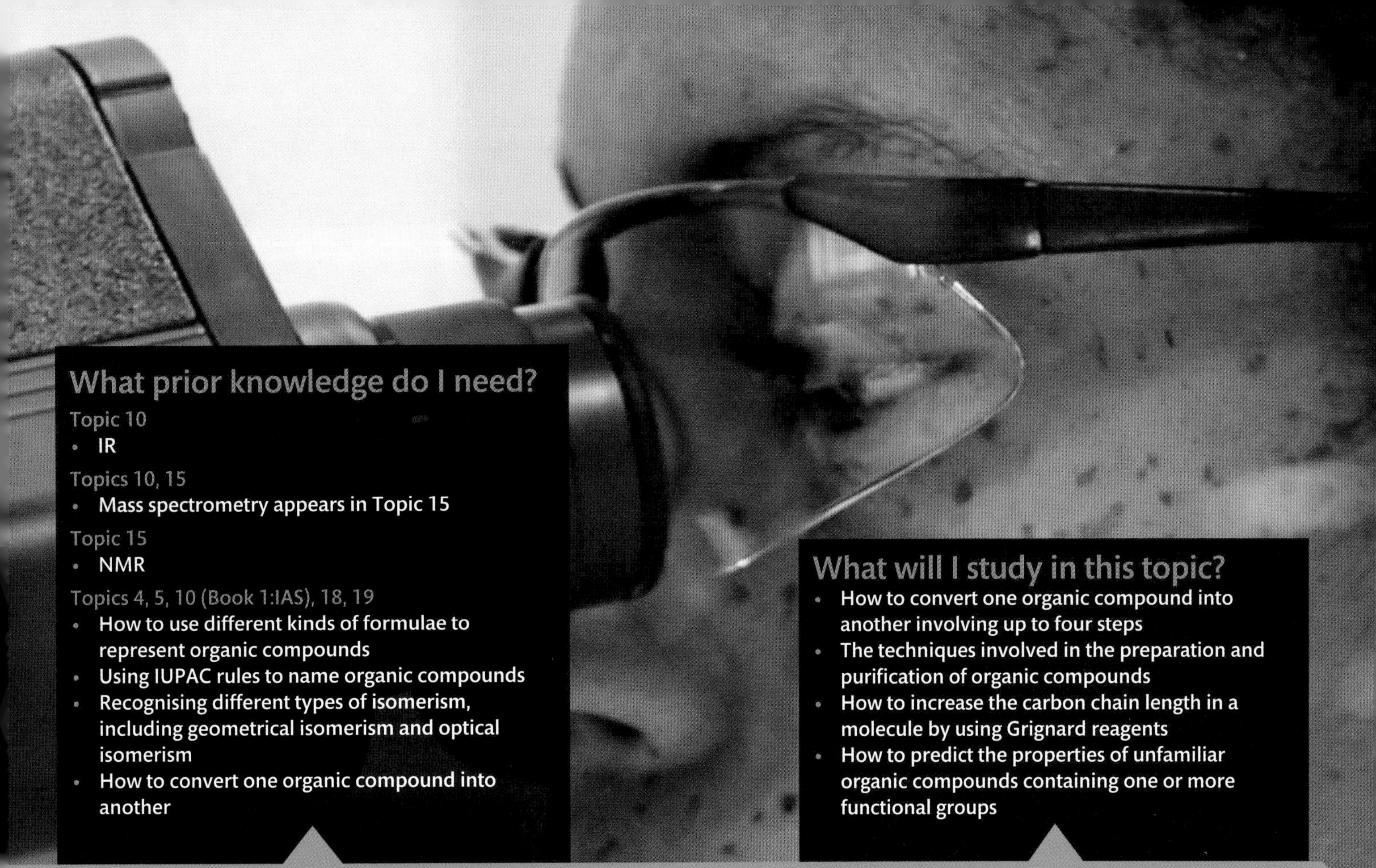

What prior knowledge do I need?

Topic 10
- IR

Topics 10, 15
- Mass spectrometry appears in Topic 15

Topic 15
- NMR

Topics 4, 5, 10 (Book 1:IAS), 18, 19
- How to use different kinds of formulae to represent organic compounds
- Using IUPAC rules to name organic compounds
- Recognising different types of isomerism, including geometrical isomerism and optical isomerism
- How to convert one organic compound into another

What will I study in this topic?

- How to convert one organic compound into another involving up to four steps
- The techniques involved in the preparation and purification of organic compounds
- How to increase the carbon chain length in a molecule by using Grignard reagents
- How to predict the properties of unfamiliar organic compounds containing one or more functional groups

What will I study later?

- How pharmaceutical companies invest heavily in research and development to design new medicines to treat a wide range of medical problems and illnesses
- How biocatalysts based on natural enzymes allow some reactions to occur at lower temperatures and pressures than other catalysts

20A 1 ORGANIC ANALYSIS

SPECIFICATION REFERENCE 20.1

LEARNING OBJECTIVES

- Be able to deduce the empirical formulae, molecular formulae and structural formulae from data drawn from combustion analysis, element percentage composition, characteristic reactions of functional groups, infrared spectra, mass spectra and NMR spectra (both ^{13}C and proton).

TRADITIONAL METHODS

The traditional method of analysis is to find the empirical formula, followed by the molecular formula and then the functional groups present.

FINDING THE EMPIRICAL AND MOLECULAR FORMULAE

You learned the method for finding the empirical formula of a compound in **Topic 1 (Book 1: IAS)**.

To find the empirical formula from the percentage composition by mass of the elements you must first divide the percentage composition of each element by its relative atomic mass. You then divide the values obtained by the smallest value. The numbers you obtain should give you an obvious whole number ratio.

This worked example will remind you of the method.

WORKED EXAMPLE 1

Find the empirical formula of the compound that has the following composition by mass:

C 40.00% H 6.67% O 53.33%

Answer

C (40.00 ÷ 12.0) = 3.33

H (6.67 ÷ 1.0) = 6.67

O (53.33 ÷ 16.0) = 3.33

Dividing all three numbers by the smallest (i.e. 3.33) gives the following ratio:

C : H : O is 1 : 2 : 1

The empirical formula = CH_2O.

If we know the molecular mass (or molar mass in g mol^{-1}) of the compound we can find the molecular formula by dividing the molecular mass by the empirical formula mass. The molecular formula is always a simple multiple of the empirical formula.

Let us assume that this compound has a molecular mass of 60.0.

The empirical formula mass = (12.0 + 2.0 + 16.0) = 30.0.

(60.0 ÷ 30.0) = 2

The molecular formula must therefore contain twice as many atoms of each element.

The molecular formula = $C_2H_4O_2$.

COMBUSTION ANALYSIS

We can use combustion analysis to find the percentage composition of a compound.

A known mass of the compound is burned in excess dry oxygen. The masses of carbon dioxide and water produced are measured.

For a compound that contains carbon, hydrogen and oxygen only we can then work out the following:

- mass of carbon dioxide → mass of carbon → % carbon
- mass of water → mass of hydrogen → % hydrogen
- % oxygen = 100 – (% carbon + % hydrogen).

WORKED EXAMPLE 2

2.90 g of an organic compound, containing carbon, hydrogen and oxygen only, was burned in excess oxygen.

6.60 g of carbon dioxide and 2.70 g of water were produced.

Calculate the percentage composition of the compound.

Answer

6.60 g of carbon dioxide contains $\frac{12.0}{44.0} \times 6.60 = 1.80$ g of carbon.

$\% \text{ carbon} = \frac{1.80}{2.90} \times 100 = 62.1\%$

2.70 g of water contains $\frac{2.0}{18.0} \times 2.70 = 0.30$ g of hydrogen.

$\% \text{ hydrogen} = \frac{0.30}{2.90} \times 100 = 10.3\%$

% oxygen = 100 – (62.1 + 10.3) = 27.6%

The percentage composition by mass of the compound is:

C 62.1% H 10.3% O 27.6%

CHARACTERISTIC REACTIONS OF FUNCTIONAL GROUPS

To determine the structural formula from the molecular formula we need to identify the functional groups in the compound. Traditionally, this is done by performing reactions in test tubes, using test reagents such as bromine water or 2,4-dinitrophenylhydrazine.

You have already seen details of these tests in **Book 1** and in **Topics 15**, **18** and **19**.

A reminder of these tests is given in **table A**.

FUNCTIONAL GROUP	TEST	OBSERVATION	NOTES
Alkene (C=C)	add $Br_2(aq)$	bromine decolourises immediately	aldehydes and ketones decolourise $Br_2(aq)$, but very slowly phenols decolourise $Br_2(aq)$ immediately and also produce a white precipitate
Halogenoalkanes (R–X), where X = Cl, Br or I	warm with NaOH(aq) and then add dilute HNO_3, followed by a few drops of $AgNO_3(aq)$	R–Cl produces a white precipitate	the precipitate is insoluble in dilute $NH_3(aq)$
		R–Br produces a cream precipitate	the precipitate is insoluble in dilute $NH_3(aq)$, but soluble in concentrated $NH_3(aq)$
		R–I produces a yellow precipitate	the precipitate is insoluble in both dilute $NH_3(aq)$ and concentrated $NH_3(aq)$
Hydroxy group (-OH)	add solid PCl_5	misty fumes	both alcohols and carboxylic acids, but not phenols, produce misty fumes
Primary alcohol (RCH_2OH)	add acidified $K_2Cr_2O_7$ solution and warm	solution turns from orange to green	the organic product of this reaction produces a silver mirror with Tollens' reagent
Secondary alcohol (R_2CHOH)	add acidified $K_2Cr_2O_7$ solution and warm	solution turns from orange to green	the organic product does not produce a silver mirror with Tollens' reagent
Carbonyl group R or H, R or H >C=O	add 2,4-dinitrophenylhydrazine	orange precipitate	both aldehydes and ketones produce an orange precipitate
Aldehydes R or H, H >C=O	add Tollens' reagent and warm	silver mirror forms	Fehling's or Benedict's solutions can also be used ketones do not react with Tollens' reagent, Fehling's solution or Benedict's solution
CH_3 >C=O or CH_3 >CHOH	add an alkaline solution of iodine and warm	yellow precipitate forms	this is known as the triiodomethane (or iodoform) test
Carboxylic acids (R–COOH)	add $NaHCO_3(aq)$ or $Na_2CO_3(aq)$ and warm if necessary	bubbles of gas	
Phenol C_6H_5–OH	add bromine water	bromine immediately decolourises and a white precipitate forms	phenylamine produces the same observation however, phenol is soluble in NaOH(aq) but insoluble in dilute HCl; phenylamine is insoluble in NaOH(aq) but soluble in dilute HCl

table A Tests for functional groups.

MODERN ANALYTIC TECHNIQUES

Chemists use a combination of techniques to identify organic compounds and determine their structures, including:

- mass spectra
- infrared spectra
- NMR spectra (both ^{13}C and proton).

Mass spectrometry allows us to determine the accurate relative molecular mass of a compound. The fragmentation patterns also allow us to suggest possible groups that may be present in the compound. For example, a peak with an m/z value of 15 could indicate that a CH_3 group is present.

Infrared spectroscopy allows us to identify particular functional groups and features in an organic compound. For example, an absorption between 1740 and 1720 cm^{-1} indicates the presence of a C=O group in an aldehyde.

^{13}C NMR spectroscopy allows us identify the number of different environments for the carbon atoms in a compound. For example, the spectrum for ethanol, CH_3CH_2OH, has two peaks. This shows that there are two different environments for the carbon atoms.

The ^{13}C NMR spectrum of 1-methylethyl propanoate, $CH_3CH_2COOCH(CH_3)_2$, has five peaks. This shows that there are five different environments for the carbon atoms. If you are wondering why there are only five peaks but six carbon atoms, it is because the carbons in the two methyl groups at the right-hand end of the molecule are in identical environments.

Proton NMR spectroscopy allows us to identify groups of hydrogen atoms in a compound. For example, if there are three hydrogen atoms in the same environment, this indicates that a $-CH_3$ group is present. Splitting patterns also allow us to identify the number of hydrogen atoms on an adjacent carbon atom.

None of these methods gives a complete answer, but together they provide good evidence to allow you to identify the molecular structure.

Mass spectrometry and infrared spectroscopy were introduced in **Topic 10 (Book 1: IAS)**. NMR spectroscopy was introduced in **Topic 15**.

It may be useful to re-read these two topics so that you fully understand the two worked examples.

WORKED EXAMPLE 3

Compound X has the following percentage composition by mass:

C 62.07% H 10.34% O 27.59%

Compound X produces an orange precipitate with 2,4-dinitrophenylhydrazine, but it does not react with either Fehling's solution or Tollens' reagent.

Fig A shows the mass spectrum of compound X.

Fig B shows the infrared spectrum of compound X.

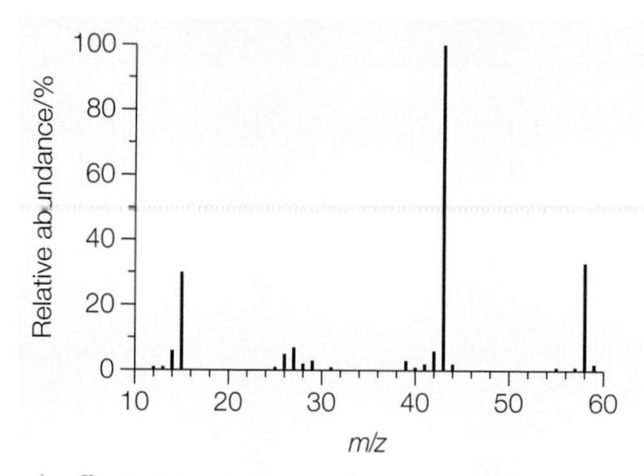

fig A Mass spectrum of compound X.

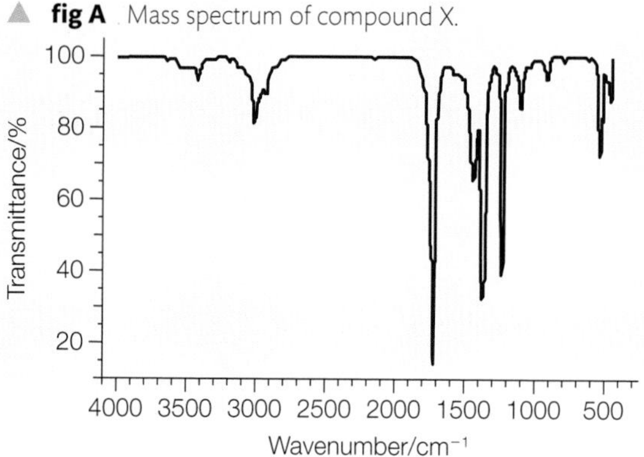

fig B Infrared spectrum of compound X.

The proton NMR spectrum of compound X has a single peak with a chemical shift of 2.2 ppm. The peak area corresponds to six carbon atoms.

Use all this information to determine the structure of compound X.

Answer

Calculation of the empirical formula:

	Divide by A_r	*Divide by smallest*
C	(62.07 ÷ 12.0) = 5.17	(5.17 ÷ 1.72) = 3
H	(10.34 ÷ 1.0) = 10.34	(10.34 ÷ 1.72) = 6
O	(27.59 ÷ 16.0) = 1.72	(1.72 ÷ 1.72) = 1

The empirical formula is C_3H_6O.

The molecular ion peak in the mass spectrum has a value of 58. The relative molecular mass of compound X is therefore 58.

The empirical formula mass is (12.0 x 3) + (1.0 x 6) + 16.0 = 58.0.

Therefore the molecular formula of compound X is C_3H_6O.

The absorption at 1720 cm^{-1} indicates that a C=O group is present. This could be in either an aldehyde or a ketone.

Compound X produces an orange precipitate with 2,4-dinitrophenylhydrazine, showing that it is a carbonyl compound (i.e. it is either an aldehyde or a ketone).

However, it does not react with either Fehling's solution or Tollens' reagent, showing that it must be a ketone.

The proton NMR spectrum shows that there are six hydrogen atoms in identical environments, indicating that there are two CH_3 groups attached to the third carbon atom. The chemical shift of 2.2 ppm indicates hydrogen atoms in the environment: H–C–C=O.

So the structure of compound X is CH_3COCH_3 (propanone).

WORKED EXAMPLE 4

(a) An organic compound **Y** contains carbon, hydrogen, nitrogen and oxygen.

0.132 g of **Y** is burned completely in oxygen to produce 0.176 g of carbon dioxide, 0.072 g of water and 24.0 cm^3 of nitrogen (molar volume of nitrogen under the conditions of the experiment = 24.0 $dm^3\ mol^{-1}$).

The relative molecular mass of compound **Y** is 132.

Determine the molecular formula of **Y**.

(b) A sample of **Y** is hydrolysed by heating under reflux with concentrated hydrochloric acid for several hours. The reaction mixture is then cooled and neutralised. The only organic product, **Z**, of this reaction has the molecular formula $C_2H_5O_2N$.

One mole of **Z** reacts with either one mole of hydrochloric acid or one mole of sodium hydroxide.

Z can exist as a zwitterion.

Use all the information above to deduce the structures of compounds **Z** and **Y**.

Answer

(a) Step 1: *calculate mass of oxygen*

$$\text{mass of carbon} = \frac{12.0}{44.0} \times 0.176 = 0.048 \text{ g}$$

$$\text{mass of hydrogen} = \frac{2.0}{18.0} \times 0.072 = 0.008 \text{ g}$$

mass of nitrogen = (24.0 ÷ 24 000) × 28.0 = 0.028 g

mass of oxygen = 0.132 g − (0.008 + 0.048 + 0.028) g

= 0.048 g

Step 2: *calculate empirical formula of* **Y**

moles of carbon = (0.048 ÷ 12.0) = 0.004

moles of hydrogen = (0.008 ÷ 1.0) = 0.008

moles of nitrogen = (0.028 ÷ 14.0) = 0.002

moles of oxygen = (0.048 ÷ 16.0) = 0.003

Empirical formula of **Y** is $C_4H_8N_2O_3$.

Empirical formula mass

= (12.0 x 4) + (1.0 x 8) + (14.0 x 2) + (16.0 x 3)

= 132

The empirical formula mass = the relative molecular mass

Therefore, the molecular formula of Y is $C_4H_8N_2O_3$.

(b) **Z** reacts with HCl, so it contains an amine group ($-NH_2$).

Z reacts with NaOH, so it contains a carboxylic acid group (−COOH).

There is one $-NH_2$ group per molecule and one −COOH per molecule (one mole of **Z** reacts with both one mole of HCl and one mole of NaOH).

Z can form a zwitterion, so it must be an amino acid.

The structure of **Z** is H_2NCH_2COOH.

Z is produced by hydrolysis of **Y**. **Y** must therefore be an amide.

Since there is only one organic product, the structure of **Y** must be $H_2NCH_2CONHCH_2COOH$.

CHECKPOINT

SKILLS ANALYSIS

1. Compound **T** does not immediately decolourise bromine water. It also does not react with an aqueous solution of sodium carbonate. When excess PCl_5 is added it produces one mole (1 mol) of hydrogen chloride gas per mole of **T**. It produces an orange precipitate when added to 2,4-dinitrophenylhydrazine.

   T has a molecular formula of $C_3H_6O_2$.

   (a) Write the structural formulae of three possible isomers of **T**. Justify your answer.

   (b) The mass spectrum of **T** has peaks at *m/z* values of 29 and 59. Use this information to deduce the identity of **T**.

2. The proton NMR spectrum of an organic compound is shown below.

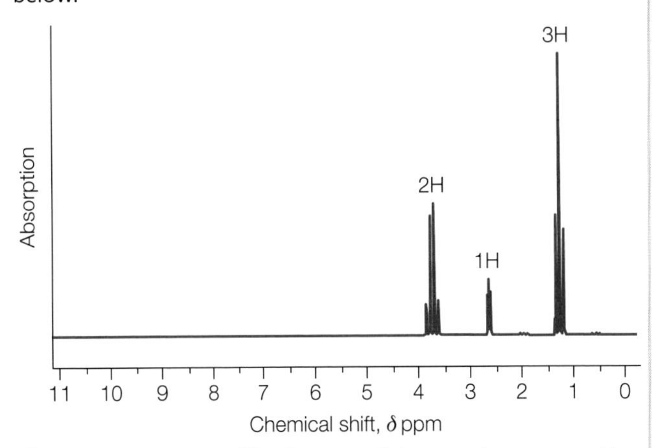

The percentage composition by mass of the organic compound is:

C 52.17% H 13.04% O 34.78%

It has a relative molecular mass of 46.

Use this information to identify the compound.

20A 2 ORGANIC SYNTHESIS

SPECIFICATION REFERENCE 20.2 20.3(i) 20.3(ii)

LEARNING OBJECTIVES

- Understand methods of increasing the length of the carbon chain in a molecule by the use of magnesium to form Grignard reagents and the reactions of the latter with carbon dioxide and with carbonyl compounds in dry ether.
- Be able to use knowledge of organic chemistry to solve problems such as:
  (i) predicting the properties of unfamiliar compounds containing one or more of the functional groups included in the specification and explain these predictions
  (ii) planning reaction schemes of up to four steps, recalling familiar reactions and using unfamiliar reactions given sufficient information.

WHAT IS ORGANIC SYNTHESIS?

The term 'synthesis' (the plural is 'syntheses') refers to making something new from what already exists. In organic chemistry it usually means using a familiar compound to make an unfamiliar compound. You have already learned many examples of converting one compound into another – for example, converting a halogenoalkane into an alcohol. In this section we will consider converting one compound into another compound that cannot be achieved in one step, but needs two or more steps. To keep things relatively simple, in this book the number of steps will be limited to four.

The ability to plan a reaction scheme to achieve this synthesis is not easily learned. It is not like recalling a simple mathematical expression and putting the numbers into a calculator to obtain the answer. Instead, it relies on a thorough knowledge of a range of chemical reactions and the ability to select the ones that are appropriate to the problem.

Here is a simple example of a two-step synthesis – how to convert bromoethane to ethyl propanoate. You could start by considering the reactions of bromoethane, and then consider reactions in which ethyl propanoate is a product. You could consider the question as a puzzle in which you need to find the identity of X in this sequence:

bromoethane → X → ethyl propanoate

If you know a product of a reaction of bromoethane that is the same as a compound that can be converted into ethyl propanoate, then you only need to include the reagents and conditions to achieve each of these two steps. When the synthesis involves three steps it is more difficult because you need to identify two intermediate compounds in the synthesis. If there are four steps, then we have reached the limit of what is reasonable to expect at this point in your study of chemistry!

In the example above, X is ethanol, and the steps are:

Step 1: bromoethane → ethanol
reflux with aqueous potassium hydroxide (hydrolysis)

Step 2: ethanol → ethyl propanoate
heat with propanoic acid using an acid catalyst (esterification).

PREPARING FOR PLANNING A SYNTHESIS

One thing that will help you learn how to plan a synthesis is to put together an outline of all the reactions you need to know. You could leave out reactions that are less useful in synthesis, such as combustion, polymerisation and chemical tests. You could try to produce a diagram (like a spider diagram) that would include all the reactions, but that would probably be too complicated to easily use. It might be better to produce a series of smaller ones. For example, **fig A** shows reactions based on alcohols.

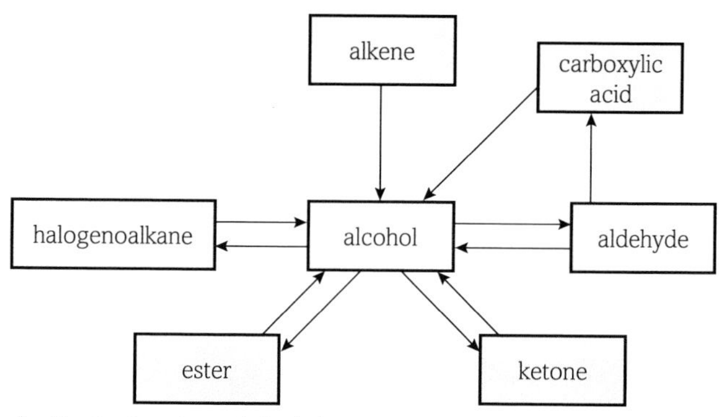

fig A Reactions of alcohols.

Where the arrows point in both directions, this means that you need to know both conversions. Note that reactions from **Book 1** are included. If you copy this onto a sheet of paper you can add the formulae of reagents and any special conditions on the arrows. You could use different colours to emphasise different features.

EXTENDING A CARBON CHAIN

It is possible that one of the steps in your synthesis will involve extending an existing carbon chain by one or more carbon atoms. You have already learned three ways of doing this:

1 Reacting a halogenoalkane with a cyanide ion forms a nitrile with one more carbon atom than the halogenoalkane.
2 The addition of hydrogen cyanide to a carbonyl compound.
3 The alkylation of benzene, which introduces an alkyl group into a benzene ring.

A different way, which is more versatile, involves the use of Grignard reagents.

GRIGNARD REAGENTS

Victor Grignard was a French chemist who devised a new way to synthesise organic compounds. His method involved the use of organometallic compounds containing magnesium, which are now known as Grignard reagents. They are made by heating under reflux the chosen halogenoalkane with magnesium in a solvent of dry ether. Halogenoarenes such as bromobenzene are also used. Bromoalkanes are used in preference to other halogenoalkanes. Grignard reagents contain magnesium covalently bonded to both the alkyl group and the halogen. A general equation for the formation of a Grignard reagent is:

$$\text{R—Br} + \text{Mg} \rightarrow \text{R—Mg—Br}$$

Grignard reagents react with water, so they are both made and used in a solvent of dry ether.

▲ **fig B** Victor Grignard (1871–1935) won the Nobel Prize in Chemistry for his work in organic synthesis.

GRIGNARD REACTIONS

Once a Grignard reagent has been made it can then be converted into a range of organic compounds, depending on the reagent used. In this book we will consider only two of these reactions: with carbon dioxide and with carbonyl compounds. They can be summarised as follows.

1	with carbon dioxide	$RMgBr \rightarrow RCOOH$	carboxylic acid
2a	with methanal	$RMgBr \rightarrow RCH_2OH$	primary alcohol
2b	with an aldehyde R′CHO	$RMgBr \rightarrow RR'CHOH$	secondary alcohol
2c	with a ketone R′COR″	$RMgBr \rightarrow RR'R''COH$	tertiary alcohol

R, R′ and R″ represent different alkyl groups, although they could be the same. Note that both carbon dioxide and methanal increase the carbon chain by one carbon atom. Both aldehydes and ketones produce a branch, of variable length, on the existing carbon chain.

Having made the appropriate Grignard reagent, the second reagent (carbon dioxide or the carbonyl compound) is added. After the reaction is complete, a dilute acid is added to obtain the desired organic product.

EXAMPLES OF GRIGNARD REACTIONS

FORMATION OF 3-METHYLBUTANOIC ACID

$$CH_3\text{—}\underset{\underset{CH_3}{|}}{CH}\text{—}CH_2\text{—}COOH$$

This is an example of reaction 1, described above. The COOH part comes from carbon dioxide, so the starting compound must supply the $(CH_3)_2CHCH_2$ part, which means using $(CH_3)_2CHCH_2Br$, i.e. 1-bromo-2-methylpropane. The equations for the reactions that occur in this example are:

Step 1: formation of the Grignard reagent

$$(CH_3)_2CHCH_2Br + Mg \rightarrow (CH_3)_2CHCH_2MgBr$$

Step 2: reaction with chosen reagent

$$(CH_3)_2CHCH_2MgBr + CO_2 \rightarrow (CH_3)_2CHCH_2COOMgBr$$

Step 3: hydrolysis using a dilute acid

$$(CH_3)_2CHCH_2COOMgBr + H_2O \rightarrow (CH_3)_2CHCH_2COOH + Mg(OH)Br$$

The inorganic product of Step 3 will react with the dilute acid. You do not need to know details of the mechanisms.

EXAM HINT

Grignard reactions are a lot safer than using hydrogen cyanide and potassium cyanide to add a carbon atom to a molecule (see **Section 20A.3**). Grignard reactions also give a wider range of chemical additions.

FORMATION OF PROPAN-1-OL

This and the other examples are abbreviated as reaction schemes, rather than being shown as three steps.

Propan-1-ol, $CH_3CH_2CH_2OH$, is a primary alcohol. So, methanal is the source of the CH_2OH part, and the starting compound is bromoethane, which is the source of the CH_3CH_2 part.

$$CH_3CH_2Br \xrightarrow{\text{Mg/ether}} CH_3CH_2MgBr \xrightarrow[\text{then } H_2O/H^+]{\text{HCHO}} CH_3CH_2CH_2OH$$

FORMATION OF PENTAN-2-OL

Pentan-2-ol, $CH_3CH(OH)CH_2CH_2CH_3$, is a secondary alcohol, so ethanal is the source of the $CH_3CH(OH)$ part, and the starting compound is 1-bromopropane, which is the source of the $CH_3CH_2CH_2$ part.

$$CH_3CH_2CH_2Br \xrightarrow{\text{Mg/ether}} CH_3CH_2CH_2MgBr \xrightarrow[\text{then } H_2O/H^+]{CH_3CHO} CH_3CH(OH)CH_2CH_2CH_3$$

FORMATION OF 2-METHYLPROPAN-2-OL

2-Methylpropan-2-ol, $(CH_3)_3COH$, is a tertiary alcohol, so propanone is the source of the $(CH_3)_2COH$ part, and the starting compound is bromomethane, which is the source of the CH_3 part.

$$CH_3Br \xrightarrow{\text{Mg/ether}} CH_3MgBr \xrightarrow[\text{then } H_2O/H^+]{CH_3COCH_3} (CH_3)_3COH$$

PREDICTING THE CHEMICAL PROPERTIES OF COMPOUNDS

As we have seen in **Section 20A.1**, chemical reagents can be used to show which functional groups are present in a compound.

Functional groups can influence the behaviour of compounds that contain them. For example:

- some compounds will act as nucleophiles and others as electrophiles
- some will be susceptible to addition reactions and others to substitution reactions
- some will be easily oxidised and others are more easily reduced.

We can show this by looking at two compounds with the molecular formula C_7H_8O: phenylmethanol and 4-methylphenol (see **fig C**).

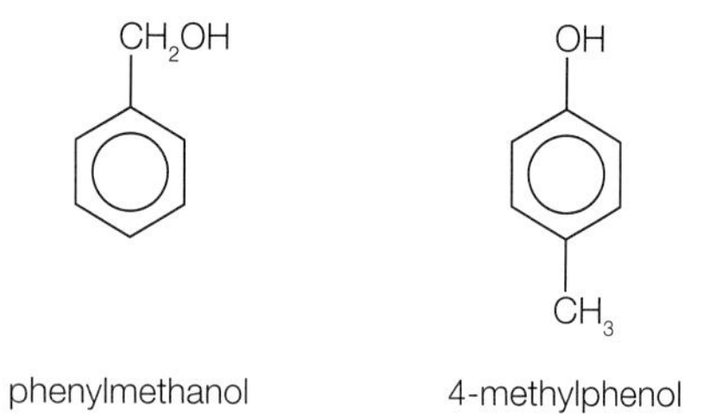

fig C

Both compounds are slightly soluble in water. Phenylmethanol forms a neutral solution, but aqueous 4-methylphenol is slightly acidic. This is because one of the lone pairs of electrons on the oxygen atom is incorporated into the delocalised electrons in the ring. This makes the O–H bond more polar (see **Topic 18**). However, 4-methylphenol is not acidic enough to react with sodium hydrogen carbonate to form carbon dioxide, distinguishing it from carboxylic acids.

Phenylmethanol is a primary alcohol because it contains a $-CH_2OH$ group. 4-Methylphenol is a phenol because it has an –OH group attached to the benzene ring.

Phenylmethanol will form an ester when heated with a carboxylic acid in the presence of concentrated sulfuric acid. This is because it is an alcohol (**Topic 15**):

$$CH_2OH \; + \; CH_3COOH \longrightarrow CH_3OOCCH_3 \; + \; H_2O$$

4-Methylphenol will not form an ester with carboxylic acids.

EXAM HINT

You can test your knowledge of organic synthesis and analysis by setting yourself a synthesis task. For example, think about how to convert a three-carbon alkene into a five-carbon secondary alcohol. Then try to **predict** the changes in the NMR spectra from reactant to product.

LEARNING TIP

Practise choosing reagents for a Grignard synthesis. First, identifying the product type (acid or alcohol) tells you the type of compound to react with the Grignard reagent. Then, work out how many carbon atoms are needed in the halogenoalkane used to form the Grignard reagent. See how you get on with the questions in this section!

EXAM HINT

You have seen in **Topic 10 (Book 1: IAS)** that primary alcohols are easily oxidised by heating with a mixture of potassium dichromate(VI) and dilute sulfuric acid. There is a colour change from orange to green in this reaction.

You may have learned that phenols are not easily oxidised. So you may think that 4-methylphenol will not be oxidised by heating with the same mixture of potassium dichromate(VI) and dilute sulfuric acid. You may decide to use this reagent to distinguish between the two compounds.

Unfortunately, this test will not work. Although the phenol group will not be oxidised by the mixture of potassium dichromate(VI) and dilute sulfuric acid, the methyl group attached to the benzene ring *will* be oxidised. So a colour change from orange to green will be observed.

This is a good warning to be very careful when choosing reagents for distinguishing between two compounds. The best advice is to use only the reactions described in **Book 1** and **Book 2**.

4-Methylphenol will decolourise bromine water:

$$CH_3C_6H_4OH + 2Br_2 \longrightarrow CH_3C_6H_2Br_2OH + 2HBr$$

However, phenylmethanol does not react with bromine water.

MULTI-STEP SYNTHESES

You may be asked to plan to a reaction scheme that shows all the intermediate compounds as well as the reagents and conditions required for each step. There are several ways you can approach this:

- by checking if the carbon chain length has increased or decreased
- by looking at the final product and 'working back' to the starting compound
- by looking at the starting compound and thinking of the types of reactions it can undergo.

Fig D shows how working back can be used to plan the synthesis of one compound from another. In this case the starting compound is bromoethane and the final compound is ethanoic acid:

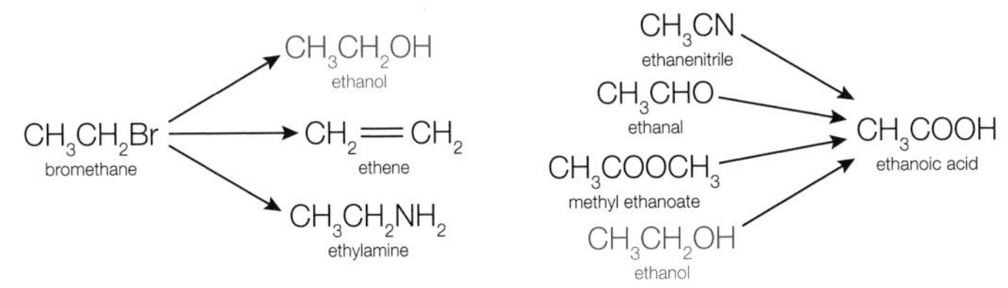

▲ **fig D**

You can plan the reaction scheme using the following stages.

- Begin by writing the formulae of the compounds that can be readily converted into ethanoic acid. These include ethanenitrile, ethanol, methyl ethanoate and ethanol.
- Now look at the starting compound, bromoethane, to see if it can be converted into one of these compounds. If necessary, write the formulae of the compounds that bromoethane can easily be converted into. These include ethanol, ethene and ethylamine.
- You should now be able to see a clear two-step route from bromoethane to ethanoic acid. In this case, the route is to convert bromoethane into ethanol, and then to convert ethanol into ethanoic acid.
- Finally, you need to add the reagents and conditions required for each step. To convert bromoethane into ethanol you need to add aqueous sodium hydroxide and heat under reflux. To convert ethanol into ethanoic acid you need to add potassium dichromate(VI) and dilute sulfuric acid and heat under reflux.

You can set out your final answer as shown in **fig E**:

$$\underset{\text{bromethane}}{CH_3CH_2Br} \xrightarrow[\text{heat under reflux}]{\text{NaOH(aq)}} \underset{\text{ethanol}}{CH_3CH_2OH} \xrightarrow[\text{heat under reflux}]{K_2Cr_2O_7\text{(aq)}/H_2SO_4\text{(aq)}} \underset{\text{ethanoic acid}}{CH_3COOH}$$

▲ **fig E**

You can use the same process for three- and four-step syntheses, although it will be a bit more complex.

WORKED EXAMPLE 1

Deduce the reagents and conditions, and identify the two intermediates, in the three-step synthesis of the following compound from benzene:

$C_6H_5NHCOCH_3$

Answer

Working backwards

The final compound is an *N*-substituted amide (a secondary amide). It can be made by the reaction between an amine and an acyl chloride, at room temperature. In this case, the reactants are phenylamine, $C_6H_5NH_2$, and ethanoyl chloride, CH_3COCl.

Phenylamine can be made by heating nitrobenzene, $C_6H_5NO_2$, under reflux with tin and concentrated hydrochloric acid.

Working forwards

Benzene, C_6H_6, is converted into nitrobenzene by heating under reflux at 50–60 °C with a mixture of concentrated nitric acid and concentrated sulfuric acid.

The three-step synthesis is therefore:

C_6H_6 —(conc HNO_3/conc H_2SO_4; heat under reflux 50–60 °C)→ $C_6H_5NO_2$ —(Sn/conc HCl; heat under reflux)→ $C_6H_5NH_2$ —(CH_3COCl; room temperature)→ $C_6H_5NHCOCH_3$

WORKED EXAMPLE 2

Deduce the reagents and conditions, and identify the three intermediates, in the four-step synthesis of ethanol, CH_3CH_2OH, from methanol, CH_3OH.

Answer

The first point to note is the increase in carbon chain length from 2 to 3. You have seen three methods for increasing the carbon chain:

- reaction between a halogenoalkane and KCN
- reaction between an alkene and HCN
- the use of Grignard reagents.

We are going to look at how to do this conversion without the use of a Grignard reagent.

Working backwards

There are several ways of making ethanol. For example:

- hydrolysing chloroethane, CH_3CH_2Cl, or bromoethane, CH_3CH_2Br
- reducing ethanal, CH_3CHO
- reducing ethanoic acid, CH_3COOH.

Ethanoic acid can be made by hydrolysing ethanenitrile, CH_3CN.

Converting a halogenoalkane into a nitrile is one method of increasing the carbon chain. CH_3CN can be made by reacting CH_3Cl with KCN.

Working forwards

CH_3OH can be converted into CH_3Cl by reaction with PCl_5.

A possible four-step synthesis is therefore:

CH_3OH —(dry PCl_5; room temperature)→ CH_3Cl —(KCN; heat under reflux)→ CH_3CN —(HCl(aq); heat under reflux)→ CH_3COOH —($LiAlH_4$ in dry ether; room temperature)→ CH_3CH_2OH

CHECKPOINT

1. Draw the structure of the organic product obtained in a Grignard synthesis starting from 1-bromo-2,2-dimethylpropane and propanal.
2. Draw the structures of the two compounds needed in a Grignard synthesis of 3,3-dimethylhexan-4-ol.
3. Deduce the reagents and conditions, and identify all intermediates, in each of the following conversions:
   (a) CH_3Br to $CH_3CH_2NH_2$ (two steps)
   (b) $CH_2{=}CH_2$ to $CH_3CH_2CH_2OH$ (four steps).

20A 3 HAZARDS, RISKS AND CONTROL MEASURES

SPECIFICATION REFERENCE 20.3(iii) 20.3(iv)

LEARNING OBJECTIVES

- Be able to use knowledge of organic chemistry to solve problems such as:
  (i) selecting suitable practical procedures for carrying out reactions involving compounds with functional groups
  (ii) identifying appropriate control measures to reduce risk based on data for hazards.

CONVERTING ONE FUNCTIONAL GROUP INTO ANOTHER

This is a summary of some of the most useful organic conversions that you have learned on your International AS/A Level course.

CONVERSION	EQUATION	REAGENT	CONDITIONS
Alkene to halogenoalkane	$CH_2{=}CH_2 + HX \rightarrow CH_3CH_2X$	hydrogen halide	mix the gases at room temperature
Halogenoalkane to alcohol	$RX + NaOH \rightarrow ROH + NaX$	aqueous sodium hydroxide	heat under reflux
Halogenoalkane to nitrile	$RX + KCN \rightarrow RCN + KX$	alcoholic potassium cyanide	heat under reflux
Halogenoalkane to amine	$RX + 2NH_3 \rightarrow RNH_2 + NH_4X$	aqueous ammonia	heat under pressure
Alcohol to chloroalkane	$ROH + PCl_5 \rightarrow RCl + HCl + POCl_3$	phosphorus(V) chloride	room temperature
Alcohol to bromoalkane	$ROH + HBr \rightarrow RBr + H_2O$	50% concentrated sulfuric acid and potassium bromide	warm
Alcohol to iodoalkane	$3ROH + PI_3 \rightarrow 3RI + H_3PO_3$	red phosphorus and iodine	heat under reflux
Primary alcohol to aldehyde	$RCH_2OH + [O] \rightarrow RCHO + H_2O$	potassium dichromate(VI) and dilute sulfuric acid	add the reagent to hot alcohol and allow the aldehyde to distil off as it is formed
Primary alcohol to carboxylic acid	$RCH_2OH + 2[O] \rightarrow RCOOH + H_2O$	potassium dichromate(VI) and dilute sulfuric acid	heat under reflux
Secondary alcohol to ketone	$RCH(OH)R' + [O] \rightarrow RCOR' + H_2O$	potassium dichromate(VI) and dilute sulfuric acid	heat under reflux
Aldehyde to primary alcohol	$RCHO + 2[H] \rightarrow RCH_2OH$	lithium aluminium hydride in dry ether	room temperature
Ketone to secondary alcohol	$RCOR' + 2[H] \rightarrow RCH(OH)R'$	lithium aluminium hydride in dry ether	room temperature
Aldehyde and ketone to 2-hydroxynitrile	$RCHO + HCN \rightarrow RCH(OH)CN$ $RCOR' + HCN \rightarrow RR'C(OH)CN$	potassium cyanide in dilute sulfuric acid	10–20 °C
Carboxylic acid to primary alcohol	$RCOOH + 4[H] \rightarrow RCH_2OH + H_2O$	lithium aluminium hydride in dry ether	room temperature
Carboxylic acid to ester	$RCOOH + R'OH \rightarrow RCOOR' + H_2O$	alcohol and concentrated sulfuric acid	heat
Acyl chloride to carboxylic acid	$RCOCl + H_2O \rightarrow RCOOH + HCl$	water	room temperature
Acyl chloride to ester	$RCOCl + R'OH \rightarrow RCOOR' + HCl$	alcohol	room temperature
Acyl chloride to primary amide	$RCOCl + 2NH_3 \rightarrow RCONH_2 + NH_4Cl$	aqueous ammonia	room temperature
Acyl chloride to secondary amide (*N*-substituted amide)	$RCOCl + R'NH_2 \rightarrow RCONHR' + HCl$	amine	room temperature

CONVERSION	EQUATION	REAGENT	CONDITIONS
Nitrile to primary amine	$RCN + 4[H] \rightarrow RCH_2NH_2$	lithium aluminium hydride in dry ether	room temperature
Nitration of benzene	$C_6H_6 + HNO_3 \rightarrow C_6H_5NO_2 + H_2O$	concentrated nitric acid and concentrated sulfuric acid	heat under reflux between 50 and 60 °C
Sulfonation of benzene	$C_6H_6 + H_2SO_4 \rightarrow C_6H_5SO_3H + H_2O$	fuming sulfuric acid	40 °C
Bromination of benzene	$C_6H_6 + Br_2 \rightarrow C_6H_5Br + HBr$	liquid bromine with iron (to form iron(III) bromide)	dry and room temperature
Friedel–Crafts alkylation of benzene	$C_6H_6 + RX \rightarrow C_6H_5R + HX$	halogenoalkane	dry, in the presence of anhydrous aluminium halide
Friedel–Crafts acylation of benzene	$C_6H_6 + RCOCl \rightarrow C_6H_5COR + HCl$	acyl chloride	dry, in the presence of anhydrous aluminium chloride

SAFETY IN CHEMISTRY LABORATORIES

Incidents that cause harm to people are rare in school and college chemistry laboratories. One of the reasons for this is that all laboratories must consider the hazards of carrying out chemistry experiments and use safe methods of working.

This applies to all chemistry experiments, whether you are given the instructions for doing them, or whether you plan your own. When you plan an organic synthesis you must consider the hazards associated with the reactants, the substance you are synthesising and with any intermediate compounds formed.

HAZARDS, RISKS AND CONTROL MEASURES

The hazards of chemical substances relate to the inherent properties of each substance and the way in which it will be used. Most people would consider that water (H_2O) is completely safe and has no hazards. In most situations this is the case. However, consider a beaker of boiling water. The steam coming from the boiling water, and the boiling water itself, could both cause harmful effects if they came into contact with your skin.

Now consider a substance that most people would consider dangerous. You probably used dilute hydrochloric acid years ago in experiments involving marble chips. You will have been told to wear eye protection when using the acid, because of the harm it could do if it got into your eyes.

The **hazard** exists because hydrochloric acid is corrosive.

The **risk** is that hydrochloric acid may get into your eyes and cause harm.

The control measure is to wear eye protection to reduce the risk of the acid getting into your eyes.

This simple example should help you to understand the difference between hazard and risk – these two words are often thought, wrongly, to have the same meaning.

HAZARD WARNING SYMBOLS

A long time ago, bottles containing certain chemicals were labelled with the word POISON. This early attempt to prevent harm to laboratory workers was well intentioned, but some people may have thought that bottles without such a label contained harmless substances.

More recently, symbols (sometimes called pictograms) have been used as labels for bottles to identify the actual hazard of the substance inside. The actual symbols have changed over the years, and you may still see older ones used, especially those in squares with an orange background, such as these:

The symbols in current use are red diamond shapes. **Table A** shows some of the more common ones, along with a simple description of their meanings.

SYMBOL	MEANING	
	Health hazard	includes warnings on skin rashes, eye damage and ingestion
	Corrosive	can cause skin burns and permanent eye damage
	Flammable	can catch fire if heated or comes into contact with a flame
	Acute toxicity	can cause life-threatening effects, even in small quantities

table A

In some cases, the substance may have more than one symbol, especially when it is an aqueous solution. For example, you are likely to use hydrochloric acid in three different concentrations:

LEARNING TIP

Become familiar with the hazards of the substances you use in an experiment, and consider how any risks of using them may be reduced.

- when used in a titration, it will have a concentration of approximately 0.1 mol dm^{-3}
- as a general laboratory reagent, it will probably have a concentration of approximately 1 or 2 mol dm^{-3}
- a bottle labelled 'concentrated hydrochloric acid' may have a concentration of more than 10 mol dm^{-3}.

These very different concentrations have different risks.

There are several other hazard symbols, including those that you are more likely to find in biology or physics laboratories.

CONTROL MEASURES

It is the responsibility of whoever is in charge of the laboratory to identify risks and hazards, and to prescribe appropriate control measures. This responsibility is shared with the student doing the experiment, who should follow the guidance. There are several organisations that provide advice and support to schools and colleges, and it is likely that the health and safety information you see in your laboratory has come from one of these sources.

As a student you may be asked to plan an experiment or a synthesis, including identifying the hazards and control measures needed, so you need to be familiar with this information.

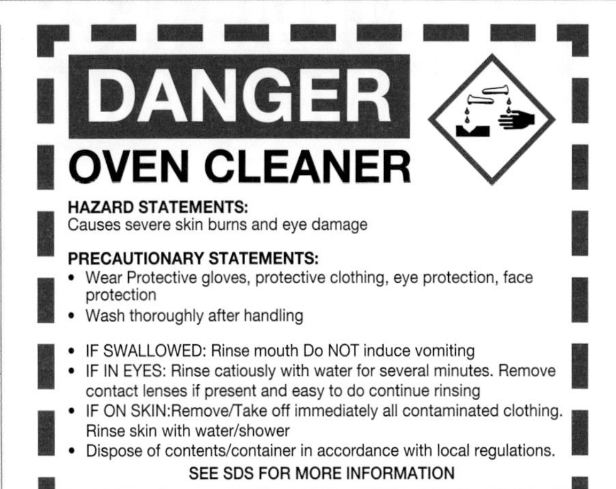

fig A You will also find examples of hazards, risks and control measures in your home.

APPARATUS

You also need to consider risks associated with apparatus. For example, mercury thermometers are often replaced by spirit thermometers or digital thermometers, which have much lower risks. Heating may be done using an electrical heating mantle instead of a Bunsen burner. You may use glass apparatus with ground glass joints, as these are safer to set up than those that use corks or bungs. You also have to consider where to support apparatus with a clamp and stand.

CHECKPOINT

SKILLS SELF-REGULATION, RESPONSIBILITY

1. Suggest some control measures you should use in an experiment to convert bromoethane into ethanol.
2. Use your chemical knowledge to suggest why spilling sodium hydroxide solution onto your skin may cause more harm than hydrochloric acid of the same concentration.

SUBJECT VOCABULARY

hazard a property of a substance that could cause harm to a user
risk the possible effect that a substance may cause to a user, and this will depend on factors such as concentration and apparatus. The level of risk is controlled using control measures

20A 4 PRACTICAL TECHNIQUES IN ORGANIC CHEMISTRY: PART 1

SPECIFICATION REFERENCE 20.5(i) 20.5(vi) 20.5(vii)

LEARNING OBJECTIVES

- Understand the following techniques used in the preparation and purification of organic compounds:
  (i) refluxing
  (ii) distillation
  (iii) steam distillation.

DIFFERENT PRACTICAL TECHNIQUES

In the preparation and purification of any organic compound, there are likely to be several techniques used. Each preparation or synthesis will use one or more of the techniques in this section and the next. Most of them have been covered in **Book 1**, but all are included in these sections for completeness.

HEATING UNDER REFLUX

Some reactions involving organic compounds are slow at room temperature, so the obvious thing to do is heat the reaction mixture. However, as many organic compounds are volatile, there is a risk that they will escape from the reaction mixture during the heating process. The normal way to prevent this happening is to heat the reaction mixture in a flask fitted with a reflux condenser (**fig A**). This technique is also known as refluxing. All the vapours rising from the reaction mixture during heating enter the condenser and change back into liquids. They then return to the flask so that the unreacted compounds can react.

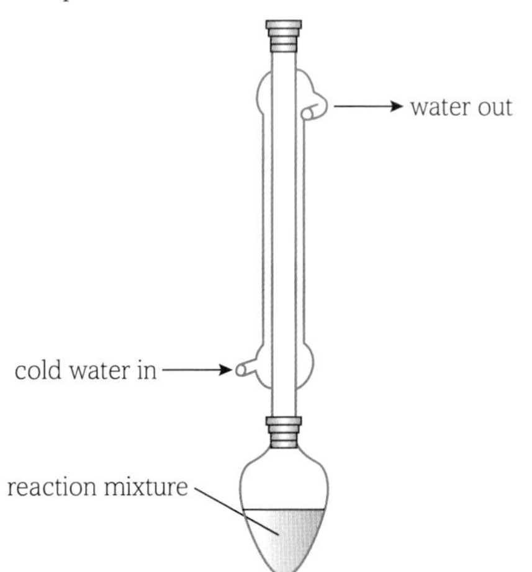

▲ **fig A** Apparatus for heating under reflux.

The flask can be heated using hot water or oil in a beaker heated by a Bunsen burner or by using an electric mantle. The use of anti-bumping granules in the reaction mixture helps to make the boiling smooth. A gentle flow of cold water enters at the bottom of the condenser.

METHODS OF SEPARATION

SIMPLE DISTILLATION

This is a separation technique used to obtain a liquid product from a reaction mixture that has a boiling temperature much lower than the other substances in the reaction mixture.

Distillation of an impure liquid involves heating it in a flask connected to a condenser (**fig B**). The liquid with the lowest boiling temperature evaporates or boils off first and passes into the condenser first, and so can be collected in the receiver separately from another liquid that may evaporate later. The purpose of the thermometer is to monitor the temperature of the vapour as it passes into the condenser. If the temperature remains steady, this indicates that one compound is being distilled. If after a while the temperature begins to rise, this indicates that a different compound is being distilled.

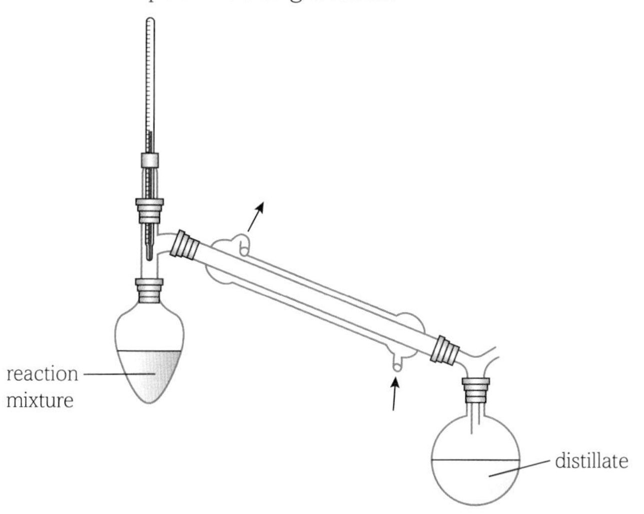

▲ **fig B** Apparatus for simple distillation.

The advantages of using simple distillation, rather than fractional distillation, are that it is easier to set up and is quicker. The disadvantage is that it does not separate the liquids as well as fractional distillation. It should only be used if the boiling temperature of the liquid being purified is very different from the other liquids in the mixture, ideally a difference of more than 25 °C.

EXAM HINT

You may be asked to **draw** distillation apparatus (simple or fractional). If so, make sure that you **draw** the bulb of the thermometer level with the side-arm linked to the condenser.

STEAM DISTILLATION

Steam distillation is a technique used to separate an insoluble liquid from an aqueous solution. It involves passing steam into a reaction mixture that contains an aqueous solution and a liquid that forms a separate layer (**fig C**). The agitation of the liquid caused by the steam bubbling through the mixture ensures that both the insoluble liquid and the aqueous solution are on the surface of the mixture and so can form part of the liquid that evaporates. The advantage of this process is that the insoluble liquid is removed from the reaction mixture at a temperature below its normal boiling temperature.

An example is phenylamine, a liquid with a boiling temperature of 184 °C. During steam distillation of a reaction mixture containing phenylamine, a liquid containing phenylamine and water distils at a temperature of 98 °C. An advantage of doing this is that the temperature at which phenylamine distils is much lower than its boiling temperature, so there is less chance of decomposition.

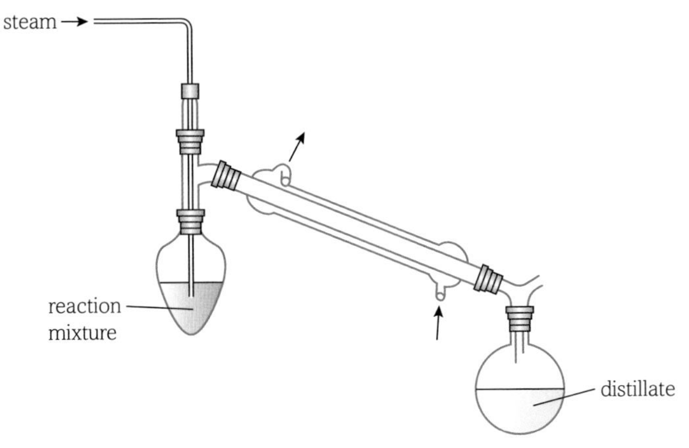

▲ **fig C** Apparatus for steam distillation.

FRACTIONAL DISTILLATION

Fractional distillation uses the same apparatus as simple distillation, but with a fractionating column between the heating flask and the still head (**fig D**). The column is usually filled with glass beads or pieces of broken glass. These act as surfaces on which the vapour leaving the column can condense, and then be evaporated again as more hot vapour passes up the column. Effectively, the vapour experiences several repeated distillations as it passes up the column, which provides a better separation.

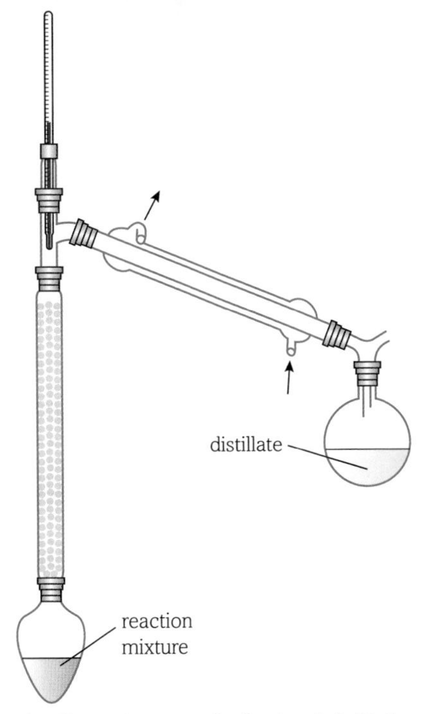

▲ **fig D** Apparatus for fractional distillation.

Fractional distillation takes longer than simple distillation. It is best used when the difference in boiling temperatures is small, and when there are several compounds to be separated from a mixture.

LEARNING TIP

Remember that there are three types of distillation, and that you need to consider which one to use in a particular case.

CHECKPOINT

SKILLS SELF-DIRECTION, CREATIVITY

1. Explain briefly why the condenser is vertical in refluxing but nearly horizontal in the three methods of distillation.
2. Suggest, with a reason, which fractionating column provides the better separation - very small glass beads or larger pieces of glass.

20A 5 PRACTICAL TECHNIQUES IN ORGANIC CHEMISTRY: PART 2

LEARNING OBJECTIVES

- Understand the following techniques used in the preparation and purification of organic compounds:
  (i) purification by washing, including with water and sodium carbonate solution
  (ii) solvent extraction
  (iii) recrystallisation
  (iv) drying
  (v) melting temperature determination
  (vi) boiling temperature determination.

MORE METHODS OF SEPARATION

SOLVENT EXTRACTION

As the name suggests, this method involves using a solvent to remove the desired organic product from the other substances in the reaction mixture. Several solvents can be used, but the choice depends mainly on these features:

- The solvent added should be immiscible (i.e. does not form a mixture) with the solvent containing the desired organic product.
- The desired organic product should be much more soluble in the added solvent than in the reaction mixture.

The method is this:

- Place the reaction mixture in a separating funnel (**fig A**), and then add the chosen solvent – it should form a separate layer.
- Place the stopper in the neck of the funnel and gently agitate the contents of the funnel for a short while (put a finger on the stopper, invert, open the tap, agitate in a circular motion, close the tap and return the funnel to its normal position).
- Allow the contents to settle into two layers.
- Remove the stopper and open the tap to allow the lower layer to drain into a flask, and then do the same to allow the upper layer to drain into a separate flask.

▲ **fig A** Apparatus for separating immiscible liquids.

If a suitable solvent is used and the method is followed correctly, most of the desired organic product will have moved into the added solvent. It is better to use the solvent in small portions rather than in a single volume (e.g. four portions of 25 cm^3 rather than one portion of 100 cm^3) because this is more efficient. Using more portions of solvent, but with the same total volume, removes more of the desired organic product.

The desired organic product has been removed from the reaction mixture, but is now mixed with the added solvent. Therefore, simple distillation or fractional distillation must now be used to separate the desired organic product from the solvent used.

WASHING

You might assume that washing is a technique used to remove impurities from a solid. Although this is correct, it is also used to remove impurities from a liquid. Washing could involve the use of water or an organic solvent. It could also involve the use of sodium carbonate solution, to remove any excess acid that may be present. Whichever liquid is used, it must be chosen carefully so as to dissolve the impurities but as little as possible of the substance being purified.

An impure solid is stirred in some of the solvent, then the mixture is filtered. If the solid is already in a filter funnel, the solvent could be added on top of the solid.

A liquid would be mixed with a solvent chosen so that it will dissolve little, if any, of the liquid to be purified. The mixture is then shaken in a separating funnel. After allowing the two liquid layers to separate, the tap is opened to allow each layer to drain into a separate container.

DRYING

No special technique is needed to dry an organic solid – it just needs to be left in a warm place or in a desiccator with a suitable drying agent (**fig B**).

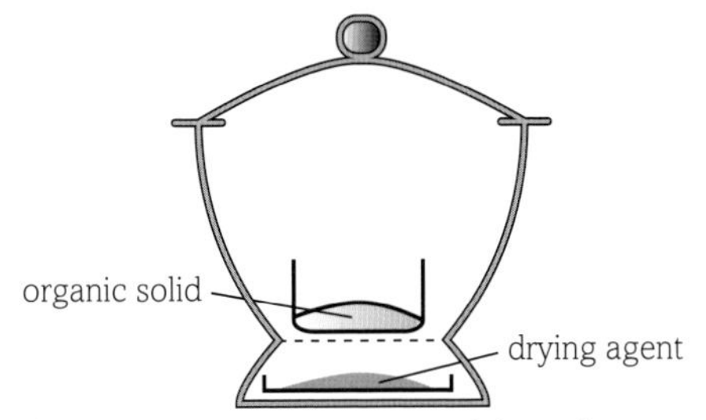

▲ **fig B** Drying an organic solid in a desiccator

Many organic liquids are prepared using inorganic reagents, which are often used in aqueous solution. A liquid organic product may partially, or even completely, dissolve in water. Therefore, water may be an impurity that needs to be removed by a drying agent. Obviously, one important feature of a drying agent is that it does not react with the organic liquid.

Several drying agents are available, but the most common ones are anhydrous metal salts – usually calcium chloride, magnesium sulfate and sodium sulfate. The property that these compounds have in common is that they form hydrated salts. So, when they come into contact with water in an organic liquid, they absorb the water as water of crystallisation.

The drying agent is added to the organic liquid and the mixture is gently agitated or shaken, and then left for a period of time. Before use, a drying agent is powdery, but after absorbing water it looks more crystalline. If a bit more drying agent is added and remains powdery, then this is an indication that the liquid is dry. The liquid also goes from cloudy to clear when water is removed.

The drying agent is removed either by decantation (pouring the organic liquid off the solid drying agent) or by filtration.

FILTRATION

It is likely that after preparation, an organic solid will need to be filtered at some stage. This always happens as part of recrystallisation, so the two pieces of apparatus (Buchner and Hirsch funnels) that are normally used will be considered under the next heading. The use of a vacuum pump means that these methods are described as filtration under reduced pressure.

RECRYSTALLISATION

When an organic solid has been prepared, it is likely to need purification. A traditional way of removing impurities is the technique of recrystallisation. The principle behind this technique is that a solid compound is dissolved in a suitable solvent that can dissolve all or most of any impurities but very little of the compound being purified. The steps used in a typical purification are:

- Add the impure solid to a conical flask.
- Add some of the chosen solvent and warm until the mixture nears the boiling temperature of the mixture.

- If there is still some undissolved solid, add further solvent and warm until the mixture boils again.
- Continue adding further solvent and heating until all of the soluble solid has dissolved.
- If insoluble impurities are present, then hot filtration could be done using fluted filter paper in a heated funnel.
- Allow the liquid to cool until crystals of the organic solid have formed.
- More crystals can be obtained by cooling the solution below room temperature in an ice bath.
- The mixture is then filtered to remove soluble impurities using a Buchner funnel or a Hirsch funnel (**fig C**).
- The crystals are washed with a small amount of ice-cold solvent and then dried in a desiccator or warm oven.

LEARNING TIP

The impure solid needs to be dissolved in the minimum volume of hot solvent. This is to make sure that the pure solid will crystallise when the solution is cooled.

Hot filtration is used to make sure that the pure solid does not crystallise at this stage.

Crystals of the pure solid form when the solution is cooled. This is because the solid is less soluble in the cold solvent than in the hot solvent.

The crystals are washed with a small amount of ice-cold solvent. This is to remove any soluble impurities, and to make sure that very little of the pure solid dissolves.

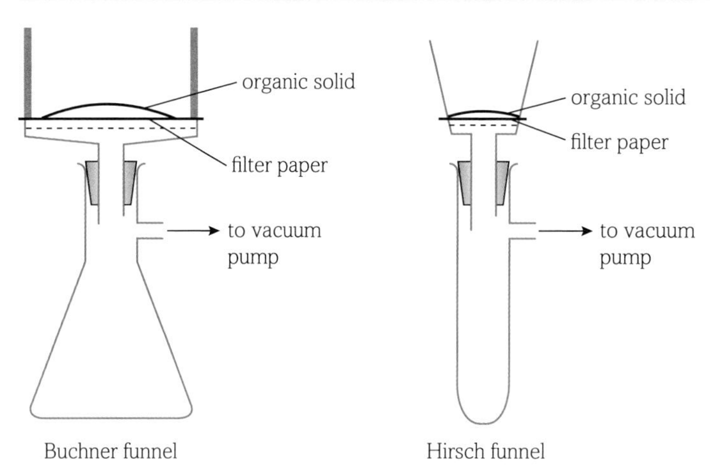

▲ **fig C** Apparatus for filtration under reduced pressure.

LEARNING TIP

Study the information about recrystallisation carefully. Many students do not understand why there might be two filtrations needed. Some throw away the substance they have just purified!

TESTING FOR PURITY

Having prepared an organic compound, a simple test can be performed to give an indication of whether the compound is pure. If the compound is a solid, then its melting temperature can be measured. If it is a liquid, then its boiling temperature can be measured.

LEARNING TIP

A pure solid has a sharp melting temperature rather than melting over a range of temperatures. This can be used as a test for purity.

DETERMINATION OF MELTING TEMPERATURE

For solids, impurities reduce the melting temperature. If you measure the melting temperature of your organic compound, then you can compare it with the known value of the pure compound. This will enable you to estimate how pure it is.

The traditional way to measure a melting temperature of a solid is to place some of the solid in a small capillary tube attached to the bulb of a thermometer and then place the assembly in a liquid that has a boiling temperature above that of the melting temperature of the solid (**fig D**).

In practice, this apparatus is often replaced by various electrical devices.

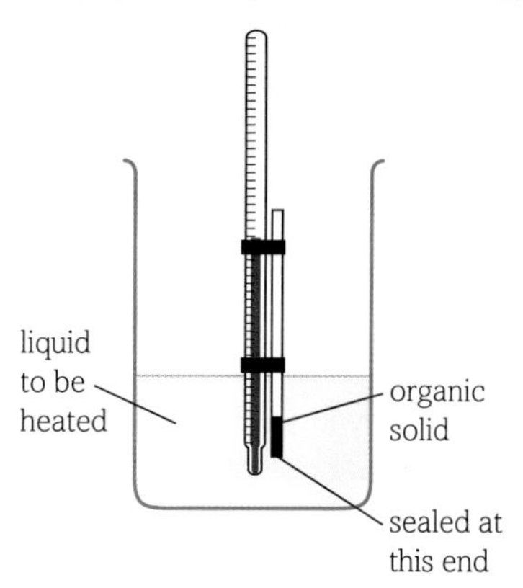

▲ **fig D** Apparatus for determination of melting temperature.

DETERMINATION OF BOILING TEMPERATURE

For liquids, impurities increase the boiling temperature.

The boiling temperatures of pure organic compounds have been carefully measured and are widely available in data books and online. If you measure the boiling temperature of your organic compound, you can compare the value you obtain with the known value of the pure compound and then estimate how pure the compound is.

The apparatus used depends on the volume of liquid available, and whether it is toxic or flammable. The apparatus used for simple distillation can be used in most cases.

A word of caution. This test may not be enough proof because you may not be able to measure the boiling temperature of your organic compound accurately enough – your thermometer might read too low or too high. So, even if your measured boiling temperature exactly matches the one in the data book or online, you may wrongly assume that your compound is pure.

It is also worth remembering that different organic compounds can, by coincidence, have the same boiling temperature. For example, both 1-chloropentane and 2-methylpropan-1-ol boil at 108 °C.

CHECKPOINT

SKILLS: CRITICAL THINKING, SELF-DIRECTION

1. Explain why the solvent extraction method will not work if the solvent dissolves the impurities.
2. Explain why, in recrystallisation, insoluble impurities are removed from a hot solution, not a cooled one.

20 THINKING BIGGER

CHALLENGING ORGANIC SYNTHESES

The following article describes the catalytic oxidation of a specific C–H bond in an organic molecule. Specific chemical transformations like this are essential in the manufacture of drug molecules.

IRON CATALYST SELECTIVELY OXIDISES UNREACTIVE C–H BONDS, UNAIDED

US chemists have cracked one of the toughest challenges in organic synthesis: how to attack a complex molecule's unreactive carbon–hydrogen bonds, without resorting to wasteful synthetic aids like protecting and directing groups.

Christina White and Mark Chen from the University of Illinois, USA, say their iron catalyst – which with hydrogen peroxide selectively and predictably oxidises C–H bonds into more reactive C–OH groups – will fundamentally alter the way complex molecules such as drugs are made in the laboratory.

The C–H bonds dotted around molecules are mostly left alone in organic syntheses as they are often very unreactive. Selectively attacking one C–H bond from a complex molecule filled with many other functional groups (in other words, a typical pharmaceutical) is especially difficult. Many C–H bonds, if not the entire molecule, are likely to be disrupted.

▲ **fig A** The Chen-White catalyst with hydrogen peroxide selectively hydroxylated the most electron-rich C-H bond in the complex antimalarial, artemisinin.

Though platinum and palladium catalysts are able to hydroxylate, aminate, and alkylate C–H bonds selectively, they rely on nearby activating or directing groups (which enhance a C–H bond's reactivity, or guide a catalyst to one C–H bond in particular) and protecting groups. But these synthetic crutches mean extra steps in a reaction. Only natural enzymes are able to pick out C–H bonds from a complex molecule without help, but their elaborate protein binding pockets mean they only work on specific substrates.

Yet White and Chen's electrophilic iron catalyst, with hydrogen peroxide, selectively oxidised lone C–H bonds from complex molecules with ease. The new method was demonstrated successfully on several complex products, including the antimalarial artemisinin. Here, one C–H bond out of five was selectively oxidised, in yield higher than that achieved by a previous enzymatic transformation. Particularly impressive was that White and Chen laid down a series of rules predicting the catalyst's selectivity: depending on the electron-richness of the C–H bond, or whether it was physically hindered from the catalyst's approach.

"This is an absolutely beautiful piece of work," commented Alan Goldman, who works on C–H activation at Rutgers University, New Jersey, USA. "The ability to functionalise C–H bonds selectively is the ultimate challenge in organic synthesis, with immense potential value."

Work in progress

The mechanism of reaction is not clear, but White suggests it involves an iron-oxo or iron-peroxo intermediate, with an oxygen atom eventually transferring from iron to form C–OH. Such chemistry has been attempted before, but usually the catalyst decomposes to generate free iron – which with hydrogen peroxide generates hydroxide radicals that wreak havoc on organic molecules. White thinks the key to success this time was making the catalyst's ligand, which surrounds the central iron, more rigid. This means that the catalyst decomposes more slowly, although it still stops working after four or five cycles.

Modifying pharmaceuticals to check if derivatives have greater activity has often meant reworking the entire synthesis from scratch, says White. Now it should be easier to simply install OH functional groups near the end of the synthesis.

"The reaction adds a new drawer to the synthetic chemist's hydroxylation toolbox. One particular application in the pharmaceutical industry will likely reside in the preparation of drug metabolites," commented Steven Brickner, who researches antibacterials for pharmaceuticals giant Pfizer.

From an article in *Chemistry World*, 'Step change for organic synthesis' by Richard Van Noorden, 1 November 2007.
https://www.chemistryworld.com/news/step-change-for-organic-synthesis/3003773.article

SCIENCE COMMUNICATION

1 The chemistry in this extract is very detailed and will be difficult for a general audience to understand. Write a short article that communicates the important ideas from this extract to the general public.

2 Why do you think such a small chemical transformation is so important to the pharmaceutical industry?

CHEMISTRY IN DETAIL

3 The organic molecule shown in **fig A** is the antimalarial drug artemisinin. Work out its molecular formula from the skeletal formula shown.

4 How many chiral carbons are there in a molecule of artemisinin?

5 You have 1 kg of artemisinin. Assuming a 54% conversion for the oxidation reaction, calculate the mass of product to 3 s.f.

6 Identify the ester functional group in artemisinin.

7 What is the shape of the iron(II) complex?

ACTIVITY

Artemisinin is a naturally occurring molecule derived from the plant *Artemisia annua*. Carry out your own research on another plant-based medicine. Address the following:

(i) Is the medicine extracted from plant material or is it synthesised in the laboratory?

(ii) Are molecular derivatives made, and if so why?

THINKING BIGGER TIP

Think about how important the position and type of functional group can be to the behaviour of a biological molecule. For example, **fig B** shows how small the differences are between the molecules (a) oestrogen (a female sex hormone) and (b) testosterone (a male sex hormone).

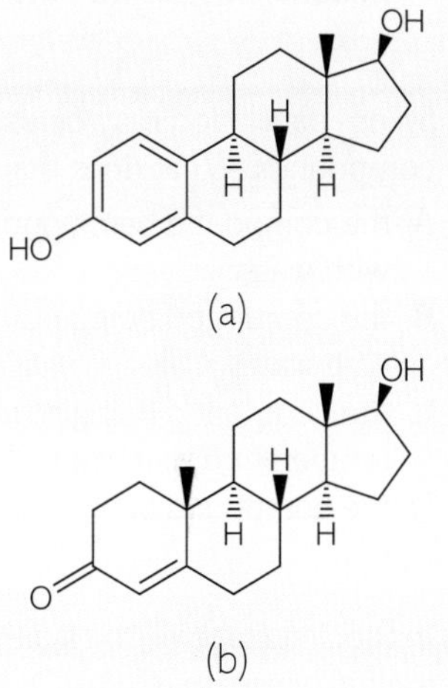

▲ **fig B** (a) Oestrogen and (b) testosterone

20 EXAM PRACTICE

1 Organic solids are often purified by recrystallisation. What is the basis for this technique?

A the impurities must be insoluble in the solvent used
B the impurities must react with the solvent used
C the impurities crystallise first when the hot solution is cooled
D the cooled solution is saturated with the desired material but not with the impurities. [1]

(Total for Question 1 = 1 mark)

2 Steam distillation may be used in the purification of some compounds. What does this technique depend on?

A the compound being purified forms a single layer with water
B the compound being purified forms two layers with water
C the compound being purified has a lower boiling temperature than water
D the compound being purified is flammable. [1]

(Total for Question 2 = 1 mark)

3 In the first stage of the synthesis of methyl 3-nitrobenzoate, methyl benzoate ($C_6H_5COOCH_3$) is prepared by the reaction of benzoic acid with methanol in the presence of concentrated sulfuric acid.

When the reaction is complete, the sulfuric acid is neutralised by adding aqueous sodium carbonate. What is the simplest way of obtaining the impure methyl benzoate from this mixture?

A refluxing
B solvent extraction
C filtration
D recrystallisation. [1]

(Total for Question 3 = 1 mark)

4 This question concerns the reaction scheme:

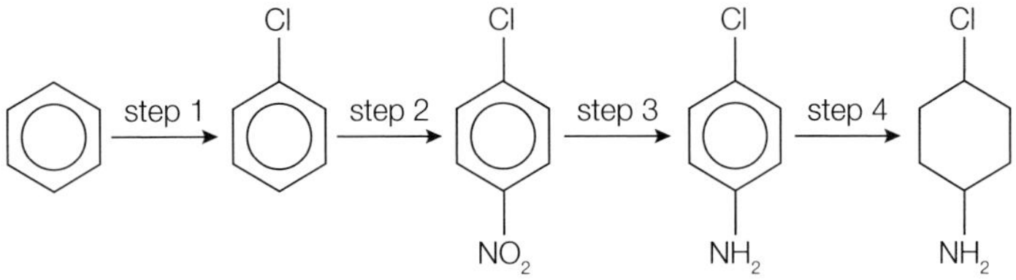

benzene —step 1→ chlorobenzene —step 2→ 1-chloro-4-nitrobenzene (Cl, NO_2) —step 3→ 4-chloroaniline (Cl, NH_2) —step 4→ 4-chlorocyclohexylamine (Cl, NH_2)

(a) Which step is most likely to need tin and concentrated hydrochloric acid?

A Step 1 **B** Step 2 **C** Step 3 **D** Step 4 [1]

(b) Which step is most likely to need a mixture of concentrated nitric acid and concentrated sulfuric acid?

A Step 1 **B** Step 2 **C** Step 3 **D** Step 4 [1]

(c) Which step is most likely to need a catalyst of aluminium chloride?

A Step 1 **B** Step 2 **C** Step 3 **D** Step 4 [1]

(Total for Question 4 = 3 marks)

5 Look at this reaction sequence:

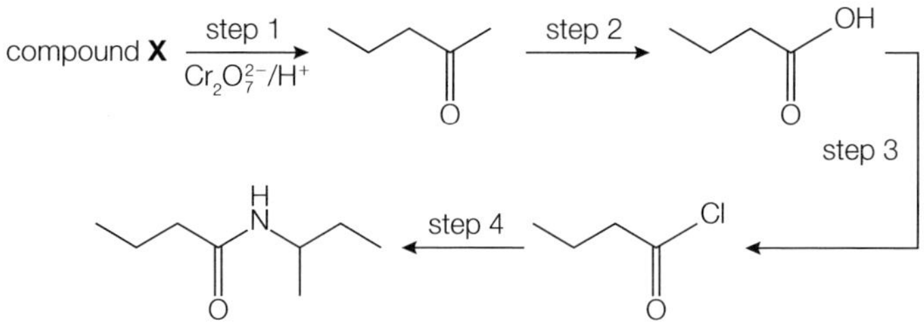

compound **X** —step 1 ($Cr_2O_7^{2-}/H^+$)→ pentan-2-one (skeletal) —step 2→ butanoic acid (OH, O) —step 3→ butanoyl chloride (Cl, O) —step 4→ N-substituted amide (H, N, O)

(a) What is compound **X**?

A pentan-1-ol
B pentan-2-ol
C 2-methylbutan-1-ol
D 2-methylbutan-2-ol [1]

(b) Which substances are required for step 2?

A acidified potassium dichromate(VI)
B iodine in alkali, followed by hydrochloric acid
C sodium hydroxide solution followed by hydrochloric acid
D ammoniacal silver nitrate (Tollens' reagent) [1]

(c) Which is the reagent for step 3?

A aqueous chlorine
B chlorine gas
C hydrochloric acid
D phosphorus(V) chloride [1]

(d) Which is the reagent for step 4?

A $CH_3CH_2CH_2CH_2NH_2$
B $CH_3CH(NH_2)CH_2CH_3$
C $CH_3CH_2CH_2CONH_2$
D $CH_3CH(CONH_2)CH_3$ [1]

(Total for Question 5 = 4 marks)

6 Two organic compounds, **P** and **Q**, have the same molecular formula, C_4H_8O.

Both compounds contain a carbonyl group.

(a) Describe what you would see when 2,4-dinitrophenylhydrazine is added to either of these compounds. [1]

(b) Outline a test you could perform to show that **P** is a ketone and that **Q** is an aldehyde. [3]

(c) (i) Give the structural formulae of the two possible isomers of **Q** that are aldehydes. [1]

(ii) Name the technique you would use to purify the product of the test with 2,4-dinitrophenylhydrazine. [1]

(iii) Other than by spectroscopic techniques, explain how you would use the purified product to identify compound **Q**. [2]

(Total for Question 6 = 8 marks)

7 Procaine can be synthesised from benzoic acid. The simplified reaction route is:

benzoic acid (COOH) —step 1→ 4-nitrobenzoic acid (COOH, NO_2) —step 2, PCl_5, heat under reflux→ (COCl, NO_2) —step 3→ ($COOCH_2CH_2N(C_2H_5)_2$, NO_2) —step 4→ procaine ($COOCH_2CH_2N(C_2H_5)_2$, NH_2)

(a) Suggest the two reagents needed for step 1. [2]

(b) Suggest why the reaction mixture in step 2 needs to be heated under reflux. [2]

(c) Give the structural formula for the organic reagent needed in step 3. [1]

(d) Name the type of reaction taking place in step 4. [1]

(Total for Question 7 = 6 marks)

8 A student produced a sample of aspirin by the esterification of 9.40 g of 2-hydroxybenzoic acid with excess ethanoic anhydride:

2-hydroxybenzoic acid + $(CH_3CO)_2O$ (ethanoic anhydride) → aspirin + CH_3COOH

After purification by recrystallisation, 7.77 g of aspirin was obtained.

(a) Calculate the percentage yield obtained. [4]

(b) (i) Outline how you would purify a solid such as aspirin by recrystallisation, using water as the solvent. [5]

(ii) Explain what effect recrystallisation has on the final yield. [2]

(Total for Question 8 = 11 marks)

9 Biodiesel can be made from vegetable oils and methanol. The equation for the reaction is:

$$3CH_3OH + CH_2OOCR / CHOOCR' / CH_2OOCR'' \xrightarrow[50\,°C,\ 98\%\ yield]{NaOH(s)} CH_2OH / CHOH / CH_2OH + RCOOCH_3 + R'COOCH_3 + R''COOCH_3$$

(a) Describe two hazards when carrying out this reaction. For each one, state the precaution you would take to minimise the risk. [4]

(b) Suggest two environmental benefits of carrying out this reaction. [2]

(Total for Question 9 = 6 marks)

10 Paracetamol is a widely used painkiller. It can be made from phenol as shown below.

phenol (OH) —step 1→ 4-nitrophenol (OH, NO_2) —step 2→ 4-aminophenol (OH, NH_2) —step 3→ paracetamol (OH, $NHCOCH_3$)

(a) State the reagent(s) used in each of steps 1, 2 and 3. [3]

(b) In the nitration of phenol in step 1, two compounds are produced:

2-nitrophenol   4-nitrophenol

These compounds can be separated by steam distillation, since 2-nitrophenol is volatile in steam but 4-nitrophenol is not.

Describe briefly how 2-nitrophenol is obtained by steam distillation, and give one advantage of steam distillation over normal distillation. [4]

(c) The paracetamol obtained in the above synthesis can be purified by recrystallisation. In this process, the impure paracetamol is dissolved in the minimum volume of hot water and the mixture is then filtered while still hot. The filtrate obtained is cooled, and the resulting crystals are filtered and dried.

The table shows the solubility in water of paracetamol at various temperatures.

Temperature/°C	5	10	20	95
Solubility/g per 100 g	0.82	0.94	1.3	5.2

(i) Explain the purpose of each of the two filtrations in the recrystallisation process. [2]

(ii) Explain, using the temperatures given in the table, which pair of temperatures will give the highest yield of paracetamol. [2]

(iii) Name the technique that could be used in a school laboratory to check the purity of the recrystallised paracetamol. [1]

(Total for Question 10 = 12 marks)

MATHS SKILLS

For you to be able to develop your skills, knowledge and understanding in Chemistry you will need to have developed your mathematical skills in a number of key areas. This section gives more explanation and examples of some key mathematical concepts you need to understand. Further examples relevant to your International A Level Chemistry studies are given throughout this book and Book 1.

USING LOGARITHMS

CALCULATING LOGARITHMS

Many formulae in science and mathematics involve powers. Consider the equation:

$$10^x = 62$$

The value of x obviously lies between 1 and 2, but how can you find a precise answer? The term logarithm means 'index' or 'power'. Logarithms enable you to solve such equations. You can take the logarithm to base 10 of each side of the equation using the log button of a calculator.

WORKED EXAMPLE 1

$10^x = 62$

$\log_{10}(10^x) = \log_{10}(62)$

$x = 1.792392...$

You can calculate logarithms using any number as the base by using the $\log_x(y)$ button.

WORKED EXAMPLE 2

$2^x = 7$

$\log_2(2^x) = \log_2(7)$

$x = 2.807355...$

Many equations relating to the natural world involve powers of e. These are called exponentials. The logarithm to base e is referred to as the natural logarithm and written ln.

USING LOGARITHMIC PLOTS

An earthquake measuring 8.0 on the Richter scale is much more than twice as powerful as an earthquake measuring 4.0 on the Richter scale, because the units used for measuring earthquakes are logarithmic. Logarithmic scales in charts and graphs can accommodate enormous increases or decreases in one variable as another variable changes.

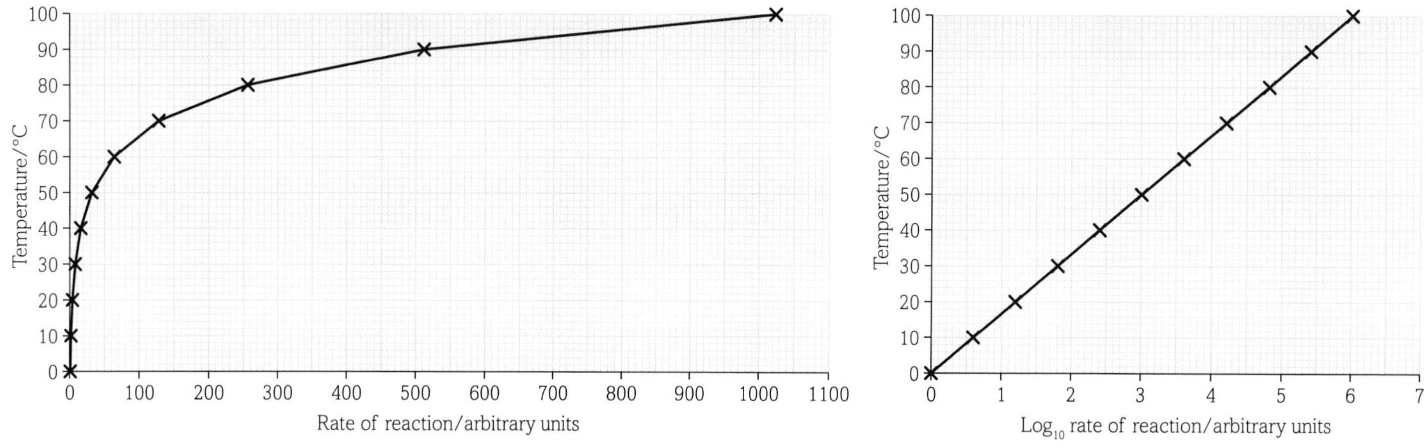

▲ **fig A** Logarithmic scales are useful when representing a very large range of values, such as in the case of rates of reaction.

GRAPHS

UNDERSTANDING THAT $y = mx + c$ REPRESENTS A LINEAR RELATIONSHIP

Two variables have a linear relationship if they increase at a constant rate in relation to one another. If you plot a graph with one such variable on the *x*-axis and the other on the *y*-axis, you get a straight line.

Any linear relationship can be represented by the equation $y = mx + c$, where the gradient of the line is m and the value at which the line crosses the *y*-axis is c. An example of a linear relationship is that between degrees Celsius and degrees Fahrenheit, which can be represented by the equation $F = 9/5C + 32$, where C is temperature in degrees Celsius and F is temperature in degrees Fahrenheit.

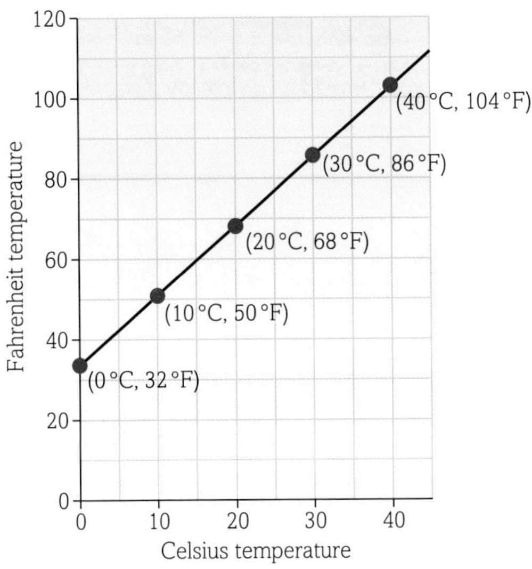

fig B Linear relationship between Fahrenheit and Celsius temperature scales

USING THE GRADIENT AS A MEASURE OF A RATE OF CHANGE

Sir Isaac Newton drew tangents to curves on graphs to find the rates of change of the variables as part of his journey towards discovering calculus, a fascinating branch of mathematics. He stated that the gradient of a curve at a given point is exactly equal to the gradient of the tangent to the curve at that point.

To find the gradient at a point on a curve:

1 Use a ruler to draw a tangent to the curve at that point.

2 Calculate the gradient of the tangent using the equation for a linear relationship. This gradient is equal to the gradient of the curve at the point of the tangent.

3 Include the unit with your answer.

APPLYING YOUR SKILLS

You will often find that you need to use more than one maths technique to answer a question. In this section you will look at two example questions and consider which maths skills are required and how to apply them.

WORKED EXAMPLE 3

The pH of an aqueous solution is related to the hydrogen ion concentration by the following equation:

$$pH = -lg[H^+]$$

where the hydrogen ion concentration, $[H^+]$, is measured in mol dm^{-3}.
Calculate the pH of an aqueous solution of 0.001 mol dm^{-3} HCl. Assume the acid is completely dissociated. Give your answer to two decimal places.

$$pH = -lg[H^+]$$

$$pH = -lg(0.001)$$
$$lg(0.001) = -3$$
$$pH = 3.00$$

WORKED EXAMPLE 4

A scientist observes an interesting by-product forming during a chemical reaction:

reactant 1 + reactant 2 → desired product + by-product

She decides to investigate how the concentration of this by-product changes during the reaction by plotting a graph of concentration against time. She also plots the change in concentration of the desired product and reactant 1.

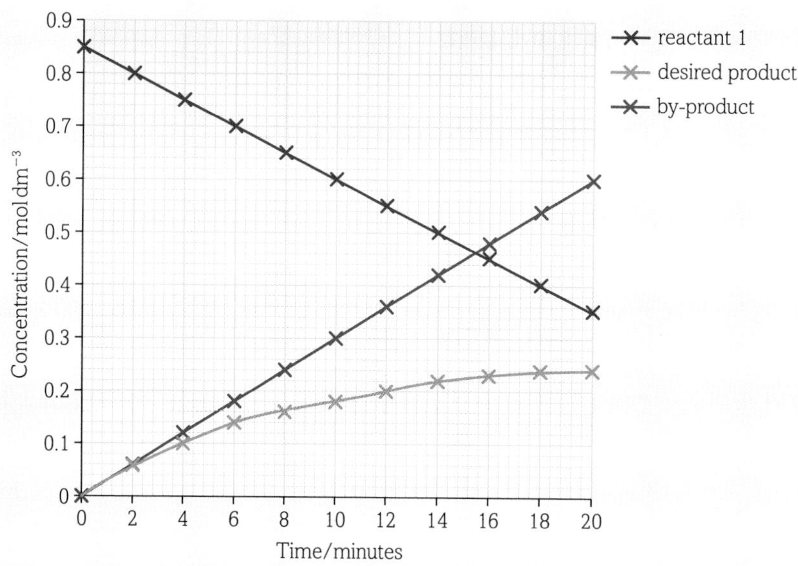

fig C Concentration of reactant and products over the course of a reaction.

(a) *What was the concentration of reactant 1 at the start of the reaction?*

(b) *Calculate the rate of change of concentration of the by-product after 10 minutes. Give your answer in mol dm^{-3} s^{-1}.*

(c) *Calculate the rate of change of concentration of the desired product after 10 minutes.*

(a) To answer this question you need to find the value for the concentration of reactant 1 when the time is equal to zero. That is, you need to find the intercept with the *y*-axis of the line for the concentration of reactant 1. From the graph you can see that this is 0.85 mol dm^{-3}.

(b) In order to calculate the rate of change of by-product concentration during the reaction you need to find the gradient of its line. The line is straight, meaning that there is a linear relationship between the concentration and time. To find the rate of change you just have to find the gradient of the line by dividing the change in concentration of by-product by the time taken for the change. Since the line is straight you can choose any two values.

Take care to use the correct units. Concentration is in mol dm^{-3} but time is in minutes, so you must convert time to seconds to get the correct answer.

concentration after 8 minutes = 0.24 mol dm^{-3}

concentration after 12 minutes = 0.36 mol dm^{-3}

change of concentration = 0.12 mol dm^{-3}

time taken for change = 4 × 60 = 240 s

$$\text{rate of change} = \frac{\text{change in concentration of by-product}}{\text{time taken for change}}$$

$$= \frac{0.12\ \text{mol dm}^{-3}}{240\ \text{s}} = 0.0005\ \text{mol dm}^{-3}\ \text{s}^{-1}$$

(c) To find the rate of change of the desired product is a little trickier because the line is curved. To do this you need to draw a tangent to the curve and then calculate the gradient of the tangent. A simple way to do this is to draw a triangle with the tangent as its hypotenuse, centred on the 10-minute point as asked (**fig D**).

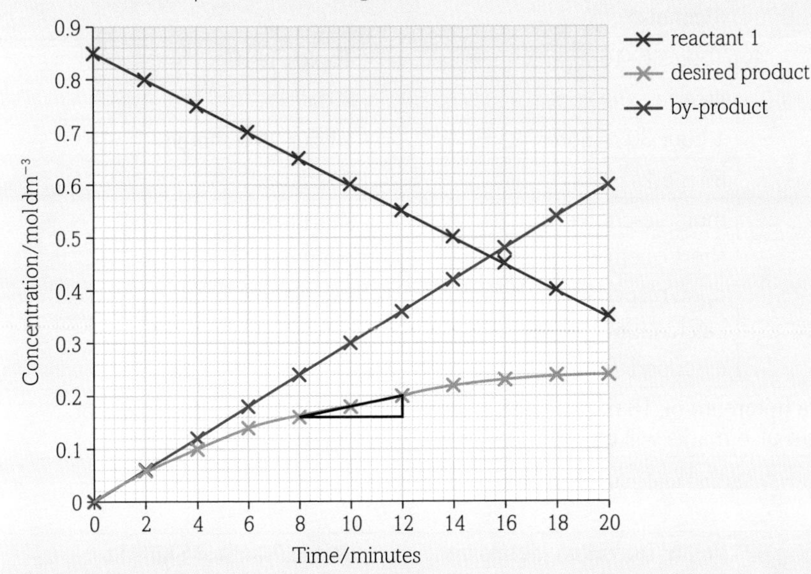

▲ **fig D** Finding the rate of change of concentration of the desired product by drawing a tangent.

change in concentration between 8 and 12 minutes = 0.04 mol dm^{-3}

time taken for change = 4 × 60 = 240 s

$$\text{rate of change} = \frac{\text{change in concentration of desired product}}{\text{time taken for change}}$$

$$= 0.04\ \text{mol dm}^{-3}/240\ \text{s} = 0.000\,166\,6... = 0.0002\ \text{mol dm}^{-3}\ \text{s}^{-1}$$

PREPARING FOR YOUR EXAMS

IAS AND IAL OVERVIEW

The Pearson Edexcel International Advanced Subsidiary (IAS) in Chemistry and the Pearson Edexcel International Advanced Level (IAL) in Chemistry are modular qualifications. The IAS can be claimed on completion of the International Advanced Subsidiary (IAS) units. The International Advanced Level can be claimed on completion of all the units (IAS and IAL units).

- International AS students will sit three exam papers. The IAS qualification can either be standalone or contribute 50% of the marks for the International Advanced Level.
- International A Level students will sit six exam papers, the three IAS papers and three IAL papers.

The tables below give details of the exam papers for each qualification.

IAS Papers	Unit 1: Structure, Bonding and Introduction to Organic Chemistry	Unit 2: Energetics, Group Chemistry, Halogenoalkanes and Alcohols*	Unit 3: Practical Skills in Chemistry I
Topics covered	Topics 1–5	Topics 6–10	Topics 1–10
% of the IAS qualification	40%	40%	20%
Length of exam	1 hour 30 minutes	1 hour 30 minutes	1 hour 20 minutes
Marks available	80 marks	80 marks	50 marks
Question types	multiple-choice short open open response calculation	multiple-choice short open open response calculation extended writing	short open open response calculation
Mathematics	For both Unit 1 and Unit 2, a minimum of 18 marks will be awarded for mathematics at Level 2 or above. For Unit 3, a minimum of 6 marks will be awarded for mathematics at Level 2 or above.		

* This paper may contain some synoptic questions which require knowledge and understanding from Unit 1.

IAL Papers	Unit 4: Rates, Equilibria and Further Organic Chemistry**	Unit 5: Transition Metals and Organic Nitrogen Chemistry†	Unit 6: Practical Skills in Chemistry II
Topics covered	Topics 11–15	Topics 16–20	Topics 11–20
% of the IAL qualification	20%	20%	10%
Length of exam	1 hour 45 minutes	1 hour 45 minutes	1 hour 20 minutes
Marks available	90 marks	90 marks	50 marks
Question types	multiple-choice short open open response calculation extended writing	multiple-choice short open open response calculation extended writing	short open open response calculation
Mathematics	For Unit 4, a minimum of 22 marks will be awarded for mathematics at Level 2 or above. For Unit 5, a minimum of 18 marks will be awarded for mathematics at Level 2 or above. For Unit 6, a minimum of 6 marks will be awarded for mathematics at Level 2 or above.		

** This paper may contain some synoptic questions which require knowledge and understanding from Units 1 and 2.

† This paper may contain some synoptic questions which require knowledge and understanding from Units 1, 2 and 4.

EXAM STRATEGY

ARRIVE EQUIPPED

Make sure you have all of the correct equipment needed for your exam. As a minimum you should take:

- pen (black ink or ballpoint pen)
- pencil (HB)
- rule (ideally 30 cm)
- rubber (make sure it's clean and doesn't smudge the pencil marks or rip the paper)
- calculator (scientific).

MAKE SURE YOUR ANSWERS CAN BE READ

Your handwriting does not have to be perfect, but the examiner must be able to read it!

PLAN YOUR TIME

Note how many marks are available on the paper and how many minutes you have to complete it. This will give you an idea of how long to spend on each question. Be sure to leave some time at the end of the exam for checking answers. A rough guide of a minute a mark is a good start, but short answers and multiple-choice questions may be quicker. Longer answers might require more time.

UNDERSTAND THE QUESTION

Always read the question carefully and spend a few moments working out what you are being asked to do. The command word used will give you an indication of what is required in your answer.

Be scientific and accurate, even when writing longer answers. Use the technical terms you've been taught.

Always show your working for any calculations. Marks may be available for individual steps, not just for the final answer. Also, even if you make a calculation error you may be awarded marks for applying the correct technique.

PLAN YOUR ANSWER

In questions marked with an asterisk (*), marks will be awarded for your ability to structure your answer logically, showing how the points that you make are related or follow on from each other, where appropriate. Read the question fully and carefully (at least twice!) before beginning your answer.

MAKE THE MOST OF GRAPHS AND DIAGRAMS

Diagrams and sketch graphs can earn marks, often more easily and quickly than written explanations, but they will only earn marks if they are carefully drawn.

- If you are asked to read a graph, pay attention to the labels and numbers on the x- and y-axes. Remember that each axis is a number line.
- If asked to draw or sketch a graph, always ensure you use a sensible scale, and label both axes with quantities and units. If plotting a graph, use a pencil and draw small crosses or dots for the points.
- Diagrams must always be neat, clear and fully labelled.

CHECK YOUR ANSWERS

For open-response and extended writing questions, check the number of marks that are available. If three marks are available, have you made three distinct points?

For calculations, read through each stage of your working. Substituting your final answer into the original question can be a simple way of checking that the final answer is correct. Another simple strategy is to consider whether the answer seems sensible. Make sure you use the correct units.

SAMPLE EXAM ANSWERS

QUESTION TYPE: MULTIPLE-CHOICE

Which substance has non-polar molecules?

A *hydrogen fluoride* ☐

B *water* ☐

C *carbon dioxide* ☐

D *ammonia* ☐ [1]

This question is about bond polarity. Electronegative atoms have the ability to draw the electrons in a covalent bond towards them, so that the electron density is unevenly distributed between the two atoms. This results in a polar bond or dipole being set up. All of these molecules contain bonds between atoms with very different electronegativities, and hence polar bonds. However, since carbon dioxide is symmetrical overall, then the molecule itself is non-polar.

Question analysis

- Multiple-choice questions may require simple recall, as in this case, but sometimes a calculation or some other form of analysis will be required.
- In multiple-choice questions you are given the correct answer along with three incorrect answers (called distractors). You need to select the correct answer and put a cross in the box of the letter next to it.
- The three distractors supplied will feature the answers that you are likely to arrive at if you make typical or common errors. For this reason, multiple-choice questions aren't as easy as you might at first think. If possible, try to answer the question before you look at any of the answers.
- If you change your mind, put a line through the box with the incorrect answer () and then mark the box for your new answer with a cross (☒) .
- If you have any time left at the end of the paper, multiple-choice questions should be put high on your list of priorities for checking answers.

Student answer

D *ammonia* ☒

The student has selected the wrong answer. They haven't remembered that symmetrical molecules can still contain bonds between very electronegative atoms and be non-polar, and just had a guess, or they haven't realised that the lone pair of electrons on the nitrogen atom means that ammonia is not symmetrical.

COMMENTARY

This is an incorrect answer because:

- The student has shown a lack of understanding of this topic.

QUESTION TYPE: SHORT OPEN

Complete the electronic configuration for the copper atom Cu and the Cu^{2+} ion.

Cu: $1s^2\ 2s^2\ 2p^6$

Cu^{2+} $1s^2\ 2s^2\ 2p^6$ [2]

You need to know the electronic configurations of the first row of transition metal elements. Generally, these follow a pattern, but there are a couple of exceptions, and copper is one of them. The points you need to remember are:

- Copper atoms have a full 3d sub-shell and only one 4s electron as this gives a more stable electron arrangement than if the 4s orbital contained two electrons, as is the case for the majority of the transition metals.
- For all of the transition metals the 4s electrons are lost before the 3d electrons.

QUESTION ANALYSIS

- Short open questions usually require simple short answers, often one word. Generally, they will be simple recall of the chemistry you have been taught.
- The command word will tell you what you need to do. Here the command word is 'complete', so you simply need to complete the electronic configurations in the space provided.

Student answer

Cu: $1s^2\ 2s^2\ 2p^6\ 3s^2\ 3p^6\ 3d^9\ 4s^2$

Cu^{2+}: $1s^2\ 2s^2\ 2p^6\ 3s^2\ 3p^6\ 3d^9$

The student has the electronic configuration of the Cu^{2+} ion correct. However, they have clearly forgotten that copper atoms have 10 electrons in the 3d sub-shell and only one 4s electron.

COMMENTARY

This is a poor answer because:

- The student has forgotten the electronic configuration of the copper atom.

QUESTION TYPE: OPEN RESPONSE

Chlorine and bromine are elements in Group 7 of the Periodic Table. Both elements possess a number of different oxidation numbers and therefore are involved in many redox reactions.

Chlorine dioxide reacts with cold, dilute aqueous sodium hydroxide. The equation for the reaction is:

$$2ClO_2(aq) + 2OH^-(aq) \rightarrow ClO_2^-(aq) + ClO_3^-(aq) + H_2O(l)$$

Using oxidation numbers, explain why the chlorine in ClO_2 has undergone disproportionation. [3]

The command word here is 'describe', so you will need to give an account of the results shown. You do not need to explain the scientific reasons behind any patterns. It is important to note the command word carefully in graph-based questions, as you may be asked to analyse, explain or compare and contrast results, each of which requires a very different approach.

Question analysis

- The command word in this question is 'explain'. This requires a justification/exemplification of a point, which in this case is that chlorine has undergone disproportionation. In some cases of this type of question, a mathematical explanation could also be used.

Student answer

In ClO_2, chlorine has an oxidation state of +4. When it reacts, its oxidation state changes to +3 in ClO_2^- and +5 in ClO_3^-. This means that its oxidation state has both increased and decreased in the same reaction. That is to say, it has been reduced and oxidised at the same time.

COMMENTARY

This is an excellent answer because:

- The answer is logically structured.
- The student has calculated all three oxidation states correctly and shown that they understand the term disproportionation.

QUESTION TYPE: EXTENDED WRITING

** The equilibrium reaction between carbon dioxide and water is important for maintaining the pH of the body through buffering.*

The equation for the reaction is:

$$CO_2 + H_2O \rightleftharpoons HCO_3^- + H^+$$

State what is meant by a buffer solution and explain how a buffer solution works by referring to this equation. [6]

It helps when composing answers to extended writing questions to think about the number of marks that are likely to be awarded to each part of the question. The first part requires a simple statement of what a buffer solution is, so only one mark is likely to be awarded for this. The second requires an explanation of how a buffer works, making reference to this particular reaction. One mark will be available for making reference to the reaction, and the other four marks will be for an explanation of how the buffer works.

Question analysis

- In questions marked with an asterisk (*), marks will be awarded for your ability to structure your answers logically, showing how the points that you make are related or follow on from each other, where appropriate.
- Four marks are available for making valid points in response to the question. In mark schemes, these are referred to as 'indicative marking points'. To gain all four marks, your answer needs to include six indicative marking points. The remaining two marks are awarded for structuring your answer well, with clear lines of reasoning and linkages between points.
- It is vital to plan out your answer before you write it down. There is always space given on an exam paper to do this, so just jot down the points that you want to make before you answer the question in the space provided. This will help to ensure that your answer is coherent and logical and that you don't end up contradicting yourself. However, once you have written your answer, go back and cross these notes out so that it is clear they do not form part of the answer.

Student answer

A buffer solution is a solution that resists changes in pH when a small amount of acid or alkali is added. So, if an acid is added, the pH won't change very much. If an alkali is added, the pH won't change very much. Buffers can do this because they contain a chemical that can react with any acid added. They also contain another chemical that can react with any alkali added, so effectively these are removed, and that means that the pH doesn't change very much.

The first sentence of the student's answer is a satisfactory statement of what a buffer solution is, so one mark can be awarded. The second sentence is a repeat of the information in the first using different words, so gains no further credit. The student has made no reference to this particular reaction, so cannot be given the second mark. The rest of the answer is far too vague, so no marks can be awarded for the explanation.

COMMENTARY

This is a poor answer because:

- The student should have stated what happens when both acid and alkali are added to this solution and how this will not affect the pH significantly. When acid is added, the H^+ ions will react with HCO_3^- ions. When alkali is added, the OH^- ions will react with CO_2 molecules. The pH of the solution is determined by the ratio of the concentrations of HCO_3^- to CO_2 (i.e. $[HCO_3^-]/[CO_2]$. If both of these two concentrations are relatively large, then this ratio will not change significantly when either H^+ or OH^- ions are added. Hence the pH remains fairly constant.

QUESTION TYPE: CALCULATION

A phosphate buffer system operates in human cells.

The reaction involved is:

$$H_2PO_4^- \rightleftharpoons H^+ + HPO_4^{2-}$$

Calculate the pH of a 0.10 mol dm^{-3} solution of $H_2PO_4^-$.

$K_a = 6.2 \times 10^{-8}$ mol dm^{-3} [2]

For weak acids the position of equilibrium is over to the left-hand side, and so the hydrogen ion concentration cannot be assumed to be the same as the original acid concentration as is the case with strong acids. To calculate the pH of a weak acid you need to first write the expression for the equilibrium constant and then rearrange that to work out the hydrogen ion concentration. Remember, you can assume that the concentration of $H_2PO_4^-$ is 0.1 mol dm^{-3}. The expression pH = −log [H$^+$] can then be used to calculate the pH.

Question analysis

- The command word here is 'calculate'. This means that you need to obtain a numerical answer to the question, showing relevant working. If the answer has a unit, this must be included.
- Always have a go at calculation questions. You may get some small part correct that will gain credit.
- The important thing with calculations is to show your working clearly and fully. The correct answer on the line will gain all of the available marks. However, an incorrect answer can gain all but one of the available marks if your working is shown and is correct. Show the calculation that you are performing at each stage and not just the result. When you have finished, look at your result and see if it is sensible.
- Take an approved calculator into every exam and make sure that you know how to use it!

Student answer

$K_a = [H^+]^2/0.1 = 6.2 \times 10^{-8}$

$[H^+] = \sqrt{(6.2 \times 10^{-8} \times 0.1)} = 7.87 \times 10^{-5}$

$pH = -\log [H^+] = 4.1$

The student has correctly used the equilibrium expression for the reaction to calculate the H^+ concentration. For this they obtained the first of the two available marks. They then went on to use the expression pH = −log [H$^+$] to calculate the pH and gain the second of the two available marks.

COMMENTARY

This is a good answer because:

- The student has completed all steps correctly and has shown their working.

QUESTION TYPE: SYNOPTIC

A student carrying out a titration wants to identify the piece of apparatus that contributes most to measurement uncertainties in the experiment.

The percentage measurement uncertainty is marked on the 25 cm^3 pipette as ±0.06 cm^3, on the 250 cm^3 volumetric flask as ±0.2 cm^3 and on the burette as ±0.05 cm^3. The average titre was 25.50 cm^3.

Deduce the most significant source of measurement uncertainty in this procedure. (4)

Paper 3 contains mainly questions that examine your knowledge of practical chemistry. However, it also includes synoptic questions that can draw on knowledge from any part of the specification. This question is expecting that you can calculate the percentage uncertainty associated with the apparatus used in a titration experiment. These are calculations that you should complete when using titration apparatus.

QUESTION ANALYSIS

- A synoptic question is one that addresses ideas from different areas of chemistry in one context. In your answer you need to use these different ideas and show how they combine to explain the chemistry in the context of the question.
- Questions in Paper 3 may draw on any of the topics in this specification. This question is assessing understanding of experimental methods.

Student answer

Percentage uncertainty of pipette = 0.06 × 100/25 = 0.24%

Percentage uncertainty of volumetric flask = 0.2 × 100/250 = 0.08%

Percentage uncertainty of burette = 0.05 × 100/25.50 = 0.2%

The pipette has the greatest percentage uncertainty.

The student has calculated the percentage uncertainty correctly for each piece of apparatus. However, they have forgotten that the percentage for the burette needs to be doubled because this reading is taken twice and therefore the error will be doubled. This has lost them one mark. This answer scores three out of the four available marks.

COMMENTARY

This is a good answer because:

- The student has used the correct formula to calculate the percentage uncertainty. There is only one error, and that is not doubling the percentage for the burette. This is penalised only once.

COMMAND WORDS

The following table lists the command words used across the IAS/IAL Science qualifications in the external assessments. You should make sure you understand what is required when these words are used in questions in the exam.

COMMAND WORD	THIS TYPE OF QUESTION WILL REQUIRE STUDENTS TO:
ADD/LABEL	Requires the addition of labelling to stimulus material given in the question, for example, labelling a diagram or adding units to a table.
ASSESS	Give careful consideration to all the factors or events that apply and identify which are the most important or relevant. Make a judgement on the importance of something, and come to a conclusion where needed.
CALCULATE	Obtain a numerical answer, showing relevant working. If the answer has a unit, this must be included.
COMMENT ON	Requires the synthesis of a number of factors from data/information to form a judgement. More than two factors need to be synthesised.
COMPARE AND CONTRAST	Looking for the similarities **and** differences between two (or more) things. Should not require the drawing of a conclusion. Answer must relate to both (or all) things mentioned in the question. The answer must include at least one similarity and one difference.
COMPLETE/RECORD	Requires the completion of a table/diagram/equation.
CRITICISE	Inspect a set of data, an experimental plan or a scientific statement and consider the elements. Look at the merits and/or faults of the information presented and back judgements made.
DEDUCE	Draw/reach conclusion(s) from the information provided.
DERIVE	Combine two or more equations or principles to develop a new equation.
DESCRIBE	Give an account of something. Statements in the response need to be developed, as they are often linked, but do not need to include a justification or reason.
DETERMINE	The answer must have an element that is quantitative from the stimulus provided, or must show how the answer can be reached quantitatively.
DEVISE	Plan or invent a procedure from existing principles/ideas.
DISCUSS	Identify the issue/situation/problem/argument that is being assessed within the question. Explore all aspects of an issue/situation/problem. Investigate the issue/situation/problem etc. by reasoning or argument.

COMMAND WORD	THIS TYPE OF QUESTION WILL REQUIRE STUDENTS TO:
DRAW	Produce a diagram either using a ruler or drawing freehand.
ESTIMATE	Give an approximate value for a physical quantity or measurement or uncertainty.
EVALUATE	Review information then bring it together to form a conclusion, drawing on evidence including strengths, weaknesses, alternative actions, relevant data or information. Come to a supported judgement of a subject's qualities and its relation to its context.
EXPLAIN	An explanation requires a justification/exemplification of a point. The answer must contain some element of reasoning/justification; this can include mathematical explanations.
GIVE/STATE/NAME	All of these command words are really synonyms. They generally all require recall of one or more pieces of information.
GIVE A REASON/REASONS	When a statement has been made and the requirement is only to give the reasons why.
IDENTIFY	Usually requires some key information to be selected from a given stimulus/ resource.
JUSTIFY	Give evidence to support (either the statement given in the question or an earlier answer).
PLOT	Produce a graph by marking points accurately on a grid from data that is provided and then drawing a line of best fit through these points. A suitable scale and appropriately labelled axes must be included if these are not provided in the question.
PREDICT	Give an expected result or outcome.
SHOW THAT	Prove that a numerical figure is as stated in the question. The answer must be to at least 1 more significant figure than the numerical figure in the question.
SKETCH	Produce a freehand drawing. For a graph, this would need a line and labelled axes with important features indicated; the axes are not scaled.
STATE WHAT IS MEANT BY	When the meaning of a term is expected but there are different ways in which these can be described.
SUGGEST	Use your knowledge and understanding in an unfamiliar context. May include material or ideas that have not been learned directly from the specification.
WRITE	When the questions ask for an equation.

GLOSSARY

absolute potential difference the potential difference between a metal and a solution of its ions

addition–elimination reaction when two molecules join together, followed by the loss of a small molecule

adsorption the adhesion of atoms, molecules or ions to the surface of a solid, or the process that occurs when reactants form weak bonds with a solid catalyst

amphoteric (substance) a substance that can act both as an acid and as a base

amphoteric behaviour the ability of a species to react with both acids and bases

analyser a material that allows plane-polarised light to pass through it

aromatic the original meaning was a description of the smell of certain organic compounds. The new meaning is a description of the bonding in a compound – delocalised electrons forming pi (π) bonding in a hydrocarbon ring

asymmetric an asymmetric carbon atom in a molecule is one that is joined to four different atoms or groups

autocatalysis when a reaction product acts as a catalyst for the reaction

basicity the extent to which a base can donate a lone pair of electrons to the hydrogen atom of a water molecule

bidentate ligand a molecule or ion that forms two dative bonds with a metal ion

bimolecular (mechanism) a mechanism in which two species are reacting in the rate-determining step

buffer capacity a measure of the amount of acid or base required to change significantly the pH of food or of a solution of an acid and a base

buffer solution a solution that *minimises* the change in pH when a *small* amount of either acid or base is added

^{13}C NMR the use of NMR spectroscopy to detect ^{13}C nuclei within the molecules of a substance, in order to determine the structure

chemical environments the chemical environments of carbon atoms in a molecule are related to whether the carbon atoms are identically, or differently, positioned within the molecule

chemical shift (of a proton or group of protons) a number (in the units ppm) that indicates its behaviour in a magnetic field relative to tetramethylsilane. It can be used to identify the chemical environment of the carbon atoms or of the hydrogen atoms (protons) attached to it

chiral a chiral atom in a molecule is one that allows it to exist as non-superimposable forms. It can also refer to the molecule itself

complementary colours colours opposite each other on a colour wheel

complex a species containing a metal ion joined to ligands

complex ion a complex with an overall positive or negative charge

condensation polymerisation the formation of a polymer, usually by the reaction of two different monomers, when another small molecule is also formed

conjugate acid when a base accepts a proton, the species formed is the conjugate acid of the base

conjugate acid–base pair either a base and its conjugate acid or an acid and its conjugate base

conjugate base when an acid donates a proton, the species formed is the conjugate base of the acid

conjugated system where single and double bonds alternate, allowing the electrons in the p-orbitals of the atoms to overlap and form a delocalised electron cloud

coordination number the number of dative (coordinate) bonds in the complex

coupling reaction a reaction in which two organic molecules or ions join together to form one new molecule

dative (coordinate) bond a covalent bond formed between the central metal atom or ion and a ligand, in which both of the bonding electrons are supplied by the ligand

deprotonation the removal of one or more hydrogen ions (protons) from a complex ion

derivatives compounds formed from other compounds, especially when the properties of the derivatives can be used to identify the original compound

desorption the process that occurs when products leave the surface of a solid catalyst

disproportionation a reaction in which an element is both oxidised and reduced at the same time

dissociated molecules are said to be dissociated when they have split to form ions

electromagnetic spectrum the range of all wavelengths and frequencies of all the types of radiation

electromotive force (emf) the measured potential difference of a cell when no current is flowing

(first) electron affinity (of an element) the energy change when each atom in a mole of atoms in the gaseous state gains an electron to form a –1 ion

electrophile an atom, molecule or ion that can accept a pair of electrons

electrophilic a reaction mechanism in which the attacking species is an electrophile

enantiomers isomers that are related as object and mirror image

enthalpy change of hydration, $\Delta_{hyd}H$ the enthalpy change when one mole of an ion in its gaseous state is completely hydrated by water

enthalpy change of solution, $\Delta_{sol}H$ the enthalpy change when one mole of an ionic solid dissolves in water to form an infinitely dilute solution

entropy a property of matter that is associated with the degree of disorder, or degree of randomness, of particles (i.e. atoms, molecules or ions), and also with the distribution of the quanta of energy between the particles

equilibrium constant a number that expresses the relationship between the amounts of products and reactants present at equilibrium in a reversible chemical reaction at a given temperature

equivalence point the point in a titration when the acid and base have reacted together in the exact proportions as dictated by the stoichiometric equation

equivalent protons hydrogen atoms in the same chemical environment

experimental lattice energy the lattice energy calculated from a Born–Haber cycle

half-life (of a reaction) the time taken for the concentration of the reactant to fall to one-half of its initial value

halogen carrier a compound such as AlX_3 or FeX_3, where X is a halogen, that can catalyse the halogenation of arenes. They produce a stronger electrophile, such as Br^+

hazard a property of a substance that could cause harm to a user

heterogeneous catalyst a catalyst that is in a different phase from the reactants

heterogeneous reaction a reaction in which at least one of the reactants and/or products is in a different phase to the others

hexadentate ligand a molecule or ion that forms six dative bonds with a metal ion

high resolution mass spectrometry (HRMS) a type of mass spectrometry that can produce M_r values to several decimal places, usually four or more

homogeneous catalyst a catalyst that is in the same phase as the reactants

homogeneous reaction a reaction in which all the reactants and products are in the same phase

hydrolysis the breaking of a compound by water into two compounds

infinitely dilute solution a solution in which there is so much water that adding any more does not cause a further enthalpy change

instantaneous reaction rate the gradient of a tangent drawn to the line of the graph of concentration against time. The instantaneous rate varies as the reaction proceeds (except for a zero order reaction)

integration trace this shows the relative numbers of equivalent protons (i.e. in the same chemical environment)

ion-dipole interaction the attraction in aqueous solution between ions and polar water molecules

ionic product of water, K_w the product of the concentration of the hydrogen ions and the hydroxide ions, both measured in mol dm^{-3}

isoelectric point (of an amino acid) the pH of an aqueous solution in which it is neutral

$K_w = [H^+(aq)][OH^-(aq)]$

kinetically stable the reaction does not take place, or is very slow

ligand a species that uses a lone pair of electrons to form a dative bond with a metal ion

ligand exchange when one ligand in a complex ion is replaced by a different ligand

mobile phase in chromatography the liquid that moves through the stationary phase and transports the components

monodentate ligand a molecule or ion that forms one dative bond with a metal ion

multidentate ligand a molecule or ion that forms several dative bonds with a metal ion

multiplets the different splitting patterns observed (singlets, doublets, triplets or quartets) in a high resolution ^1H NMR spectrum

nuclear magnetic resonance spectroscopy (NMR) a technique used to find the structures of organic compounds. It depends on the ability of nuclei to resonate in a magnetic field

nucleophilic addition a type of mechanism in which a molecule containing two atoms or groups is added across a polar double bond (usually C=O) and the attacking species in the first step is a nucleophile

optical activity a substance shows optical activity if it rotates the plane of polarisation of plane-polarised light

order (of a reactant species) the power to which the concentration of the species is raised in the rate equation

overall order (of a reaction) the sum of all the individual orders

partial pressure (of a gas in a mixture of gases) the pressure that the gas would exert if it alone occupied the volume of the mixture

partially dissociated only a small fraction of the acid molecules have dissociated

peak a peak in a ^1H NMR spectrum shows the presence of hydrogen atoms (protons) in a specific chemical environment

peptide bond the bond formed by a condensation reaction between the carbonyl group of one amino acid and the amino group of another amino acid

pH (of an aqueous solution) the reciprocal of the logarithm to the base 10 of the hydrogen ion concentration measured in moles per cubic decimetre: $pH = -\lg[H^+]$. This definition is difficult to remember, so either of the two equations given on page 83 can be used to define pH

$pK_a = -\lg K_a$

plane-polarised light monochromatic light that has oscillations in only one plane

polarimeter the apparatus used to measure the angle of rotation caused by a substance

polarisation the distortion of the electron density of a negative ion (anion)

polariser a material that converts unpolarised light into plane-polarised light

polypeptide a condensation polymer formed from many amino acids

promotion (or **excitation**) when an electron moves from a lower energy level to a higher energy level

protein a polypeptide that has folded into a specific shape in order to have a specific function

proton acceptor a base is a proton acceptor

proton donor an acid is a proton donor

proton NMR the use of NMR spectroscopy to detect ^1H nuclei within the molecules of a substance, in order to determine the structure

racemic mixture an equimolar mixture of two enantiomers that has no optical activity

rate-determining step (of a reaction) the slowest step in the mechanism for the reaction

rate equation an equation expressing the mathematical relationship between the rate of reaction and the concentrations of the reactants

(overall) **rate of reaction** the change in concentration of a species divided by the time it takes for the change to occur. All reaction rates are positive

reaction quotient a measure of the relative amounts of products and reactants present during a reaction at any given time

retention time (of a component) in HPLC and GC, the time taken from injection to detection

risk the possible effect that a substance may cause to a user, and this will depend on factors such as concentration and apparatus. The level of risk is controlled using control measures

six-fold coordination complexes in which there are six ligands forming coordinate bonds with the transition metal ion

(chemical) **species** an atom, a molecule or an ion that is taking part in a chemical reaction

splitting pattern the appearance of a peak as a small number of small sub-peaks very close to each other

spontaneous process a process that takes place without continuous intervention by us

square planar this shape contains a central atom or ion surrounded by four atoms or ligands in the same plane and with bond angles of 90°

standard electrode potential the emf measured when a half-cell is connected to a standard hydrogen electrode under standard conditions

standard enthalpy change of atomisation (of an element) the enthalpy change measured at a stated temperature, usually 298 K, and 100 kPa when one mole of gaseous atoms is formed from an element in its standard state

standard lattice energy (of an ionic compound) the energy change measured at a stated temperature, usually 298 K, and 100 kPa when one mole of an ionic compound is formed from its ions in the gaseous state

stationary phase in chromatography the stationary phase is the liquid or solid that does not move

substitution a reaction in which an atom, or group of atoms, in a molecule is replaced by another atom, or group of atoms

theoretical lattice energy the lattice energy calculated using the principles of electrostatics

thermodynamically feasible reaction a reaction that should take place without any intervention by us, if we consider the enthalpy and entropy changes involved

transition metal an element that forms one or more stable ions with incompletely filled d-orbitals

unimolecular (mechanism) a mechanism in which only one species is reacting in the rate-determining step

unpolarised light light that has oscillations in all planes at right angles to the direction of travel

zwitterion a molecule containing positive and negative charges but which has no overall charge

Key

Atomic (proton number)
Atomic symbol
Name
Relative atomic mass

Group

Period	1 (1)	2 (2)	(3)	(4)	(5)	(6)	(7)	(8)	(9)	(10)	(11)	(12)	3 (13)	4 (14)	5 (15)	6 (16)	7 (17)	8 (18)
1								1 **H** Hydrogen 1.0										2 **He** Helium 4.0
2	3 **Li** Lithium 6.9	4 **Be** Beryllium 9.0											5 **B** Boron 10.8	6 **C** Carbon 12.0	7 **N** Nitrogen 14.0	8 **O** Oxygen 16.0	9 **F** Fluorine 19.0	10 **Ne** Neon 20.2
3	11 **Na** Sodium 23.0	12 **Mg** Magnesium 24.3											13 **Al** Aluminium 27.0	14 **Si** Silicon 28.1	15 **P** Phosphorus 31.0	16 **S** Sulfur 32.1	17 **Cl** Chlorine 35.5	18 **Ar** Argon 39.9
4	19 **K** Potassium 39.1	20 **Ca** Calcium 40.1	21 **Sc** Scandium 45.0	22 **Ti** Titanium 47.9	23 **V** Vanadium 50.9	24 **Cr** Chromium 52.0	25 **Mn** Manganese 54.9	26 **Fe** Iron 55.8	27 **Co** Cobalt 58.9	28 **Ni** Nickel 58.7	29 **Cu** Copper 63.5	30 **Zn** Zinc 65.4	31 **Ga** Gallium 69.7	32 **Ge** Germanium 72.6	33 **As** Arsenic 74.9	34 **Se** Selenium 79.0	35 **Br** Bromine 79.9	36 **Kr** Krypton 83.8
5	37 **Rb** Rubidium 85.5	38 **Sr** Strontium 87.6	39 **Y** Yttrium 88.9	40 **Zr** Zirconium 91.2	41 **Nb** Niobium 92.9	42 **Mo** Molybdenum 95.9	43 **Tc** Technetium (98)	44 **Ru** Ruthenium 101.1	45 **Rh** Rhodium 102.9	46 **Pd** Palladium 106.4	47 **Ag** Silver 107.9	48 **Cd** Cadmium 112.4	49 **In** Indium 114.8	50 **Sn** Tin 118.7	51 **Sb** Antimony 121.8	52 **Te** Tellurium 127.6	53 **I** Iodine 126.9	54 **Xe** Xenon 131.3
6	55 **Cs** Caesium 132.9	56 **Ba** Barium 137.3	57 **La*** Lathanum 138.9	72 **Hf** Hafnium 178.5	73 **Ta** Tantalum 180.9	74 **W** Tungsten 183.8	75 **Re** Rhenium 186.2	76 **Os** Osmium 190.2	77 **Ir** Iridium 192.2	78 **Pt** Platinum 195.1	79 **Au** Gold 197.0	80 **Hg** Mercury 200.6	81 **Ti** Thallium 204.4	82 **Pb** Lead 207.2	83 **Bi** Bismuth 209.0	84 **Po** Polonium (209)	85 **At** Astatine (210)	86 **Rn** Radon (222)
7	87 **Fr** Francium (223)	88 **Ra** Radium (226)	89 **Ac*** Actinium (227)	104 **Rf** Rutherfordium (261)	105 **Db** Dubnium (262)	106 **Sg** Seaborgium (266)	107 **Bh** Bohrium (264)	108 **Hs** Hassium (277)	109 **Mt** Meitnerium (268)	110 **Ds** Darmstadtium (271)	111 **Rg** Roentgenium (272)	112 **Cn** Copernicium (285)	113 **Nh** Nihonium (286)	114 **Fl** Flerovium (289)	115 **Mc** Muscovium (289)	116 **Lv** Livermorium (293)	117 **Ts** Tennessine (293)	118 **Og** Oganesson (294)
				58 **Ce** Cerium 140.1	59 **Pr** Praseodymium 140.9	60 **Nd** Neodymium 144.2	61 **Pm** Promethium 144.9	62 **Sm** Samarium 150.4	63 **Eu** Europium 152.0	64 **Gd** Gadolinium 157.2	65 **Tb** Terbium 158.9	66 **Dy** Dysprosium 162.5	67 **Ho** Holium 164.9	68 **Er** Erbium 167.3	69 **Tm** Thulium 168.9	70 **Yb** Ytterbium 173.0	71 **Lu** Lutetium 175.0	
				90 **Th** Thorium 232.0	91 **Pa** Protactinium (231)	92 **U** Uranium 238.1	93 **Np** Neptunium (237)	94 **Pu** Plutonium (242)	95 **Am** Americium (243)	96 **Cm** Curium (247)	97 **Bk** Berkelium (245)	98 **Cf** Californium (251)	99 **Es** Einsteinium (254)	100 **Fm** Fermium (253)	101 **Md** Mendeleevium (256)	102 **No** Nobelium (254)	103 **Lr** Lawrencium (257)	

INDEX

A

B

C

D

E

F

G

H

I

Q

R

S